SOILS

AN INTRODUCTION TO SOILS
AND PLANT GROWTH

PRENTICE-HALL INTERNATIONAL, INC., *London*
PRENTICE-HALL OF AUSTRALIA, PTY. LTD., *Sydney*
PRENTICE-HALL OF CANADA, LTD., *Toronto*
PRENTICE-HALL OF INDIA PRIVATE LIMITED, *New Delhi*
PRENTICE-HALL OF JAPAN, INC., *Tokyo*

SOILS

AN INTRODUCTION TO SOILS
AND PLANT GROWTH

THIRD EDITION

ROY L. DONAHUE

JOHN C. SHICKLUNA

LYNN S. ROBERTSON

Professors of Soil Science
Michigan State University
East Lansing, Michigan

PRENTICE-HALL, INC., ENGLEWOOD CLIFFS, NEW JERSEY

Humbly and affectionately dedicated
to our college professor

JETHRO OTTO VEATCH

Professor Emeritus of Soil Science
Michigan State University

Roy L. Donahue

John C. Shickluna

Lynn S. Robertson

13-821876-5

Library of Congress Card Catalog Number: 73–159573

Current printing (last digit):

10 9 8

Printed in the United States of America

PREFACE

This book is written to be read and studied by students of all ages who are interested in learning more about soils, soil ecology, and plant environment as an introductory subject. It is not intended to be a reference work for research soil scientists. Because it is, in fact, "an introduction," as the title states, this book does not include all that is known on any aspect of the subject. The subject matter has been carefully selected to present most of the simpler principles and the corresponding examples of practices of soil science, soil ecology, and plant environment.

The principal objectives of this textbook are itemized in the following ten categories:

1. To help the student develop intellectual curiosity leading toward an understanding of the basic principles underlying soil science and soil ecology, and their relationship to the growth of plants and their environment.
2. To assist the reader in organizing, evaluating, and applying scientific information about soils.
3. To convey an appreciation of the dynamic character of soils and the need for continuing and imaginative soil research.
4. To try to instill into the student a scientific attitude by stressing the need for continually weighing new findings in terms of past observations and experiences.
5. To discourage the tendency to accept established practices and published information without subjecting them to critical study.
6. To provide the student with a practical approach to typical present-day soil problems.
7. To encourage the reader to enjoy studying the soil as a cultural subject and to appreciate the importance of soils, soil ecology, and plant environment to all human activity.

8. To challenge the reader to initiate research on one or more of the many unsolved problems of the relationships among soils, soil ecology, and plant environment.
9. To make the student more readily aware of problems of environmental pollution and their control.
10. To broaden the horizon of the reader toward soil characterization in the Tropics as a prelude to more effectively assisting developing countries, and for developing countries to assist themselves more effectively.

ROY L. DONAHUE

JOHN C. SHICKLUNA

LYNN S. ROBERTSON

ACKNOWLEDGMENTS

For this third edition of, "Soils: An Introduction to Soils and Plant Growth," the authors gratefully acknowledge the following assistance:

To our wives, Lola, Ann, and Betty, for encouragement and understanding.

To Dr. Charles E. Kellogg, Deputy Administrator for Soil Survey, Soil Conservation Service, United States Department of Agriculture, for cooperation in obtaining photographs and reference materials, especially for Chapters 6 and 7.

To Dr. Eugene P. Whiteside, Professor of Soil Science, Michigan State University, for critically reviewing Chapter 7.

To Dirk van der Voet, Soil Conservation Service, United States Department of Agriculture, for supplying photographs of Oxisols and Vertisols from Brazil and of organic soils from Michigan.

To Frank Shuman, Champaign, Illinois, for obtaining difficult-to-obtain photographs.

To the State Soil Scientists, Soil Conservation Service in Texas, Georgia, Virginia, Iowa, Minnesota, Florida, Kentucky, Nebraska, Colorado, Arizona, Louisiana, Vermont, Wisconsin, Kansas, Tennessee, Alaska, Utah, New York, Hawaii, and Pennsylvania, for sending soil series descriptions and photographs of soils to more accurately represent the United States.

To Dr. J. C. F. Tedrow, Professor of Soil Science, Rutgers University, New Brunswick, New Jersey, for assistance in obtaining photographs and reference materials on soil characterizations in Alaska.

To Dr. William M. Johnson, Assistant to Deputy Administrator for Soil Survey, Soil Conservation Service, United States Department of Agriculture, for courteously sending black and white photographs of the 10 new soil orders.

To Dr. Curtis L. Godfrey, Professor, In Charge of Soil Survey, Texas

A&M University, College Station, Texas, for cooperation in sending photographs and characterizations of Vertisols.

To Dr. John Box, Extension Agronomist, Texas A&M University, College Station, Texas, for assistance in obtaining photographs representing irrigation.

To Dr. Guy D. Smith, Director, Soil Survey Investigations, Soil Conservation Service, United States Department of Agriculture, for supplying advanced sheets of "Soil Taxonomy" for use in writing Chapter 6, Soil Classification.

To John Hesse, Michigan Water Resources Commission, for reviewing the new chapter on soil ecology and environmental pollution.

To Dr. Robert E. Lucas, Professor, Crop and Soil Sciences Department, Michigan State University, East Lansing, Michigan, for reviewing Chapter 22 on Soil Diagnosis.

To H. Wayne Pritchard, Director, Soil Conservation Society of America, for the courtesy to permit the use in our glossary of selections from "Resource Conservation Glossary."

To Dr. John H. Day, Soil Research Institute, Central Experimental Farm, Ottawa, Ontario, Canada, for sending photographs of the soil orders.

To Dr. Arthur L. Brundage, Professor of Animal Husbandry, Agricultural Experiment Station, Palmer, Alaska, and to Dr. Donald H. Dinkel, Plant Physiologist, University of Alaska, College, Alaska, for providing photographs on the use of plastic as a surface mulch for crop production.

To Dr. Werner Nelson, Senior Vice President, American Potash Institute, Inc., Lafayette, Indiana, for assistance and permission in the use of photographs.

To Dr. Ronald P. White, Canada Department of Agriculture, Research Station, Charlottetown, Prince Edward Island, Canada, for providing a photograph on molybdenum deficiency.

To Drs. J. F. Davis, E. C. Doll, and B. D. Knezek, Professors, Crop and Soil Sciences Department, Michigan State University, East Lansing, Michigan, for supplying relevant photographs.

To the late Ray Linville, Extension Specialist, Department of Agronomy, The Ohio State University, Columbus, Ohio, for providing a photograph of a computerized soil test report form.

To Dr. R. A. Cline, Research Scientist, Horticulture Research Institute of Ontario, Department of Agriculture and Food, Vineland Station, Ontario, Canada, for providing photograph of magnesium deficiency.

To Dr. Paul E. Rieke, Professor, Department of Crop and Soil Sciences, Michigan State University, for furnishing information on soil mixes and fertilizer use in greenhouses.

To the St. Louis Testing Laboratories, Inc., St. Louis, Missouri, for providing plant analysis nutrient evaluation forms.

To Mrs. Norma Green, a teacher of English, for editing the entire manuscript.

To secretaries Janet Botimer and Netsanet Tesfamariam for typing the manuscript.

To Dr. James A. Pomerening, Professor in the Department of Plant and Soil Science at California State Polytechnic College, for critically reviewing the book and offering very helpful suggestions for improving this third edition.

To Dr. Ivan F. Schneider, Associate Professor of Soil Science, Michigan State University, East Lansing, Michigan, for general encouragement and specific help on Chapter 6.

The authors would be grateful if other professors would tell them how this edition could be improved.

R. L. D.

J. C. S.

L. S. R.

CONTENTS

APPENDICES

INTRODUCTION

Since the second edition was published in 1965, the United States and the rest of the world have changed drastically.

Man has walked on the moon; man is rapidly polluting his environment; and man is less happy because massive and oft-repeated headlines of evil drive away headlines of virtue—and man is brainwashed into seeing only meaninglessness in an increasingly impersonalized society of mass communication.

The early 1970's for the United States will no doubt be typified by greater speed in tractorized farming, easier processes of learning, faster computerized solutions to massive and complex problems, a renewed emphasis on greater efficiency in the application of the factors of production, renewed vigor to control pollution, a greater awareness of the problems of the world, and an intensification toward writing simplified textbooks in applied science that are interesting for the average student to read, understand, and apply, but that will not insult the intelligence of the most scholarly.

In this third edition, the authors have "joined the decade of the 70's" by writing a book for the serious student (rather than primarily for the professor): by giving evidence how a computer is used to interpret soil test results more quickly; by updating all information on growth and production factors, such as fertilizers; by adding a chapter on soil ecology and pollution; by including a new chapter on soils of the Tropics; and by incorporating the latest (1971) information on the new United States system of soil taxonomy, and a new soil map of the world based upon this new taxonomy.

PART ONE

PRINCIPLES OF SOIL SCIENCE
AND PLANT GROWTH

CHAPTER 1

SOIL SCIENCE AND PLANT ENVIRONMENT

[Soil is] the surface stratum of earthy material, as far as the roots of plants reach.—E. W. HILGARD, 1860.

Soil in its traditional meaning, is the natural medium for the growth of land plants, whether or not it has developed discernible soil horizons. . . . In this sense soil has a thickness that is determined by the depth of rooting of plants.*

All sciences, including soil science, are changing as man explores the oceans, new lands, and the planets. Yet the best scientists in 1860 and the best ones 110 years later are in agreement that soil comprises the surface few feet of the earth that has been influenced by and influences plant roots.

SOIL FACTORS INFLUENCING PLANT ENVIRONMENT

From the soil, plant roots receive mechanical support, essential elements, water, and oxygen.

Soil physical properties largely determine the soil's water-supplying capacity to plants. Soil chemical properties determine the soil nutrient-supplying capacity to plants. Both physical and chemical properties determine root extension and the volume of soil that serves as a reservoir for both water and essential nutrients for plants.

---

* *Soil Taxonomy of the National Cooperative Soil Survey.* Soil Conservation Service, U.S.D.A. (Dec., 1970), pp. 1-1 to 1-2.

Some soils are in such a physical and chemical condition as to encourage plant roots to grow deeply and to extend long distances laterally. These soils are ideal because the plants growing in them will not blow over, will be drought resistant, and capable of absorbing nutrients from a large volume of soil. The growth of plants may be restricted by naturally or artificially compacted layers, infertile horizons, too much or too little soil moisture, or soluble salts in toxic quantities.

At present, 16 elements are known to be essential for the growth of crop plants. They are carbon, hydrogen, and oxygen from air and water; phosphorus, potassium, sulfur, calcium, iron, magnesium, boron, manganese, copper, zinc, molybdenum, and chlorine from the soil; and nitrogen from both air and soil.[1]

Water and air occupy pore spaces in the soil. Following a heavy and prolonged rain or irrigation, the soil pores may be almost completely filled with water for a few hours. After a day or two, some water will have moved downward in response to gravity, and the larger pores will be emptied of their water but filled with air. With a further loss of water by evaporation or transpiration, air will replace more of the space occupied by water. The next soaking rain or irrigation will repeat this process.

The important part of this air-water relationship is that there must be enough *total* pore spaces and that the pore spaces must be the proper size ranges to hold enough air and water to satisfy plant roots between cycles of rainfall or irrigation.

From the soil, plants *accumulate* nitrogen, phosphorus, and sulfur. In other words, plants nearly always contain a higher percentage of these elements than the soil in which the plants are growing. Conversely, soils almost always contain more iron, calcium, potassium, magnesium, and manganese than the plants growing in them.

---

[1] The number of elements considered essential for the growth of higher plants now varies from 16 to 20 or more, depending upon the definition of essentiality. The authors of this book are aware that Arnon limits essentiality to only those elements that are needed for higher plants to complete all life functions, and that the deficiency can be corrected by the application *only of this specific element causing the deficiency.* Other scientists, such as Nicholas, believe that an element should be considered essential if its addition enhances plant growth even though it merely substitutes for one of the 16 elements that Arnon declares to be essential. For example, since sodium can substitute in plant nutrition for some potassium, and vanadium for some molybdenum, Nicholas would consider both sodium and vanadium as essential, but Arnon would not. On the bases of the criteria used, Arnon specifies 16 elements and Nicholas 20 elements as being essential for the growth of higher plants, such as cotton and corn.

Refer to Chapters 11, 22, and 23 for further information on essential elements for plants.

OTHER ENVIRONMENTAL FACTORS INFLUENCING
PLANT GROWTH

Plant growth is influenced by many factors that are not directly related to the soil. These include the following:

1. Temperature of the air
2. Light intensity and quality
3. Day length
4. Plant diseases
5. Plant insects
6. Weeds
7. Floods
8. Violent winds

EARLY CONCEPTS OF SOIL SCIENCE AND
PLANT ENVIRONMENT

Xenophon, a Greek historian (430–355 B.C.), is credited with first recording the value of green-manuring crops when he wrote, "But then whatever weeds are upon the ground, being turned into the earth, enrich the soil as much as dung."

Cato (234–149 B.C.) wrote a practical handbook in which he recommended intensive cultivation, crop rotations, the use of legumes for soil improvements, and the value of manure in a system of livestock farming. Cato was also the first to classify land according to its relative value for specific crops. His classification included the following:

1. Land for vineyards (grapes)
2. Land for gardens
3. Willow land
4. Land for olive trees
5. Meadow land
6. Corn land
7. Timber land
8. Land for small trees
9. Land for oak trees

The usefulness of turnips for soil improvement was emphasized by Columella about 45 A.D. Liberal amounts of manure were recommended for turnips, which were to be plowed under and the land planted to corn. Also advocated by Columella were land drainage and the use of ashes, marl, clover, and alfalfa to make the soil more productive.

Then the barbarians of the north conquered Rome. Scientific agriculture and other forms of art and culture were arrested until nearly 1600.

An experiment was performed by Van Helmont (1577–1644) in Holland. He put a 5-pound willow tree in 200 pounds [2] of soil (oven-dry basis). The tree received only water for five years. At the end of this period, the soil weighed only 2 ounces less than 200 pounds, but the willow tree weighed 169 pounds and 3 ounces. Since the tree was given only water, Van Helmont reasoned that water was the "principle" of vegetation.

Although Van Helmont's experiment was advanced for the times, his reasoning was later proved false on two counts:

1. The loss of 2 ounces of soil, which he ignored, consisted of minerals such as calcium, potassium, and phosphorus, that were absorbed by the tree. If Van Helmont had burned the willow tree at the end of the experiment, he would have recovered his 2 ounces of soil minerals.
2. The willow tree consisted of carbon from the carbon dioxide of the atmosphere and oxygen from the atmosphere. The occurrence of carbon and oxygen from air as constituents of the willow tree refutes Van Helmont's conclusion that water is the "principle" of vegetation.

In 1731, Jethro Tull of Oxford concluded that cultivation was one of the prime essentials of growing plants, and that nitre, water, air, fire, and earth all contributed to increases in plant growth.

Justus von Liebig (1803–1873), a German chemist, published in 1840 *Chemistry in its Application to Agriculture and Physiology*. His thesis was that carbon for plant nutrition came from carbon dioxide of the atmosphere, hydrogen and oxygen from air and water, and nitrogen from ammonia. Phosphates were stated to be necessary for seed production and potassium for the development of grasses and cereals. He also believed that by analyzing crop plant ash he could formulate a fertilizer that would supply all essential elements for the next crop. And he developed the "law of the minimum," which stated that the growth of plants is limited by the essential element present in the least relative amount, as shown graphically in Figure 1.1.

Edmund Ruffin, a Virginia farmer-scientist, has been given credit for

---

[2] Conversion factors from English to metric units appear in Appendix C.

being the first to apply lime to correct a low productivity of soil due to soil acidity. This work was conducted from 1825 to 1845.

The oldest agricultural experiment station in the world was established in 1843 by J. B. Lawes and J. H. Gilbert at Rothamsted, England, a few miles from London. This experiment station laid the groundwork for modern research techniques in soils and plant growth.

SOIL CLASSIFICATION—A BRIEF HISTORY

The first scientific classification of soils was proposed in 1886 by the Russian, V. V. Dokuchaiev, and finalized in 1900.[3] Soils were classified into "Normal" (upland), "Transitional" (meadow, calcareous, alkali), and "Abnormal" (organic, alluvial, aeolian). The "Normal" class was further

---

[3] J. N. Afanasiev, "The Classification Problem in Russian Soil Science," Academy of Science, U.S.S.R. *Pedology,* 5, Leningrad (1927).

Note: E. W. Hilgard, working in the southern United States and later in California, between about 1860 and 1900 published his observations and chemical analyses of the relationships among soils, climate, and vegetation. Hilgard never developed a system of soil classification but the same interrelated factors were the principal ones used by Dokuchaiev in Russia as the basis for his system. However, neither man knew of the work of the other because of language differences. Dokuchaiev's work was translated from Russian into German in 1914 by Glinka and from German into English in 1927 by Marbut. The principal references are:

1. E. W. Hilgard, *Soils* (New York: The Macmillan Company, 1911), pp. 487–549.
2. K. D. Glinka, "Dokuchaiev's Ideas in the Development of Pedology and Cognate Sciences," Academy of Science, U.S.S.R., *Russian Pedological Institute,* 1, Leningrad (1927), 32 pp.

FIG. 1.1. "The law of the minimum" is illustrated here.

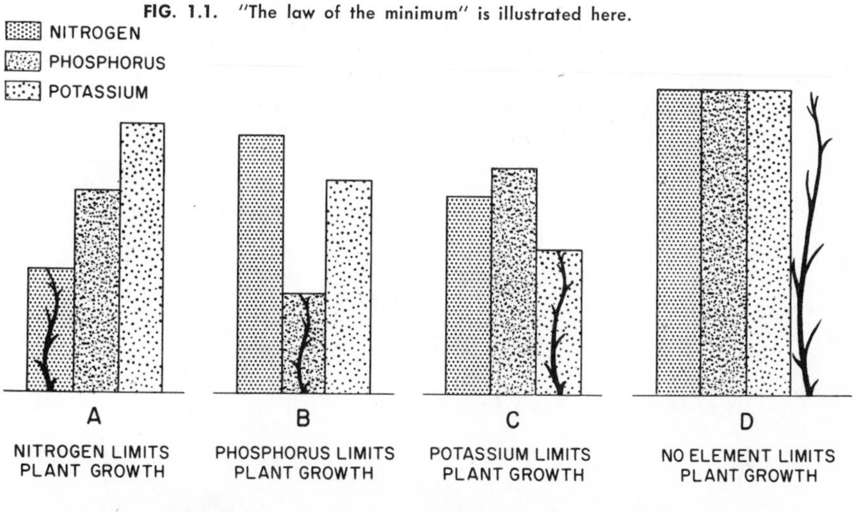

NITROGEN
PHOSPHORUS
POTASSIUM

A
NITROGEN LIMITS
PLANT GROWTH

B
PHOSPHORUS LIMITS
PLANT GROWTH

C
POTASSIUM LIMITS
PLANT GROWTH

D
NO ELEMENT LIMITS
PLANT GROWTH

divided into seven "zones" based upon climate and into "soil types" based upon soil color.

In 1912, Coffey classified the soils of the United States into five categories, as follows:

1. Arid soils
2. Dark-colored prairie soils
3. Light-colored timbered soils
4. Black swamp soils
5. Organic soils

Marbut's system of soil classification was proposed in 1927 and progressed to a 1936 version. Strong emphasis was given to "Mature" soils, similar to the "Normal" class of Dokuchaiev. Marbut's justification for emphasizing mature soils was his analogy that immature plants and animals are not classified; why try to classify immature soils? Marbut divided United States soils into two broad categories: *Pedalfer* (soil with aluminum and iron concentration) and *Pedocal* (soil with calcium carbonate concentration).[4]

In 1938, the U.S. system of soil classification was modified quite radically and the concepts of soil order, soil suborders, and soil great groups were introduced.[5] The system was revised in 1949.[6]

The present U.S. system of soil classification started in 1951 and went through seven major revisions ("approximations"). It was made "official" in 1965 and was released in 1971 as two books, *Soil Taxonomy* and *Classification of the Soils of the United States* (see Chapter 6, Soil Classification).

---

[4] C. F. Marbut, "A Scheme for Soil Classification," *First International Congress Soil Science Proceedings* 4:1–31 (1927); and C. F. Marbut, "Soils of the U.S.," in *United States Department of Agriculture, Atlas of American Agriculture,* Part 3 (1935).

[5] Mark Baldwin, Charles E. Kellogg, and James Thorp, "Soil Classification," in *Soils and Men: United States Department of Agriculture Yearbook* (1938), pp. 979–1001.

[6] James Thorp and Guy D. Smith, "Higher Categories of Soil Classification: Order, Suborder, and Great Soil Groups," *Soil Science,* Vol. 67 (1949), pp. 117–26.

A MODERN CONCEPT OF SOIL SCIENCE AND
PLANT ENVIRONMENT

Soil is a three-dimensional, dynamic, natural body occurring on the surface of the earth, which is a medium for plant growth and whose characteristics have resulted from the forces of climate and living organisms acting upon parent material, as modified by relief, over a period of time.

With the use of modern scientific methods, it is possible to study soil characteristics and the response to soil management practices in the field, in the greenhouse, and in the laboratory. From this information, soil maps can be made that permit accurate predictions of soil management responses in new areas that have soil characteristics similar to those soils where research has previously been conducted. It is not, therefore, necessary to conduct research on every acre of land to predict crop responses to soil management practices (Figure 1.2).

A close look at a handful of productive loam soil will reveal that the soil is composed of lumps held together by fine roots (Figure 1.3). Upon closer examination with a 10-power hand lens and a 100-power microscope, these lumps of soil are also observed to be held together by fungal hyphae and the vegetative parts of actinomycetes. Shiny streaks on the soil lumps were probably made by the slimy secretions of earthworms. This

FIG. 1.2. The modern concept of the soil has been developed to a large extent by Dr. Charles E. Kellogg, Deputy Administrator for Soil Survey, Soil Conservation Service, U.S. Dept. of Agr. Dr. Kellogg is shown here examining a shallow Vertisol in Central India that is used extensively for growing grain sorghum and cotton. *Photo by Roy L. Donahue.*

FIG. 1.3. A close look at a productive soil reveals many plant roots (1), cracks (2), small-animal burrows (3), peds (natural aggregates) (4), clods (man-induced aggregates) (5), sand grains, and finer silt and clay particles too small to be seen without magnification, but their presence is inferred because silt and clay (mostly clay) supply adhesive and cohesive properties that result in the formation of peds and clods. (Kenya.) *Photo by Roy L. Donahue.*

slime, as well as the gums and resins remaining as temporary end-products of organic decomposition, are good cementing agents that help living plant roots, fungi, and actinomycetes to glue the individual particles into desirable lumps.

Now imagine that you plant a seed in the handful of soil. Soon root hairs permeate the entire soil mass, following mainly along the crooked paths left by previous plant roots, earthworms, ants, fungal hyphae, and the vegetative parts of actinomycetes.

The root hairs obtain nutrients by an exchange of cations and anions between the surface of the roots and the surface of clay and humus particles. There is also a similar exchange between the root surface and the soil solution. The root uptake of calcium, for example, is accomplished by the release of two hydrogen ions from the root to either the soil solution or the surfaces of clay or humus, in exchange for a calcium ion from these sources to the root. Similarly, the plant root appears to exchange OH ions for nitrates, sulfates, phosphates, and other anions (Figure 1.4). One difference in the plant uptake of cations and anions is the fact that most cations are temporarily held on the surfaces of clay and humus, whereas most anions remain in the soil solution.

SOIL FERTILITY AND NATIVE VEGETATION

Contrary to popular belief, many soils, even when virgin, were never plentifully supplied with all essential elements. Before the common use of lime and fertilizers, farmers would adjust to differences in native fertility by choosing the best virgin land, farming it until crop yields declined, then clearing other virgin land. This is the common practice even today in many underdeveloped countries where pressure of population on land is not great and where lime and fertilizers are not readily available.

The total chemical composition of the leaves, branches, and stems of 30-year-old shortleaf pine on fertile soils in North Carolina was approximately 200 pounds of N, 15 pounds of P, 100 pounds of K, and 200 pounds of Ca per acre. Stands of the same age on soils of low fertility accumulated approximately one-half these amounts of nutrient elements. When such forested lands are cut and burned in preparation for planting crops, the nutrients in the trees become available for use by crop plants.[7]

In his book first copyrighted in 1906, Hilgard, one of our best early American soil scientists, made a shrewd observation on the relationships between native vegetation and soil fertility in the southern United States when he wrote:

> Thus in the longleaf pine uplands of the Cotton States, the scattered settlements have fully demonstrated that after two or three years cropping with corn, ranging from as much as 25 bushels per acre the first year to ten and less the third, fertilization is absolutely necessary to further paying cultivation. Should the shortleaved pine mingle with the longleaved, production may hold out from five to seven years. If oaks and hickory are super-added, as many as twelve years of good production without fertilization may be looked for by the farmer; and should the longleaved pine disappear altogether, the mingled growth of oaks and shortleaved pine will encourage him to hope for from twelve to fifteen years of fair production without fertilization.[8]

---

[7] M. S. L. Vaidya and C. W. Ralston, "Nutrient Mineral Uptake in Shortleaf Pine Plantations," *Agronomy Abstracts,* American Society of Agronomy (1962), p. 53.

[8] E. W. Hilgard, *Soils* (New York: The Macmillan Company, 1911), p. 314.

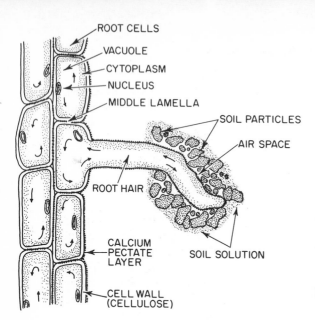

ROOT CELLS
VACUOLE
CYTOPLASM
NUCLEUS
MIDDLE LAMELLA
SOIL PARTICLES
AIR SPACE
ROOT HAIR
CALCIUM
PECTATE
LAYER
SOIL SOLUTION
CELL WALL
(CELLULOSE)

FIG. 1.4. A root hair magnified to show its intimate contact with soil particles and the soil solution from which the root obtains essential elements for growth and reproduction. Arrows within the cells indicate streaming of cytoplasm, a process by which plant food elements (plant nutrients) and water from the soil move into the plant. *Source:* Hawaiian Sugar Planters' Association.

ESSENTIAL ELEMENTS FOR PLANTS

Sixteen elements have been demonstrated to be essential for plant growth and reproduction. These essential plant food elements from air, water, soil, and fertilizers are shown in the following material.

Nitrogen was first proved to be essential for plants in the eighteenth century; seven more elements were proved essential in the nineteenth century; and thus far in the twentieth century, eight additional elements have been shown essential. The last element that was so proved was chlorine, in 1956. As a result of more refined research methods, other elements probably will be proved necessary.

Plants contain sodium, iodine, selenium, and cobalt, which have not yet been proved essential for plants but are a necessity for man and animals who eat these plants. Although sodium increases the growth of certain plants, it has not been demonstrated as essential. Silicon and aluminum also occur in all plants, but apparently they serve no useful function.

SOIL AS THE PRINCIPAL SOURCE OF PLANT NUTRIENTS

Throughout the world today, soil is the principal supply source of essential elements for plant growth. Fertilizers and manures are used only as a supplement. Green-manure crops function primarily to make the existing soil nutrients more readily available.

Soil is the source of 13 of the 16 elements essential for plant growth. Twelve of the thirteen elements originated in the parent rock from which

ESSENTIAL ELEMENTS FOR PLANTS

From Air and Water

Element	Symbol
Carbon	(C)
Hydrogen	(H)
Oxygen	(O)
Nitrogen	(N)

From Soil and Fertilizer
Major Nutrients

Element	Symbol
Nitrogen	(N)
Phosphorus	(P)
Potassium	(K)

Secondary Nutrients

Calcium	(Ca)
Magnesium	(Mg)
Sulfur	(S)

Micronutrients

Iron	(Fe)
Boron	(B)
Manganese	(Mn)
Copper	(Cu)
Zinc	(Zn)
Molybdenum	(Mo)
Chlorine	(Cl)

the soil developed. (Nitrogen was present in some parent rock in very small quantities.)

The total supply of plant nutrients in the soil is therefore of fundamental importance to economic production of plants. Although it is true that plants can be grown in pure quartz sand when essential water, air, and nutrients are supplied, economic production is seldom possible by this technique.

Table 1.1 shows the total nitrogen, phosphorus, potassium, calcium, and magnesium found in representative surface soils in the U.S. and on the moon. The data are arranged by average annual rainfall, from 0 to 50 inches.

In general, the less the rainfall the higher the percentage composition of total phosphorus, potassium, calcium, and magnesium in the surface soils. Calcium and magnesium are highest in moon rocks.[9] Nitrogen is

---

[9] L. H. Ahrens and R. V. Danchin, "Lunar Surface Rocks and Fines: Chemical Composition," *Science,* Vol. 167, No. 3914 (Jan. 2, 1970), pp. 87–88.

Note: Moon dust has been demonstrated to stimulate the growth of plants on earth.

TABLE 1.1   RANGE OF TOTAL PLANT NUTRIENTS IN THE SURFACE SOILS OF SELECTED
SOIL ORDERS IN THE UNITED STATES AND ON THE MOON *

Moon or State in U.S.	Average Annual PPT. (in.)	Soil Order	Soil Type	Horizon	Depth (in.)	Total Composition (Per Cent)				
						N	P	K	Ca	Mg
Moon †	0	Entisols?	Moon soil	Surface	0–2	0.003	0.14	0.11	7.4	4.8
Arizona	10	Aridisols	Mohave loam	1	0–6	0.03	0.10	2.26	1.45	0.09
South Dakota	20	Mollisols	Barnes silt loam	1	0–2½	0.46	0.10	1.62	1.21	0.05
				2	2½–8	0.30	0.09	1.52	0.96	0.05
Iowa	30	Mollisols	Marshall silt loam	A	0–10	0.17	0.05	1.85	0.56	0.49
Massachusetts	40	Spodosols	Becket fine sandy loam	A₁	0–6	1.04	0.06	1.71	0.64	0.09
South Carolina	50	Ultisols	Norfolk sandy loam	A₁	0–3	0.04	Trace	0.02	0.20	0.06
				A₂	0–6	0.20	Trace	Trace	0.07	0.05

* Source: C. F. Marbut, Atlas of American Agriculture, Part 3, Soils of the United States, U.S.D.A. (1935).
† "Space Science," Chemical and Engineering News (Jan. 19, 1970), pp. 20–21.

14

highest in amount in Spodosols [10] in Massachusetts because of the cool, humid climate, which favors an accumulation of organic matter on the surface. The nitrogen is a constituent of the organic matter.

Total reserves of plant nutrients, however, are not an accurate index of their rate of availability to plants. For this reason, data are given for *available* nutrients by surface soil textural class in Michigan.

Table 1.2 portrays available plant nutrients in clays, loams, and sands for phosphorus, potassium, and magnesium. Clays are mostly low in available phosphorus and are high in available potassium and magnesium. Loams test low to medium in available phosphorus, medium in potassium, and high in magnesium. Sands are shown as low to medium in available phosphorus and potassium and low to high in magnesium. The ability of

---

[10] See Chapter 6 for a discussion of Spodosols and other soil orders.

TABLE 1.2   RANGE OF AVAILABLE PLANT NUTRIENTS IN CLAY, LOAM, AND SAND SOILS IN MICHIGAN *

Nutrient and Soil Texture	Range in Pounds of Nutrients Per Acre		
	0–40	41–150	More than 150
	Percentage of Samples in Each Range		
Phosphorus (P)			
Clays	100	0	0
Loams	63.2	34.8	2.0
Sands	47.2	48.3	4.5
	Range in Pounds of Nutrients Per Acre		
	0–100	101–200	More than 200
	Percentage of Samples in Each Range		
Potassium (K)			
Clays	0	35.8	64.2
Loams	19.0	53.0	28.0
Sands	50.1	38.6	11.3
Magnesium (Mg)			
Clays	0	0	100
Loams	7.1	11.9	81.0
Sands	45.2	26.7	28.1

* Source: J. C. Shickluna, "The Relationship of pH, Available Phosphorus, Potassium, and Magnesium to Soil Management Groups," *Michigan State University Quarterly Bulletin*, Vol. 45, No. 13 (Aug., 1962), pp. 136–47.

sand soils to supply potassium and magnesium to plants is naturally low; for this reason, any medium and high tests for these elements on sand soils are usually the result of heavy applications of fertilizers.

For comparison with mineral soils, available plant nutrients are given for Michigan peat and muck soils in Table 1.3. Phosphorus is predominantly low to medium, available potassium averages mostly in the medium range, and available magnesium and calcium are mostly high.

NUTRIENT-SUPPLYING POWER OF SOILS

Before the general availability of commercial fertilizers, farmers depended chiefly on the natural nutrient-supplying power of soils, supplemented by a small amount of farmyard manure. Today, in most parts of the world, this condition still exists. However, in the United States and Canada, research and demonstration results have indicated that the lower

TABLE 1.3 RANGE OF AVAILABLE PLANT NUTRIENTS IN ORGANIC SOILS IN MICHIGAN *

Nutrient	Range in Pounds of Nutrients Per Acre			
	0–40	41–70	71–150	More than 150
	Percentage of Samples in Each Range			
Phosphorus (P)	51.2	22.2	23.5	3.1
	Range in Pounds of Nutrients Per Acre			
	0–100	101–200	201–400	More than 400
	Percentage of Samples in Each Range			
Potassium (K)	12.9	38.9	31.5	16.7
Magnesium (Mg)	3.1	4.3	8.7	83.9
	Range in Pounds of Nutrients Per Acre			
	0–2000	2000–4000	4000–6000	More than 6000
	Percentage of Samples in Each Range			
Calcium (Ca)	1.8	14.8	26.8	56.6

* *Source:* J. C. Shickluna and R. E. Lucas, "The pH, Phosphorus, Potassium, Calcium, and Magnesium Status of Organic Soils in Michigan," *Michigan State University Quarterly Bulletin,* Vol. 45, No. 3 (1963), pp. 417–25.

the nutrient-supplying power of soils for a particular element, the greater amount of that element as fertilizer that must be added to obtain maximum yields and profits (Figure 1.5).

The nutrient-supplying power of sand and silt is very low because these are composed mostly of relatively undecomposed primary minerals. By contrast, clay has a greater power to supply nutrients to plants because it is composed of secondary weathered minerals (kaolinite, montmorillonite, and illite), and hydrous oxides of iron and aluminum. The nutrient-supplying power of clays also vary depending upon the kinds.

The nitrogen, phosphorus, and sulfur in soils are constituents of organic matter. Approximately 95 per cent of the total soil nitrogen, 5 to 60 per cent of the total soil phosphorus, and 10 to 80 per cent of the total soil sulfur are present in the organic matter. These three nutrients become available to plants only after biological decomposition. The nutrient-supplying power of the soil for almost all of the nitrogen and for variable amounts of phosphorus and sulfur is therefore dependent upon the total amount of organic matter and its rate of decomposition in relation to the

FIG. 1.5. The nutrient-supplying power of soils in Illinois for phosphorus and potassium are shown here as generalized areas. Soils low in supplying-power of phosphorus or potassium must have more of the particular element added as fertilizer to obtain maximum economic yields. On a field-by-field basis, this information can be obtained from a soil test that has been properly interpreted. *Source:* D. W. Graffis, "Going It Alone," *Better Crops with Plant Food Magazine,* No. 4, 1969.

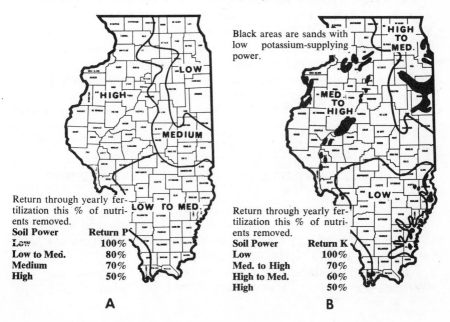

Black areas are sands with low potassium-supplying power.

Return through yearly fertilization this % of nutrients removed.

Soil Power	Return P
Low	100%
Low to Med.	80%
Medium	70%
High	50%

Return through yearly fertilization this % of nutrients removed.

Soil Power	Return K
Low	100%
Med. to High	70%
High to Med.	60%
High	50%

A   B

time that plants require these nutrients. Factors that favor organic decomposition are high temperatures, a favorable oxygen and moisture supply, and an abundant supply of nutrients that are required by bacteria.

In addition to organic matter decomposition as a source of sulfur to plants, rainwater returns to the earth about 5 or more pounds of sulfur per acre per year.

Phosphorus is taken into the plant from the soil solution in the water-soluble form. The concentration of phosphorus in the soil solution is then replenished from slowly soluble forms such as calcium phosphates, hydroxyapatite, fluorapatite, iron phosphates, and aluminum phosphate, and from the decomposition of organic matter.

Potassium occurs in the soil mostly as a constituent of primary minerals that exist in sand and silt particles. Such minerals as orthoclase and microcline release potassium slowly, but others, such as muscovite (white mica) and biotite (black mica) are important sources of potassium for plant nutrition, especially in tropical areas where temperatures and moisture are high. Some of the potassium that is released by the decomposition of minerals will remain in the soil solution and be available to plants, and some will be adsorbed on the surfaces of clay and humus particles in a readily exchangeable and available form. Of the total potassium in a soil, approximately 1 per cent will be in the exchangeable form; of this amount, from 1 to 5 per cent will be in the soil solution.

Calcium and magnesium in soils occur in both slowly and readily soluble minerals and rocks. Slowly soluble sources of calcium include oligoclase (a sodium and calcium silicate) and anorthite (a calcium alumino silicate), whereas readily soluble sources include calcite and limestone. The slowly soluble sources of magnesium are augite, biotite, and hornblende. A readily soluble source of both calcium and magnesium is dolomitic limestone. Calcium and magnesium are supplied to plants in ionic form in a manner similar to that of potassium—that is, from rocks and minerals to the soil solution and to an exchangeable form on the surfaces of clay and humus particles.

SOIL AND PLANT TESTS FOR AVAILABLE PLANT NUTRIENTS

In all 50 states and in Puerto Rico, state and commercial soil testing laboratories tested more than 3.5 million soil samples and more than 300,000 plant samples in 1968. The modal charge for a soil test was $2.00 per sample, and for a plant analysis, $5.00.

States ranking highest in numbers of soils tested with more than 100,-000 each are Illinois, 700,000; Indiana, 253,000; Wisconsin, 225,000;

Georgia, 192,000; North Carolina, 185,000; Iowa, 180,000; Minnesota, 171,000; Ohio, 150,000; Michigan, 143,000; and California, 105,000.

More than 10,000 plant analyses each were reported from California, 115,000; Ohio, 44,000; Puerto Rico, 29,960; New York, 27,700; Maine, 22,800; Pennsylvania, 13,600; and Oregon, 10,400.

AREAS OF MICRONUTRIENT DEFICIENCY IN PLANTS [11]

A report from the 50 states shows that micronutrient deficiencies for crop plants are becoming more common because crop yields per acre are increasing, soils are becoming more depleted, fertilizers are becoming increasingly higher in N, P, and K, and correspondingly lower in secondary nutrients and micronutrients, and smaller amounts of animal manures are being used on the farm.

Boron deficiency has been reported in one or more crops in 41 states, zinc deficiency in 30 states, iron and manganese deficiencies each in 25 states, molybdenum in 21, and copper deficiency in 13. The number of states indicating specific micronutrient deficiencies by crops is given in Table 1.4.

Animals throughout the United States and over the entire world are slow growing, crippled, barren, or die young when forages do not have sufficient protein or essential elements. Sometimes toxic substances, such as organic substances, excess selenium, or excess molybdenum, restrict successful livestock production. To overcome some known or unknown deficiency, animals may chew on bones or the bark of trees (Figures 1.6 and 1.7) or lick areas of natural accumulation of salts (Figure 1.8).

AREAS OF MINERAL TOXICITY TO PLANTS

Although boron deficiency has been reported on one or more crops in 41 states, primarily in areas of high rainfall, in many areas of low rainfall certain irrigation waters and a few soils contain toxic quantities of boron as water-soluble sodium borate. The permissible level of boron in irrigation waters for use on sensitive crops such as beans and on fruit and nut trees

---

[11] Kermit C. Berger, "Micronutrient Deficiencies in the United States," *Journal of Agricultural and Food Chemistry* (May–June, 1962), pp. 178–81; J. C. Shickluna, J. R. Miller, C. L. W. Swanson, G. W. Hardy, A. R. Halvorson, T. W. Scott, and R. D. Munson, *A Survey on Micronutrient Deficiencies in the U. S. A. and Means of Correcting Them,* Soil Testing Committee, Soil Science Society of America, Special Publication (March, 1965).

TABLE 1.4 EXTENT OF MICRONUTRIENT DEFICIENCIES IN PLANTS IN THE UNITED STATES *

Micronutrient and Crop	No. of States Reporting Deficiency	Micronutrient and Crop	No. of States Reporting Deficiency
Boron		*Copper*	
Alfalfa	38	Onion	7
Cruciferae	25	Small grains	4
Fruit trees	21	Corn	3
Clover	13	Fruit trees	3
Beet	12	Grasses	3
Celery	10		
		Manganese	
Iron		Bean	13
Fruit trees	11	Small grains	10
Shrubs	11	Fruit trees	9
Grasses	7	Spinach	8
Shade trees	7	Corn	5
Bean	5		
Corn	3	*Zinc*	
		Corn	20
Molybdenum		Fruit trees	12
Alfalfa	13	Nut trees	10
Cruciferae	9	Bean	7
Clover	6	Onion	4
Soybean	3	Potato	3

* Deficiencies were also observed on many other crops but with less frequency.

FIG. 1.6. Many soils are deficient in phosphorus and as a result the phosphorus content of plants is also low. Rodents that eat plants low in phosphorus require additional phosphorus, which they may get from chewing on bones. (East Texas.) *Source:* Ext. Serv., Texas A&M University.

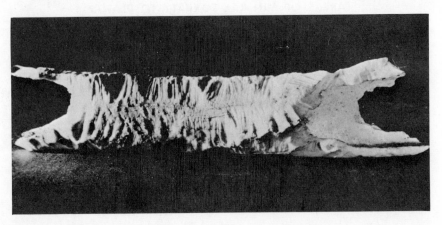

FIG. 1.7. The forage available to these cows in Minnesota does not contain sufficient phosphorus because the soil is deficient in phosphorus. The result is weakened animals that chew on bones or on the bark of trees. *Source:* Ext. Serv., U. of Minn.

FIG. 1.8. When animals do not get enough minerals from the grasses that they eat, they seek other sources. Here cattle, sheep, and goats supplement their intake of essential minerals by licking this layer of soil where salts have accumulated. (Ethiopia.) *Photo by Roy L. Donahue.*

is 0.66 parts per million (p.p.m.). On semitolerant crops (which include most crops), the limit is 1.33 p.p.m., and for use on tolerant crops, such as alfalfa, sugar beets, onions, carrots, and cabbages, the maximum limit is 2.0 p.p.m.

Lithium toxicity has more recently been reported on citrus trees in southern California. In a 1961 survey, approximately 5 per cent of the citrus acreage showed symptoms of lithium toxicity—that is, leaf margin mottling that changes to necrosis (death of mottled parts) in advanced deficiency symptoms. Citrus that contained more than 12 p.p.m. of lithium on a dry-weight basis showed symptoms of lithium toxicity. In some areas, as much as 40 p.p.m. of lithium in the leaves was reported.[12]

AREAS OF MINERAL DEFICIENCY AND TOXICITY
FOR LIVESTOCK AND MAN

It is often true that nutrients may be adequate for normal plant growth but not in sufficient concentration in the plant for adequate livestock nutrition. Likewise, a plant can absorb certain elements which it may or may not need and which do no harm to the plant but are toxic to livestock.

The forage growing in many soils in the following states does not contain sufficient phosphorus for adequate livestock nutrition:

Delaware	Montana	Oregon
Florida	Nevada	South Carolina
Idaho	New Mexico	Tennessee
Kentucky	New York	Texas
Michigan	North Carolina	Virginia
Minnesota	North Dakota	

Cobalt has not yet been added to the list of elements essential for plants, but animals require it in their nutrition. Cobalt deficiency in livestock has been reported in these states because the soils and forage plants do not contain a sufficient amount of cobalt (Figure 1.9):

Arkansas	Louisiana	New Hampshire
Delaware	Massachusetts	New Jersey
Florida	Michigan	New York

---

12 G. R. Bradford, "Lithium Toxicity in Southern California Citrus," *California Agriculture*, Vol. 15, No. 12 (1961).

North Carolina   South Carolina   Virginia
Oregon   Vermont

A deficiency of iodine in soils results in inadequate iodine in plants. Without additional iodine, livestock and man, who depend upon these plants for feed and food, may develop a swollen neck condition known as *goiter*. Many soils in the following states are deficient in iodine:

California	Michigan	Pennsylvania
(northeastern)	Minnesota	(northern)
Colorado	Montana	South Dakota
Idaho	Nevada	Washington
Indiana	North Dakota	West Virginia
Iowa	Ohio	Wisconsin
	Oregon	Wyoming

Although iodine can be used as a fertilizer on the soil or sprayed on plant foliage, the simplest cure of iodine deficiency for both livestock and man is for them to eat iodized salt.

Copper deficiency in livestock due to inadequate copper in the soil and plants has been reported in:

Florida   Oregon   Virginia

FIG. 1.9. Areas of cobalt (Co) deficiency and molybdenum (Mo) toxicity for livestock. *Source:* Joe Kubota, "Trace Element Studies Link Animal Ailments with Soils," *Soil Conservation,* Vol. 35, No. 4, Nov., 1969, p. 87.

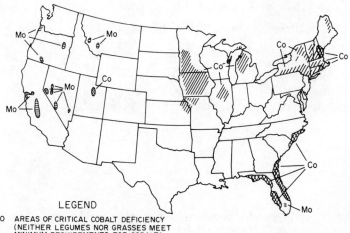

LEGEND

CO   AREAS OF CRITICAL COBALT DEFICIENCY
(NEITHER LEGUMES NOR GRASSES MEET
MINIMUM REQUIREMENTS FOR COBALT)

CO   AREAS OF SPORADIC COBALT DEFICIENCY
(LEGUMES BUT NOT GRASSES MEET
MINIMUM REQUIREMENTS FOR COBALT)

MO   GENERAL AREAS OF MOLYBDENUM TOXICITY
(LEGUMES HAVE 10 PPM OR MORE OF
MOLYBDENUM)

U.S. PLANT, SOIL AND NUTRITION LABORATORY,
ARS, USDA
AND
SOIL SURVEY INVESTIGATIONS, SCS, USDA

J. KUBOTA – 1969

Molybdenum toxicity has been officially observed in livestock in (Figure 1.9):

California	Florida	Texas
Colorado	Nevada	Utah
	Oregon	

Plants absorb selenium even though it is not essential for them; animals obtain their essential or toxic levels of selenium by eating plants.

Selenium toxicity in livestock has been observed in the northern Great Plains since 1857, but the cause was understood first in 1928. In recent years, selenium has been shown to be an essential element for lambs, calves, chickens, and turkeys, but the *essential* level and the *toxic* level are very narrow. Feeds should contain as much as 0.1 p.p.m. selenium but no more than 4 p.p.m. Soils containing as little as 1 p.p.m. of selenium may produce vegetation with more than 4 p.p.m. of selenium, which is toxic to most animals (Figure 1.10).

Although not a mineral, *nitrate* toxicity of forages was first recognized in Kansas in the 1890's when steers became dizzy after eating corn stalks grown on a former corral site where the soil was very high in nitrates.

FIG. 1.10. Map of the United States showing areas of relative selenium concentration in plants in relation to its adequacy or toxicity for animals. *Source: The Yearbook of Agriculture* (1968), U.S. Dept. of Agr., p. 366.

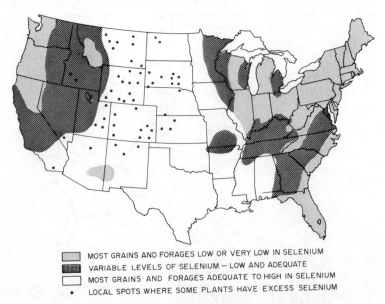

MOST GRAINS AND FORAGES LOW OR VERY LOW IN SELENIUM
VARIABLE LEVELS OF SELENIUM – LOW AND ADEQUATE
MOST GRAINS AND FORAGES ADEQUATE TO HIGH IN SELENIUM
• LOCAL SPOTS WHERE SOME PLANTS HAVE EXCESS SELENIUM

FIG. 1–10

The level of nitrates in plants tends to accumulate to toxic levels for animals and man under the following conditions:

1. When plants are stunted by drought, high temperatures, low soil phosphorus, or shading.
2. When the soil level of nitrates is excessively high, caused by large applications of animal manures or nitrogenous fertilizers.
3. When plants are young and fast-growing.
4. When certain plants are used for forage, such as sudangrass and grain sorghum.[13]

SOILS—POLLUTION—ENVIRONMENT

The affluent, "throw away" society of the United States has now reached a stage in which the waste products are ugly if *not* burned and pollute the environment with smoke if they *are* burned. Furthermore, many chemicals are so durable that they are not biodegradable; thus they degrade the environment. Pesticides, detergents and fertilizers are so relatively cheap that they may be used in excess quantities which cause pollution. Sediment from excessive and careless use of soil on farms, highways, and urban developments chokes the water courses.

The soil is also an agent of control of pollution of the environment by its capacity to serve as a filter, chemical combiner, and biological supplier of bacteria and fungi to decompose and destroy useless organic debris.

More and more effective answers must be found by massive research on the relationships among soils, pollution, and environment.

CONTINUOUS RESEARCH NEEDED

Research on the subject of soils and plant environment is continuing at an accelerated pace. The new technique of using radioactive tracer elements in the study of plant nutrition and the efficient use of fertilizers have already added greatly to the knowledge of the uptake of plant nutrients in relation to ecology and environment.

However, many unsolved problems exist that will require the best research talent. Five such problems are as follows:

1. How to reduce losses of nitrogen by denitrification and leaching.
2. How to maintain the nitrate content in plants below the level toxic to animals and man.

---

[13] Larry S. Murphy, "Nitrate Toxicity, Guidelines for Livestock Men," *Crops and Soils Magazine* (June–July, 1968), pp. 11–13.

3. How to make a larger percentage of soil and fertilizer phosphorus available to plants.
4. How to make the plentiful but slowly soluble potassium in soil minerals more readily available to plants.
5. How to maximize the use of the soil for pollution control.

SUMMARY

Soil is the original source of most nutrients for all plant and animal life. Water and air are the other original sources. Most soils need lime, fertilizers, drainage, or irrigation to produce satisfactory crops. Crops receive from the soil mechanical support, essential elements, water, and oxygen. There are at present 16 elements known to be essential for the growth of crop plants. Water and air in the soil alternately occupy the same pore spaces; after a rain, most of the pores contain water, but in a few days air gradually replaces some of the water.

The value of manure, cover crops, and legumes was recognized before the time of Christ. Early scientists and philosophers looked for the "principle" of vegetation, but centuries later scientists concluded that there were many "principles." Liebig in 1840 was among the first of the scientists to explain plant nutrition in a manner somewhat similar to the way it is known today. Then the Rothamsted Experiment Station was established in 1843 near London, England, and field plot techniques as a scientific instrument came into being.

A new system of soil classification has been developing since 1951 and was officially adopted in the United States in 1965.

Soil granules are held together by plant roots, the vegetative parts of fungi and actinomycetes, slimes from such animal life as earthworms, and waxes and resins from the partial decomposition of bacterial bodies and the remains of higher plants.

Plant roots take in nutrients by an exchange of cations and anions from the soil solution and the surfaces of clay and humus particles.

The soil is the primary source of 13 of the 16 elements essential for plant growth; therefore, the nutrient-supplying power of the soil is of prime importance. Each state has available to its citizens a soil and plant testing service. Micronutrient deficiencies in plants are becoming increasingly common. Boron and lithium may be present in water or soil in sufficient concentration to be toxic to plants. Phosphorus, cobalt, copper, and iodine may be so deficient in soils and plants that they may cause deficiency symptoms in animals or man, and molybdenum and selenium in soils and plants may effect toxicity symptoms in livestock.

Man's environment is becoming polluted and therefore less favorable for habitation. Soils and soil management may *add to this pollution* by eroded silt to fill lakes and rivers and by excessive use of fertilizers. Soils may also aid in *controlling pollution* by absorbing many pollutants and by providing a "home" for many bacteria and fungi that aid in decomposing otherwise harmful substances.

Continuous research is needed to solve new problems in soil science in relation to plant, animal, and human ecology and environment.

QUESTIONS

1. Name the essential elements that come only from the soil.
2. What was Liebig's "law of the minimum"?
3. Briefly discuss the history of soil classification.
4. Discuss the importance of the nutrient-supporting power of soils.
5. Explain how soil can be a pollutant as well as how it can be used to control pollution.

REFERENCES

Aldrich, Daniel G., Jr. (ed.), *Research for the World Food Crisis,* American Association for the Advancement of Science, Publication No. 92. Wash. D.C., 1970, 323 pp.

Alexander, Martin, *Introduction to Soil Microbiology.* New York: John Wiley & Sons, Inc., 1961.

Animal Diseases: The Yearbook of Agriculture (1956). United States Department of Agriculture.

Bartelli, L. J., A. A. Klingebiel, J. V. Baird, and M. R. Heddleson (eds.), *Soil Surveys and Land Use Planning.* Madison, Wisc.: Soil Science Society of America and American Society of Agronomy, 1966.

Bear, Firman E., *Earth, The Stuff of Life.* Norman, Okla.: University of Oklahoma Press, 1961.

———, *Soils in Relation to Crop Growth.* New York: Reinhold Publishing Corp., 1965.

Beeson, Kenneth C., *The Mineral Composition of Crops with Particular Reference to the Soil in Which They Were Grown,* Miscellaneous Publication 369, United States Department of Agriculture, 1941.

Berger, Kermit C., *Introductory Soils.* New York: The Macmillan Company, 1965.

Buckman, Harry O., and Nyle C. Brady, *The Nature and Properties of Soils.* (7th ed.). New York: The Macmillan Company, 1969.

Carter, D. L., M. J. Brown, W. H. Allaway, and E. E. Cary, "Selenium Content of Forage and Hay Crops in the Pacific Northwest," *Agronomy Journal.* Vol. 60, No. 5 (Sept.–Oct., 1968), pp. 532–34.

Clark, Raymond M., and Henry D. Foth, *Understanding Our Soils,* Professional Series Bulletin No. 35. East Lansing, Mich.: Bureau of Educational Research, Michigan State University, 1958.

Cobalt Deficiency in Soils and Forages: How It Affects Cattle and Sheep, Leaflet No. 488. United States Department of Agriculture, 1961.

Cook, R. L., *Soil Management for Conservation and Production.* New York: John Wiley & Sons, Inc., 1962.

Donahue, Roy L., *Our Soils and Their Management* (3rd ed.). Danville, Ill.: The Interstate Printers & Publishers, Inc., 1970.

Food: The Yearbook of Agriculture (1959). United States Department of Agriculture.

Hardy, Glen W., A. R. Halvorson, J. B. Jones, R. D. Munson, R. D. Rouse, T. W. Scott, and Benjamin Wolf (eds.), *Soil Testing and Plant Analysis,* Special Publication No. 2. Madison, Wisc.: Soil Science Society of America, 1967.

Hilgard, E. W., *Soils.* New York: The Macmillan Company, 1911.

Investigating the Earth, Earth Science Curriculum Project. Boston: Houghton Mifflin Company, 1967.

Jenny, Hans., "E. W. Hilgard and the Birth of Modern Soil Science," Collanda Della Rivista *Agrochimica.* Pisa, Italy: 1961.

Kellogg, Charles E., *Our Garden Soils.* New York: The Macmillan Company, 1952.

————, "Potentials for Food Production," *The Farmer's World: The Yearbook of Agriculture* (1964). United States Department of Agriculture, pp. 57–69.

Know Your Soil, Agricultural Information Bulletin No. 267. Washington: United States Department of Agriculture, Soil Conservation Service, 1963.

Leet, Don L., and Sheldon Judson, *Physical Geology* (3rd ed.). Englewood Cliffs, N.J.: Prentice-Hall, Inc., 1965.

Millar, C. E., L. M. Turk, and H. D. Foth, *Fundamentals of Soil Science* (4th ed.). New York: John Wiley & Sons, Inc., 1965.

Retzer, John L., "Soil Formation and Classification of Forested Mountain Lands in the United States," *Soil Science.* Vol. 96, No. 1 (July, 1963).

Russell, Sir E. John, *The World of the Soil* (2nd ed.). London: Collins Press, 1959.

Russell, E. Walter, *Soil Conditions and Plant Growth* (9th ed.). London: Longmans, Green & Company, Ltd., 1961.

Scarseth, George D., *Man and His Earth.* Ames, Iowa: Iowa State University Press, 1962.

Science for Better Living: The Yearbook of Agriculture (1968). United States Department of Agriculture.

Shickluna, John C., *A Survey on Micronutrient Deficiencies in the U.S.A. and Means of Correcting Them.* Madison, Wisc.: The Soil Testing Committee, Soil Science Society of America, 1965.

————, "The Relationship of pH, Available Phosphorus, Potassium, and Magnesium to Soil Management Groups," Article 45–13, *Quarterly Bulletin of the Michigan Agricultural Experiment Station.* Vol. 45, No. 13 (1962), pp. 136–47.

Soil: The Yearbook of Agriculture (1957). United States Department of Agriculture.

Soil Taxonomy of the National Cooperative Soil Survey. Soil Conservation Service, United States Department of Agriculture, Wash. D.C., Dec., 1970.

Sprague, Howard B. (ed.), *Hunger Signs in Crops* (3rd ed.). New York: David McKay Co., Inc., 1964.

CHAPTER 2

PHYSICAL PROPERTIES OF SOILS AND PLANT ENVIRONMENT

All you need do was dig a hole. If the dirt you took out went back in with room to spare, your ground would grow most anything*

The soil is a complex mechanical system. For a soil to be in good physical condition for plant growth, the air, water, and solid particles must be in the right proportions at all times. Every cubic inch of soil that is expected to support plant life must be:

1. Open enough to permit the right amount of rainwater or irrigation water to enter the soil but not so open as to allow excessive loss of water and plant nutrients by deep percolation.
2. Sufficiently retentive of moisture to supply roots with all needed water but not so retentive as to create undesirable suspended water tables.
3. Well enough aerated to permit all plant root cells to obtain oxygen at all times, but not excessively aerated to the point of preventing a continuous contact of roots with moist soil particles (Figure 2.1).

To understand how an engine functions, it is necessary to take the engine apart and study each separate piece. The same is true with a soil.

---

* Conrad Richter in *The Trees* (New York: Alfred A. Knopf, Inc., 1940). (Reproduced with permission of the publisher.)

FIG. 2.1. Before a soil can serve as a suitable environment for plant roots, it must be physically compatible. This means that at all times the air, water, and solid particles must be in the right proportion. But contrary to popular belief, not all soils, even when virgin, were physically compatible for all plants, as depicted here with the root systems of longleaf pine in Georgia and Florida. *Above:* When the soil is physically suitable, the root of longleaf pine will extend downward for perhaps 20 feet. (Shown here in northern Florida at a depth of 10 feet.) *Below:* Because of a naturally compact silty clay layer at a depth of four feet, this root of longleaf pine in southern Georgia was stopped and blunted when it reached this compact layer. *Source:* J. O. Veatch.

When a soil is taken apart and each part studied, the process is called a *mechanical analysis.*

MECHANICAL ANALYSIS

When scientists take a soil sample apart, they usually find that it is composed of:

1. Large pebbles and stones
2. Coarse sand
3. Fine sand
4. Something resembling flour (silt)
5. Lumps or clods of varying sizes, which consist of clusters of soil particles cemented by organic matter and clay
6. Plant roots
7. A dark substance spread throughout the soil mass, called *humus*
8. Dead leaves and twigs
9. Ants, earthworms, and other forms of animal life

Not all of these substances are properly a part of the mechanical analysis of the soil. For this purpose, only the mineral matter which is less than 2 millimeters in diameter is to be considered as soil. This means that, before a mechanical analysis is made, the soil sample must be screened through a 2-millimeter sieve. All the rocks, pebbles, leaves, and plant roots that do not go through the 2-millimeter screen are discarded.

In like manner, the humus is not a part of an official mechanical analysis. If the humus percentage is fairly large, it must be destroyed before the determinations are made. Treatment with hydrogen peroxide will destroy most of the humus.

After the soil sample has been screened through a 2-millimeter sieve and, if necessary, the humus destroyed, a particle size-distribution analysis can be made. The purpose of a mechanical analysis is to determine the amounts of *individual* soil grains of the various sizes. To get an accurate estimate of the percentages of each group of individual (primary) soil particles, it is necessary to disperse completely all lumps or aggregates so that they can be separated into their primary groups. If complete dispersion were not obtained, a small lump of clay the same size as a sand grain would be reported as sand in the results of the mechanical analysis.

There are several methods of mechanical analysis, but only two have wide acceptance: the *Pipette Method* and the *Bouyoucos Hydrometer*

Method. Both methods are based upon the differential rate of settling of soil particles in water, and accuracy of the methods depends upon these conditions or assumptions:

1. Soil is completely dispersed in water.
2. Soil is in a dilute suspension in water so that the soil grains can settle without bumping into or otherwise influencing each other.
3. All soil particles settle as if they were smooth and rigid spheres.
4. The rate of settling of the particles is assumed not to be influenced by the walls of the settling vessel.
5. The temperature of the soil-water suspension is constant and known.
6. All soil particles are of the same density.

SOIL SEPARATES

A mechanical analysis reports the percentages of different size groups of particles. From coarse to fine, these size groups (soil separates) and their diameter ranges, according to the system of the U.S. Department of Agriculture, are given in Table 2.1. For convenience in memorizing the size ranges, a comparison is shown of the diameter range of the soil separates with the United States monetary system (Figure 2.2).

TABLE 2.1 THE SOIL SEPARATES AND THEIR DIAMETER RANGE, COMPARED WITH THE U.S. MONETARY SYSTEM

Soil Separate	Diameter Range (mm)	Monetary System (Dollars)
Very coarse sand *	2.0–1.0	2.00–1.00
Coarse sand	1.0–0.5	1.00–0.50
Medium sand	0.5–0.25	0.50–0.25
Fine sand	0.25–0.10	0.25–0.10
Very fine sand	0.10–0.05	0.10–0.05
Silt	0.05–0.002	0.05–
Clay	< 0.002	—

* Before 1947, this soil separate was known as fine gravel. When comparing recent mechanical analysis data with that made before Jan. 1, 1938, it should be noted that before 1938, silt was defined as particles between 0.05–0.005 mm in diameter and *clay* as less than 0.005 mm. Particles less than 0.002 mm in diameter were called *colloid.*

Natural field soils are always mixtures of soil separates. The relative percentages of the seven soil separates in a field soil are almost infinite in

FIG. 2.2. Students in an introductory soils laboratory making a mechanical analysis of a soil by the Bouyoucos hydrometer method. *Source:* Agronomy Dept., U. of N.H.

possible combinations. Soils may contain from less than 1 per cent to almost 100 per cent of any soil separate. To bring order to this apparently chaotic situation, the U.S. Department of Agriculture has established limits of variations among the soil separates and has assigned a textural name to each group. For example, soils containing a large amount of sand are *sandy,* those with a high content of silt are *silty,* and soils containing a high percentage of clay are *clayey.* When the soil does not exhibit the properties of either sand, silt, or clay, it is called a *loam.* The limitations in the range of each textural name were established upon significant differences in the physical properties of each textural class.

SOIL TEXTURE

Soil texture refers to the relative percentages of sand, silt, and clay in a soil. The size of the sand grains further modifies the textural name. The more common soil textural names, listed in order of increasing fineness, are the following: [1]

Sand	Silt loam	Silty clay loam
Loamy sand	Silt	Sandy clay
Sandy loam	Sandy clay loam	Silty clay
Loam	Clay loam	Clay

For convenience in determining the textural name of a soil from the mechanical analysis, an equilateral triangle has been adapted for this

---

[1] The terms "clay," "silt," "very fine sand," "fine sand," and "coarse sand" are used as names for soil separates as well as for specific soil textural classes.

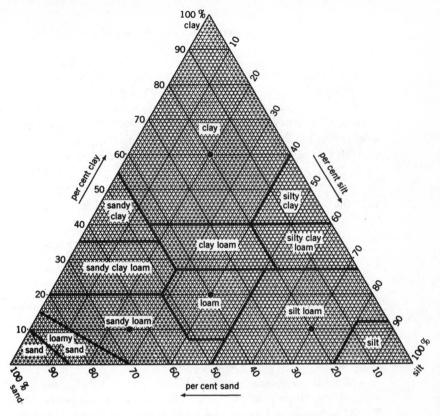

FIG. 2.3. Graphic guide for textural classification. Because of the small scale of the chart, it is not possible to show all recognized soil textures in the sandy range. For this purpose it is necessary to use verbal definitions, given in Appendix D under soil texture.

purpose. Figure 2.3 is a guide for a textural classification. The left angle represents 100 per cent sand, the right 100 per cent silt, and the top angle 100 per cent clay.

The textural name of a soil may be obtained from the results of a mechanical analysis in this way. Assume that the mechanical analysis of a soil is as follows, in percentages: sand, 40; silt, 40; and clay, 20. To find the textural name of this soil, locate the 40 along the bottom of the triangle, then proceed upward to the left along the heavy line. All points along this 40 line are 40 per cent sand. To locate the 40 per cent silt line, look along the right side of the triangle labeled "per cent silt" until you

Note: For use in the soil family name in the new United States System of Soil Taxonomy, soil textural classes have been grouped into broad particle-size classes. These are detailed in: *Soil Taxonomy of the National Cooperative Soil Survey.* Soil Conservation Service, U.S.D.A. (Dec., 1970), pp. 3-43, 3-44.

come to 40. Follow down and to the right along the heavy line until this line intersects the 40 per cent line of the sand. Now check this point of intersection by locating the 20 per cent point along the left side of the triangle (per cent clay). Follow this 20 per cent line to the right; it should intersect at the same point where the 40 per cent sand line and the 40 per cent silt line intersected. This point (at dot) is the midpoint of the textural name *loam*.

Follow the same instructions and confirm the textural names of soils with the following mechanical analyses, expressed in percentages: [2]

Sand	Silt	Clay	Textural Class
65	25	10	Sandy loam
20	20	60	Clay
20	70	10	Silt loam

COARSE FRAGMENTS

Mineral fragments in the soil larger than 2 millimeters in diameter are not strictly a part of the soil but must be recognized because they greatly influence the use of the land. In the classification and mapping of soils, the names of the large fragments are attached to and precede the textural name of the soil. For example, if a sandy loam contains a large amount of gravel, the textural name becomes *gravelly* sandy loam.

The commonly accepted classification of the coarse fragments in soils is set forth in Table 2.2.

Stoniness refers to the relative proportion of loose stones more than 10 inches in diameter in a soil, and *rockiness* refers to the relative proportion of bedrock.

ORGANIC SOILS

The preceding textural names are used for designating the mineral fractions in a soil. Under field conditions, materials are classified as *organic* under these conditions: [3]

---

[2] *Soil Classification: A Comprehensive System, Seventh Approximation,* Soil Survey Staff, Soil Conservation Service, U.S.D.A. (Aug., 1960), pp. 250–56.

[3] *Source: Supplement to Soil Classification System, Seventh Approximation: Histosols,* Soil Survey Staff, Soil Conservation Service, U.S.D.A. (Sept., 1968).

1. If the mineral fraction has 50 per cent or more of clay, there must be 30 per cent or more of organic matter.
2. If the mineral fraction has no clay, there must be 20 per cent or more of organic matter.
3. When the mineral fraction percentage is between 0 and 50, the amount of organic matter necessary to be classified as *organic* is proportionately between 20 and 30 per cent.

PARTICLE DENSITY

One assumption necessary in making a mechanical analysis is that all soil particles have the same density. Although the assumption is fairly accurate for this purpose, the differences in particle density are important in the determination of other soil constants, such as pore space.

Particle density is usually expressed as grams per cubic centimeter. Thus, if a solid soil particle were the shape of a cube, one centimeter on each edge, and it weighed 2.65 grams, it would have a particle density of 2.65.

The particle density of mineral soils over the world varies from approximately 2.60 to 2.80, with an average of approximately 2.65. Soils

TABLE 2.2   THE SHAPE, KIND, SIZE, AND NAME OF COARSE FRAGMENTS IN SOILS *

Shape of Fragment	Kind of Fragment	Size and Name of Fragment		
		Up to 3 in. in Diameter	3–10 in. in Diameter	More than 10 in. in Diameter
Rounded	Any kind	Gravelly	Cobbly	Stony
Angular	Chert	Cherty	Coarse Cherty	Stony
	Other than chert	Angular Gravelly	Angular Cobbly	Stony
		Up to 6 in. Long	6–15 in. Long	More than 15 in. Long
Thin, flat	Sandstone, limestone, or schist	Channery	Flaggy	Stony
	Slate	Slaty	Flaggy	Stony
	Shale	Shaly	Flaggy	Stony

* *Source: Soil Classification: A Comprehensive System, Seventh Approximation*, Soil Survey Staff, Soil Conservation Service, U.S.D.A. (Aug., 1960), pp. 254–56.

with a large amount of organic matter will have average particle densities of approximately 2.5 or below.

Accurate determinations of particle density may be made with a specific gravity bottle (pycnometer). An approximate method of determining the average particle density of a handful of pebbles is to:

1. Weigh the pebbles.
2. Drop the pebbles into a graduated cylinder which is partly full of water and observe the increase in volume. For example, if 300 grams of pebbles displaced 113 cubic centimeters of water, the particle density is obtained by dividing the 300 by 113, or approximately 2.65. This example merely illustrates a principle and is not suggested as an accurate method.

BULK DENSITY [4]

Bulk density is the density or weight of a given bulk (unit volume) of soil. The grams per cubic centimeter of soil, including the pores, is the bulk density.

It may be easier to visualize bulk density if a method for its determination is described. Obtain an iron cylinder, with open ends, which is slightly smaller than a pint ice cream container. One edge should be sharpened for ease of driving into the soil. Drive the cylinder into the soil so that the top is even with the soil surface. Dig out the buried cylinder, using care to slice off the bottom evenly without disturbing the soil core. Transfer the entire contents into a pint ice cream container and place in an oven to dry. (Drying this large a soil volume will take a week or more.) Weigh the soil daily after it has been in the oven for a week. When two successive daily weighings are the same, it can be assumed that the soil has reached oven dryness. The bulk density can be calculated by dividing the oven-dry weight in grams by the volume of the soil in cubic centimeters. Thus, if the oven-dry weight of the soil core is 630 grams and the volume of the same soil core is 450 cubic centimeters, the bulk density is obtained by dividing 630 by 450, or 1.4.

Tree growth on sand and loam that was loosened (less bulk density) was less than that on undisturbed soil. Loosened clay, however, supported the best growth of all treatments on clay. This indicates adequate natural pore space and aeration on sand and loam but inadequate aeration on undisturbed clay. On sand and loam, the best growth of loblolly pine was

---

[4] *Bulk density* was formerly called *volume weight* and *apparent specific gravity*.

on undisturbed (natural) soil; whereas, growth on natural clay soil was inferior to loosened clay. With increasing pressures, sand, loam, and clay exhibited decreasing growth rates (Figure 2.4).

PORE SPACE

The pore spaces in a soil consist of that portion of a given volume of soil not occupied by solids, either mineral matter or organic matter. Under field conditions, the pore spaces are occupied at all times by water and air.

The relative amounts of air and water in the pore spaces fluctuate almost hourly. During a rain, water drives air from pores, but as soon as water disappears by deep percolation, evaporation, and transpiration, air replaces the water which is lost.

The percentage of any given volume of soil occupied by pore space may be calculated from the formula:

$$\% \text{ pore space} = 100 - \frac{\text{Bulk density}}{\text{Particle density}} \times 100$$

It is easier to understand why this formula is correct by using the following line of reasoning:

$$\% \text{ solid space} = \frac{\text{Bulk density}}{\text{Particle density}} \times 100$$

and % pore space + % of solid space = 100%

so the % pore space = 100 — % solid space

and therefore $\% \text{ pore space} = 100 - \frac{\text{Bulk density}}{\text{Particle density}} \times 100$

SOIL STRUCTURE

The relative proportion of primary (individual) particles in a soil mass is known as *texture;* how these particles are grouped together into aggregates is *structure*. Natural aggregates are called *peds* and are fairly water stable; the word *clod* is restricted to an artificially formed aggregate that is usually not water stable. Plowing a clay soil when it is too wet will make the soil cloddy.

Two other terms are often confused with a ped. One is a *fragment,*

FIG. 2.4. Bulk density in relation to the growth of loblolly pine in Louisiana. Sand, loam, and clay soils were subjected to the following kinds of loosening or pressure and the resulting soil material was used to grow loblolly pine seedlings: 1. Soil loosened, 2. Soil undisturbed, 3. Soil subjected to static pressure of 3.5 kilograms per centimeter, 4. Soil subjected to static pressure of 7.0 kilograms per centimeter, 5. Soil subjected to static pressure of 10.5 kilograms per centimeter, 6. Soil puddled plus same pressure as in treatment 5. *Source:* R. Rodney Foil, La. State U.

which consists of a broken ped, and the other is a *concretion,* which is formed within the soil by the precipitation of salts dissolved in percolating waters.

Soil structural terms for peds are divided into three categories: [5]

1. Type (shape and arrangement of peds)
2. Class (size of peds)
3. Grade (degree of distinctness of peds)

Each of these three categories of soil structure will be discussed separately.

Type of Soil Structure (*Shape and Arrangement of Peds*)

There are four principal *types* of soil structure:

1. *Platy.* Peds exhibit a matted, flattened, or compressed appearance.
2. *Prismlike.* Peds exhibit a long verticle axis and are bounded by flattened sides.
3. *Blocklike.* Peds resemble imperfect cubes like baby blocks, but are usually smaller.
4. *Spheroidal.* Peds are imperfect spheres like marbles, but are usually smaller.

Class of Soil Structure (*Size of Peds*)

The five general *classes* of soil structure are:

1. Very fine or very thin
2. Fine or thin
3. Medium
4. Coarse or thick
5. Very coarse or very thick

Actual size designations of the peds vary with each *type* of soil structure.

Grade of Soil Structure (*Degree of Distinctness of the Peds*)

The four terms in common use to designate the *grade* of soil structure are:

---

[5] *Soil Classification: A Comprehensive System, Seventh Approximation,* Soil Survey Staff, Soil Conservation Service, U.S.D.A. (Aug., 1960), pp. 256–57.

1. *Structureless.* No noticeable peds. This may mean the condition exhibited by loose sand or the cementlike condition of some clay soils or cemented soils.
2. *Weak.* Peds are indistinctly formed.
3. *Moderate.* Peds are moderately well formed.
4. *Strong.* Peds are very well formed (Figures 2.5 and 2.6).

Soil structure influences many important properties of the soil, such as the rate of infiltration of water. Figure 2.7 portrays graphically the relationship between type of soil structure and the rate of infiltration. Both granular (spheroidal) and single-grain (structureless) soils have rapid infiltration rates; blocky (blocklike) and prismatic soils have moderate rates; and platy and massive soil structures result in slow infiltration rates.

SOIL CONSISTENCE [6]

Consistence refers to the attribute of cohesion and adhesion or resistance of soil to rupture or deformation. Although structure and consistence are interrelated, structure deals with the shape, size, and distinctness of natural aggregates (peds), whereas consistence has reference to the force required to rupture soil material or to the properties of a deformed soil mass.

Consistence may be described under three soil-moisture conditions, *wet, moist,* and *dry.*

Consistence Terms when Soil Is Wet

Consistence when soil is wet is determined at a moisture content slightly above the field capacity. When wet, the soil material is characterized by stickiness and plasticity.

Stickiness. Stickiness is the quality of adhesion to other objects. For field evaluation of stickiness, soil material is pressed between thumb and forefinger and its adherence is noted. Degrees of stickiness are described as follows:

Nonsticky: After release of pressure, practically no soil material adheres to thumb or finger.

Slightly sticky: After pressure, soil material adheres to both thumb and finger but comes off rather cleanly. The soil is not appreciably stretched when the digits are separated (Figure 2.8).

---

[6] *Soil Classification: A Comprehensive System, Seventh Approximation,* Soil Survey Staff, Soil Conservation Service, U.S.D.A. (Aug., 1960), pp. 257–58.

FIG. 2.5. An example of soil structure desirable for plant growth. *Type:* spheroidal. *Class:* medium. *Grade:* moderate. *Source:* Ezee Flow Division, Avco Distributing Corp.

FIG. 2.6. An example of soil structure undesirable for plant growth. *Type:* blocklike. *Class:* coarse. *Grade:* strong. *Source:* Ezee Flow Division, Avco Distributing Corp.

FIG. 2.7. Soil structure influences the infiltration rate of water, as shown. *Source: Irrigation on Western Farms,* U.S. Dept. of Interior and U.S. Dept. of Agr., Agr. Inf. Bul. No. 199, 1959.

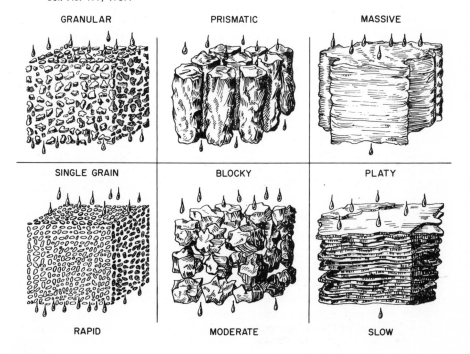

GRANULAR PRISMATIC MASSIVE

SINGLE GRAIN BLOCKY PLATY

RAPID MODERATE SLOW

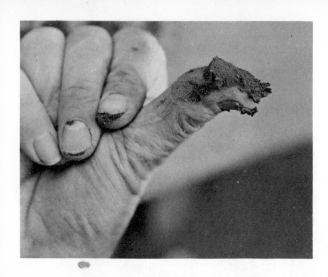

FIG. 2.8. The consistence of this wet soil is slightly sticky. *Source:* C. H. Diebold, Soil Conservation Service, U.S. Dept. of Agr.

Sticky: After pressure, soil material adheres to both thumb and finger and tends to stretch somewhat and pull apart rather than pulling free from either digit.

Very sticky: After pressure, soil material adheres strongly to both thumb and forefinger and is decidedly stretched when digits are separated.

Plasticity. Plasticity is the ability to change shape continuously under the influence of an applied stress and to retain the impressed shape on removal of the stress. For field determination of plasticity, roll the soil material between thumb and forefinger and observe whether or not a thin rod of soil can be formed. If it will be helpful to the reader of particular descriptions, state the range of moisture content within which plasticity continues: plastic when slightly moist or wetter; plastic when moderately moist or wetter; and plastic only when wet; or plastic within a wide, medium, or narrow range of moisture content. Express degree of resistance to deformation at or slightly above field capacity as follows:

Nonplastic: No wire is formable.
Slightly plastic: Wire is formable but soil mass is easily deformable.
Plastic: Wire is formable and moderate pressure is required for deformation of the soil mass.
Very plastic: Wire is formable and much pressure is required for deformation of the soil mass.

Consistence Terms when Soil Is Moist

Consistence when soil is moist is determined at a moisture content approximately midway between air dry and field capacity. At this moisture content, most soil materials exhibit a form of consistence characterized by: (1) a tendency to break into smaller masses rather than into powder; (2)

some deformation prior to rupture; (3) absence of brittleness; and (4) ability of the material after disturbance to cohere again when pressed together. The resistance decreases with moisture content, and accuracy of field descriptions of this consistence is limited by the accuracy of estimating moisture content. To evaluate this consistence, select and attempt to crush in the hand a mass that appears slightly moist.

Loose: Noncoherent.

Very friable: Soil material crushes under very gentle pressure; coheres when pressed together.

Friable: Soil material crushes easily under gentle to moderate pressure between thumb and forefinger; coheres when pressed together.

Firm: Soil material crushes under moderate pressure between thumb and forefinger; resistance is distinctly noticeable.

Very firm: Soil material crushes under strong pressure; barely crushable between thumb and forefinger.

Extremely firm: Soil material crushes only under very strong pressure; cannot be crushed between thumb and forefinger and must be broken apart bit by bit.

The term *compact* should be used only to denote a combination of firm consistence and close packing or arrangement of particles. It can be given the degrees of "very" and "extremely."

Consistence Terms when Soil Is Dry

The consistence of soil material when dry is characterized by rigidity, brittleness, maximum resistance to pressure, more or less tendency to crush to a powder or to fragments with rather sharp edges, and inability of crushed material to cohere again when pressed together. To evaluate, select an air-dry mass and break in the hand.

Loose: Noncoherent.

Soft: Soil mass is very weakly coherent and fragile; breaks to powder or individual grains under very slight pressure.

Slightly hard: Weakly resistant to pressure; easily broken between thumb and forefinger.

Hard: Moderately resistant to pressure; can be broken in the hands without difficulty but is barely breakable between thumb and forefinger.

Very hard: Very resistant to pressure; can be broken in the hands only with difficulty; not breakable between thumb and forefinger.

Extremely hard: Extremely resistant to pressure; cannot be broken in the hand.

Consistence Terms when Soil Is Cemented

Cementation of soil material refers to a brittle, hard consistence caused by some cementing substance other than clay materials, such as calcium carbonate, silica, or oxides or salts of iron and aluminum. Typically, the cementation is altered little, if at all, by moistening; the hardness and brittleness persist in the wet condition. Semi-reversible cements, which generally resist moistening but soften under prolonged wetting, occur in some soil and give rise to soil layers having a cementation that is pronounced when dry but very weak when wet. Some layers cemented with calcium carbonate soften somewhat with wetting. Unless the contrary is stated, descriptions of cementation imply that the condition is altered little, if at all, by wetting. If the cementation is greatly altered by moistening, it should be so stated. Cementation may be either continuous or discontinuous within a given horizon.

> **Weakly cemented:** Cemented mass is brittle and hard; can be broken in the hands.
>
> **Strongly cemented:** Cemented mass is brittle and hard; cannot be broken in the hand but is easily broken with a hammer.
>
> **Indurated:** Very strongly cemented; brittle, does not soften under prolonged wetting, and is so extremely hard that, for breakage, a sharp blow with a hammer is required; hammer generally rings as a result of the blow.

SOIL COLOR [7]

Soil colors are conveniently measured by comparison with the Munsell soil color chart that includes only the particular portion needed for soil colors, about one-fifth the entire range of color. The color chart consists of 175 different colored papers systematically arranged according to their Munsell notations. The arrangement is by *hue, value,* and *chroma*—the three simple variables that combine to give all colors. *Hue* is the dominant spectral (rainbow) color; it is related to the dominant wave length of light. *Value* refers to the relative lightness of color and is a function (approximately the square root) of the total amount of light. *Chroma* (sometimes called *saturation*) is the relative purity of strength of the spectral color and increases with decreasing grayness.

In the modified Munsell soil color chart, all colors on a given card

---

[7] *Soil Classification: A Comprehensive System, Seventh Approximation,* Soil Survey Staff, Soil Conservation Service, U.S.D.A. (Aug., 1960), pp. 249–53.

are of a constant hue, designated by the symbol in the upper right-hand corner of the card. Vertically, the colors become successively lighter by visually equal steps; their value increases. Horizontally, they increase in chroma to the right and become grayer to the left. The value and chroma of each color in the chart are printed immediately beneath the color. The first number is the value and the second, the chroma. As arranged in the chart, the colors form three scales: (1) radial, or from one card to the next, in hue; (2) vertical in value; and (3) horizontal in chroma.

The nomenclature for soil color consists of two complementary systems: color names and the Munsell notation of color. Alone, neither of these is adequate for all purposes. The color names are employed in all descriptions for publication and general use. The Munsell notation is used to supplement the color names wherever greater precision is needed—for example, as a convenient abbreviation in field descriptions, expression of the specific relations between colors, and statistical treatment of color data. The Munsell notation is especially useful for international correlation, since no translation of color names is needed. The names for soil colors are common terms now so defined as to obtain uniformity and yet be in accord, as nearly as possible, with past usage by soil scientists.

The Munsell notation for color consists of separate notations for hue, value, and chroma, which are combined in that order to form the color designation. The symbol for hue is the letter abbreviation of the color of the rainbow (R for red, YR for yellow-red or orange, Y for yellow) preceded by number from 0 to 10. Within each letter range, the hue becomes more yellow and less red as the numbers increase. The middle of the letter range is 5; the 0 point coincides with the 10 point of the next redder hue. Thus, 5YR is in the middle of the yellow-red hue, which extends from 10R (zero YR) to 10 YR (zero Y).

The notation for value consists of numbers from 0, for absolute black, to 10, for absolute white. Thus, a color of value 5/ is visually midway between absolute white and absolute black. One of value 6/ is slightly less dark, 60 per cent of the way from black to white and midway between values of 5/ and 7/.

The notation for chroma consists of numbers beginning at 0 for neutral grays and increasing at equal intervals to a maximum of about 20, which is never approached in soil. For absolute achromatic colors (pure grays, white, and black), which have zero chroma and no hue, the letter N (neutral) takes the place of a hue designation.

In writing the Munsell notation, the order is hue, value, chroma, with a space between the hue letter and the succeeding value number, and a virgule (/) between the two numbers for value and chroma. If expression beyond the whole numbers is desired, decimals are always used, never fractions. Thus, the notation for a color of hue 5YR, value 5, chroma 6, is 5YR 5/6, a yellowish-red. The notation for a value midway between the 5YR 5/6 and 5YR 6/6 chips is 5YR 5.5/6. The notation is decimal and capable of expressing any degree of refinement desired. Since color determinations cannot be made precisely in the field—generally no closer than half the interval between colors in the chart—expression of color should ordinarily be to the nearest color chip.

In using the color chart, a more accurate comparison is obtained by holding the soil sample above the color chips that are being compared. Rarely will the color of the sampe be perfectly matched by any color in the chart. It should be evident, however, between which colors the sample lies. The principal difficulties encountered in using the soil color chart are: (1) selecting the appropriate hue card; (2) determining colors that are intermediate between the hues in the chart; and (3) distinguishing between value and chroma where chromas are strong.

SOIL TEMPERATURE

Soil temperatures vary from perpetual frost at shallow depth in many soils in frigid Alaska to tropical Hawaii where daytime temperatures of the bare soil surface seldom fall below 100°F.

Over most of the earth, the daily soil temperature seldom changes below a depth of 20 inches. Below this depth, the soil temperature can be approximated by adding 2°F to the mean annual air temperature (°F).[8]

Temperature of the soil is a very important factor under the following situations:

1. To permit germination and growth of seeds, soil temperatures must be correct; for corn—45–85°F; for potatoes—60 to 70°F; and for sorghums and melons—above 80°F.
2. When applying anhydrous ammonia in the fall to reduce nitrification, soil temperatures should be below 50°F.
3. Freezing and thawing of bare, saturated, fine-textured soils may cause heaving of such crops as alfalfa.
4. Alternate freezing and thawing improves the structure of cloddy soils.
5. Cold soils tend to retard the absorption by the plant of phosphorus

---

[8] Guy D. Smith, Franklin Newhall, Luther H. Robinson, and Dwight Swanson, *Soil-Temperature Regimes—Their Characteristics and Predictability*, Soil Conservation Service, U.S.D.A., SCS–TP–144 (Apr., 1964), 14 pp.

from the soil. To increase phosphorus absorption, soils should be "warmed" by draining them and/or adding more phosphorus fertilizer.

6. The use of clear plastic surface soil covers in Alaska increase soil temperatures and permit a successful crop of corn and squash to be grown (Figure 2.9).

7. A black polyethylene mulch on the soil in a pineapple plantation in Hawaii was responsible for a 50 per cent increase in growth of pineapple. This increase in growth was caused by an increase in soil temperature of 2.7°F during winter and not to an increase in soil moisture.[9]

SUMMARY

To be in the proper physical condition to support luxuriant plant growth, the soil must be sufficiently open to permit water and air to circulate freely but not so open as to lower the necessary retention of water for plant growth.

---

[9] Paul C. Ekern, "Soil Moisture and Soil Temperature Changes with the Use of Black Vapor-Barrier Mulch and Their Influence on Pineapple Growth in Hawaii," *Soil Science Society of America Proceedings*, Vol. 1 (1967).

FIG. 2.9. These clear plastic strips are trapping the rays of the sun and warming the root environment, thus making the growth of this corn possible in Alaska. Source: Lee Allen, D. H. Dinkel, and Arthur L. Brundage, U. of Alaska Agr. Exp. Sta. *Agroborealis*, Vol. 1, No. 2, Sept., 1969.

To study the physical parts of a soil, it is necessary to make a mechanical analysis, using only the mineral particles less than 2 millimeters in diameter. These mineral particles are classified by size ranges into seven groups known as *soil separates*. From the relative percentages of each soil separate, a special equilateral triangle is consulted in assigning the proper textural name to a particular soil. Mineral particles larger than 2 millimeters in diameter, known as coarse fragments, are classified by shape, kind, and size.

For the mineral soil, particle density averages approximately 2.65 and bulk density, which includes pore space, 1.4.

Soil structure is classified by type (shape and arrangement of peds), class (size of peds), and grade (degree of distinctness of peds).

Soil consistence refers to the attribute of cohesion and adhesion or resistance of soil to rupture or deformation.

Soil color is best determined by comparison with the Munsell soil color chart, which consists of 175 different-colored chips arranged according to hue, value, and chroma.

Soil temperature is now considered as a soil attribute that can be approximated by adding 2°F to the mean annual air temperature.

QUESTIONS

1. Name the soil separates and their diameter ranges.
2. Name five soil textures in the order of increasing fineness.
3. Differentiate between bulk density and particle density.
4. Discuss the relationship between soil structure and infiltration.
5. Describe the Munsell color chart and its use.

REFERENCES

Bear, Firman E. (ed.), *Chemistry of the Soil* (2nd ed.). New York: Reinhold Publishing Corp., 1964.

Buckman, Harry O., and Nyle C. Brady, *The Nature and Properties of Soils* (7th ed.). New York: The Macmillan Company, 1969, 653 pp.

Dinkel, D. H., *Growing Sweet Corn in Alaska's Cool Environment*. University of Alaska Agricultural Experiment State Bulletin 39 (Nov., 1966).

Field Manual of Soil Engineering (5th ed.). Lansing, Mich.: State Highway Commission, Michigan Department of State Highways (Jan., 1970), 474 pp., $6.00.

Kilmer, V. J., and L. T. Alexander, "Methods of Making Mechanical Analysis of Soils," *Soil Science* (1949), 68:15–24.

Soil: The Yearbook of Agriculture (1957). United States Department of Agriculture.

Soil Classification: A Comprehensive System, Seventh Approximation. Soil Survey Staff, Soil Conservation Service, United States Department of Agriculture (Aug. 1960), pp. 257–58.

Soil Survey Manual, United States Department of Agriculture Handbook No. 18. United States Department of Agriculture, 1962.

CHAPTER 3

CHEMICAL AND COLLOIDAL PROPERTIES OF SOILS AND PLANT ENVIRONMENT

The crops on a field diminish or increase in exact proportion to the dimunition or increase of the mineral substances conveyed to it in manure.—JUSTUS VON LIEBIG

Approximately 100 years ago, chemists began a scientific study of the soil. For more than 50 years, the soil chemist analyzed soils and the crops growing on them for the purpose of finding out why some soils were more productive than others. The early concept consisted of the soil as a "bank" into which nutrients were stored for use by plants. As the plant grew, it drew "checks" from the "bank." Thus, if the total amount of calcium car-

FIG. 3.1. Colloidal clay in the soil acts as a large anion, that is, it adsorbs cations and holds them in exchangeable forms that are available to growing plants. The negatively charged clay, in the presence of cations in the soil solution, adsorbs the cations to satisfy the capacity of the negative bonds. In arid regions, the cations will be dominated by calcium and sodium and in humid regions, by hydrogen. Ammonium is also capable of being adsorbed on the colloidal clay, but the amount is usually small and, under certain conditions, the ammonium is oxidized to nitrate. Colloidal humus also acts in a similar manner as colloidal clay. However, certain highly weathered clays in the tropics act as cations.

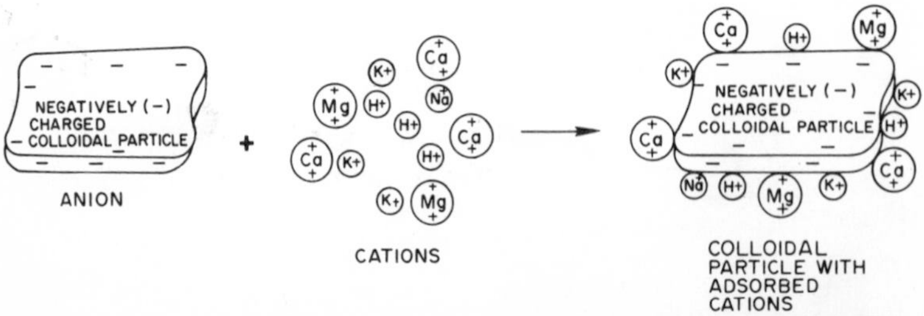

bonate in the soil was 1000 pounds per acre and if alfalfa removed 100 pounds each year, the lime would therefore be enough to last 10 years.

As research techniques improved, concepts changed. The present concept is that soils consist of solids, liquids, and gases. The chemically active fraction of the solids is confined primarily to clay and humus particles which are between 0.5 micron to 1.0 millimicron in diameter, known as the *colloidal range*. Almost all colloidal clay is now known to be crystalline, whereas humus is amorphous.

Plants feed by releasing ions and absorbing other ions. As a result, ions in the liquids and gases are in a continuously changing equilibrium with ions on the surfaces of the clay crystals and humus particles (Figure 3.1).

COLLOIDAL CLAY (CLAY MINERALS)

Nearly all clay in natural soils is colloidal, and almost all colloidal clay is crystalline. The principal exception is amorphous clay in many highly weathered soils in the tropics. Whereas sands and silts are ground and essentially unaltered primary minerals (mostly quartz), clays are composed of secondary minerals that have been built in nature from the products of chemical transformation of primary minerals. Clay minerals are mostly silica and alumina in definite spatial arrangements constituting clay crystals. Some clay minerals in the tropics may also be composed of hydrous oxides, mostly of iron and aluminum.

Clay crystals are of many kinds, but many resemble the pages of this book—that is, they are thinner in one dimension than in the other two dimensions, and the clusters of molecules are arranged in sheets. Many clay minerals are *secondary* minerals and have a close semblance to the mica minerals (*primary* minerals), such as muscovite and biotite.

If clay crystals are examined under a high-powered microscope, their crystalline structure can be seen (Figure 3.2). Most clay crystals are composed of one of two types, known as two-layer or three-layer crystals. The two-layer type consists of one layer of silicon and oxygen atoms and the other layer of aluminum and oxygen atoms, all in definite spatial arrangement. Three-layer clay crystals have two outside layers made of silicon and oxygen and a middle layer of aluminum and oxygen.

Clay minerals are usually classified into one of three groups: kaolinite, montmorillonite, and hydrous mica (the best known member of which is illite).

Kaolinite

The kaolinite group of clay materials also includes nacrite, dickite, halloysite, anauxite, and allophane; kaolinite is the most common member.

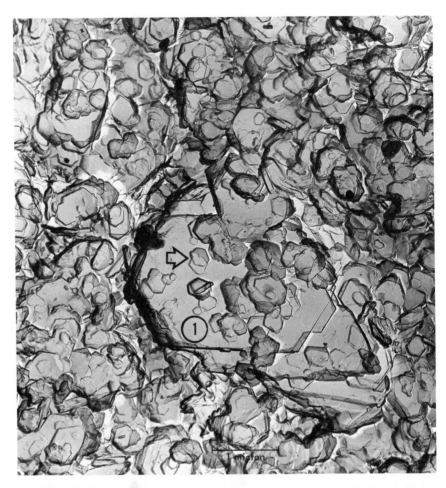

FIG. 3.2. Microscopic evidence that *primary* minerals and *secondary* (*clay*) minerals are crystalline is portrayed here. The large crystals are mica, a primary mineral (1), and the small crystals are kaolinite, a clay mineral (arrow). (Mexico.) (Magnification 26,300 times.) *Source:* Professor R. L. Sloane, U. of Arizona.

Kaolinite is a clay mineral with a crystalline structure composed of two-layer unit cells, one layer of which consists of aluminum and oxygen octahedra (eight-sided) and one layer of silicon and oxygen tetrahedra (four-sided). The silica and alumina are therefore in a 1:1 ratio. These two-layer unit cells are held together by common oxygen-hydroxyl linkage into a rigid structure that does not expand when wet. The nonexpanding lattice structure prevents the exchange of available cations; the exchange of cations is thus restricted to the broken edges of the kaolinite mineral (Figure 3.3).

Kaolinite

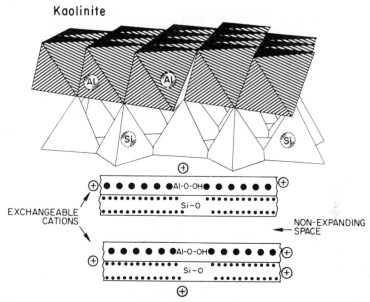

FIG. 3.3. A kaolinite clay mineral consists of a crystalline structure made of two-layer unit cells comprising one layer of aluminum-atom octahedra (eight-sided figure) and one layer of silicon-atom tetrahedra (four-sided figure) bonded by oxygen atoms, shown in a two-dimensional sketch in the lower figure and in a three-dimensional drawing in the upper figure. Spaces between the unit cells are non-expanding. *Source:* Richard Weismiller and J. L. White.

The structural formula for kaolinite is

$$(OH)_8Al_4Si_4O_{10}$$

The oxide formula for kaolinite is

$$4H_2O \cdot 2Al_2O_3 \cdot 4SiO_2$$

Montmorillonite

The montmorillonite group of clay minerals includes beidellite, saponite, and nontronite; montmorillonite is the most common member.

Montmorillonite is a clay mineral that has a unit crystalline structure of one layer of silicon and oxygen tetrahedra, one layer of aluminum and oxygen octrahedra, and another layer of silicon and oxygen tetrahedra, all linked to another such unit by oxygen and hydroxyl bonds. The silica and alumina are therefore in a 2:1 ratio. Between each two units of silica-alumina-silica is space that expands when wet. In this space, as well as on the edges, cations such as calcium, magnesium, sodium, potassium, and ammonium are held in an exchangeable form, which is available to plants (Figure 3.4).

Montmorillonite

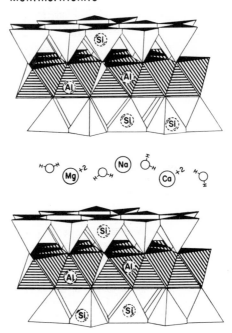

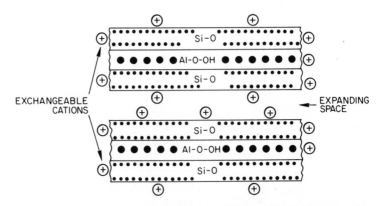

FIG. 3.4. A montmorillonite clay mineral consists of three-layer unit cells comprising two silicon-atom tetrahedra (Si) on each side of an aluminum-atom octahedra (Al) bonded by oxygen atoms (not shown). Spaces between the unit cells are expanding and permit the presence of varying amounts of water (H—O—H) and exchangeable cations such as calcium (Ca), magnesium (Mg), and sodium (Na). *Above:* Three-dimensional drawing of montmorillonite. *Below:* Two-dimensional sketch of montmorillonite, showing the expanding space between the three-layer unit cells. *Source:* Richard Weismiller and J. L. White.

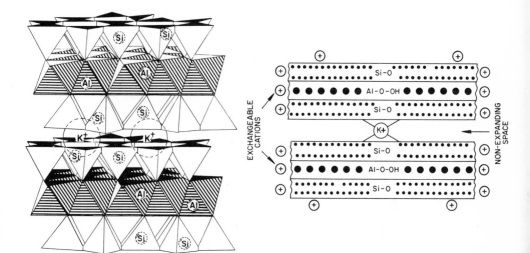

FIG. 3.5. An illite clay mineral is similar in its three-layer unit cell arrangement to the structure of montmorillonite but with two exceptions: 1. Illite has a non-expanding lattice structure, similar to kaolinite, 2. Illite fixes large amounts of potassium in a form unavailable to plants. *Left:* Three-dimensional drawing of illite. *Right:* Two-dimensional sketch of illite, showing the non-expanding lattice where potassium (K) is held in a form not available to plants. *Source:* Richard Weismiller and J. L. White.

The structural formula for montmorillonite is
$$(OH)_4Al_4Si_8O_{20} \cdot nH_2O$$

The oxide formula for montmorillonite is
$$3H_2O \cdot 2Al_2O_3 \cdot 8SiO_2$$

Illite

Illite is a hydrous mica and is closely related to vermiculite. Illite is similar in structure to montmorillonite by having a 2:1 silica:alumina ratio in its structure. However, it does not expand when wet (as is true also for kaolinite); and illite has a special characteristic of rapidly fixing large numbers of potassium ions (K^+) between the nonexpanding plates (Figure 3.5).

The structural formula for illite is
$$(OH)_4 K_y(Al_4, Fe_4, Mg_{4-6}) (Si_{8-y}Al_y)O_{20}$$
where y varies from 1 to 5.

The oxide formula for illite is
$$4H_2O \cdot K_2O \cdot 4Al_2O_3 \cdot 16SiO_2$$

In the natural weathering processes, the clay minerals are continuously but slowly being transformed by ion substitution. For this reason, clays in the soil order of Entisols, representing the least weathered group of soils, are dominated by kaolinite and illite; Aridisols by illite and montmorillonite; Mollisols by montmorillonite; and Oxisols by kaolinite, goethite, hematite, bauxite, gibbsite, boehmite, and amorphous iron oxides.

ORGANIC SOIL COLLOIDS (HUMUS)

Organic soil colloids are a temporary end-product of the decomposition of plants and animals—temporary because the organic soil colloids are themselves slowly decomposing. *Humus* is a term used to designate organic matter which is colloidal in size. In chemical composition, humus contains approximately 30 per cent each of lignin, protein, and polyuronides (complex sugars plus uronic acid).

Humus is amorphous, dark brown to black, nearly insoluble in water but soluble in dilute alkali solutions. It contains approximately 5 per cent nitrogen and 60 per cent carbon. Humus has a cationic exchange capacity many times greater than that of colloidal clay.

CATIONIC EXCHANGE

Place a filter paper in a funnel, add several grams of soil, then pour a solution of ammonium acetate through the soil. Catch the filtrate and test it for calcium, magnesium, potassium, sodium, and hydrogen. Filtrates from most soils will contain at least traces of all of these cations. Where did the cations come from when only ammonium was added?

The NH_4^+ replaced calcium, magnesium, potassium, sodium and hydrogen ions on the surfaces of the clay crystals and humus particles. These cations were released to the soil solution and were moved down into the filtrate. This mechanism is known as *cationic exchange.*

Cationic exchange takes place when any cation is added to the soil, such as Ca^{++} when lime is used, K^+ from potassium fertilizers, and NH_4^+ from anhydrous ammonia, ammonium phosphate, or ammonium sulfate.

Cation exchange is a common and very important chemical soil reaction. The exchange of cations takes place almost entirely on the surfaces of clay crystals, humus particles, and, to some extent, with ferric hydroxide. This is so because these surfaces have a net negative charge and therefore attract positive ions (cations).

The exchange of cations in the soil takes place between

1. Cations in the soil solution and those on the surfaces of clay crystals and humus

2. Cations released by plant roots and those on the surfaces of clay crystals and humus
3. Cations on the surface of either two clay crystals, two humus particles, or a clay crystal and a humus particle

Research in Florida, as well as that in many other states, has repeatedly shown that the capacity of a soil to exchange cations is the best single index of potential soil fertility.

Representative exchange capacities of soils throughout the nation are shown in Table 3.1. The variation in exchange capacity is from 1 for a fine sand to more than 100 for a mucky fine sand. Between these values, the cationic exchange capacity increases with increasing amounts of clay and humus in the soil.

TABLE 3.1 THE CATIONIC EXCHANGE CAPACITY OF REPRESENTATIVE SURFACE SOILS IN THE UNITED STATES

Soil	State	Cationic Exchange Capacity (Milliequivalents per 100 gm of Soil) *
Charlotte fine sand	Florida	1.0
Ruston fine sandy loam	Texas	1.9
Gloucester loam	New Jersey	11.9
Grundy silt loam	Illinois	26.3
Gleason clay loam	California	31.6
Susquehanna clay	Alabama	34.3
Davie mucky fine sand	Florida	100.8

* A milliequivalent is one-thousandth of an equivalent weight, and equivalent weight is defined as a weight equal to the weight of 1 gram of hydrogen. Therefore, the equivalent weight of hydrogen is 1.00 and the milliequivalent weight of hydrogen is 0.001 gram. The equivalent weight of calcium would be

$$\frac{\text{Atomic weight}}{\text{Valence}} = \frac{40}{2}$$

Equivalent weight of $Ca^{++} = 20$

Milliequivalent weight $= \frac{20}{1000} = 0.020$ gm

SOIL TEXTURE AND EXCHANGE CAPACITY

In general, the more clay of a given type there is in a soil, the higher the cationic exchange capacity. This fact is clearly demonstrated in Table 3.2. For sand soils, the exchange capacity is between 1 and 5 milliequivalents

TABLE 3.2 THE RELATIONSHIP BETWEEN SOIL TEXTURE AND
CATIONIC EXCHANGE CAPACITY

Soil Texture	Cationic Exchange Capacity (In Milliequivalents per 100 gm of Soil) (Normal Range)
Sands	1–5
Fine sandy loams	5–10
Loams and silt loams	5–15
Clay loams	15–30
Clays	Over 30

per 100 grams of oven-dry soil. The value of greatest frequency (mode) for fine sandy loams is 5–10 milliequivalents; for loams and silt loams, 5–15; for clay loams, 15–30; and for clay soils, over 30.

For general purposes, it can be assumed that kaolinite has a cationic exchange capacity of approximately 8 milliequivalents; illite, 30 milliequivalents; montmorillonite, 100 milliequivalents, and humus (colloidal organic matter), 200 milliequivalents per 100 grams of material, on an oven-dry basis.

ORGANIC MATTER AND EXCHANGE CAPACITY

Several studies have demonstrated that the cationic exchange capacity of soils is due as much, if not more, to soil organic matter as to the percentage of clay. In fact, there was a fairly definite increase in the exchange capacity of sandy Florida soils of 2 milliequivalents for each 1 per cent increase in organic matter.

Research in New York on Honeoye silt loam compared the exchange capacity of each horizon of the profile with that of the extracted soil organic matter. The percentage of the total exchange capacity due to organic matter varied from 51 for the A_{22} horizon to 22 for the B_3 horizon. The most important fact is that less than 4 per cent organic matter in the A_{22} horizon is responsible for more than 50 per cent of the total cationic exchange capacity.

SOIL FERTILITY AND EXCHANGE CAPACITY

The cation exchange complex in the soil is always saturated with cations. In humid regions, the most common cations present are hydrogen, calcium, magnesium, and potassium. When no other cations are present, hydrogen saturates the complex; the result is a very strongly acid and infertile soil. The most fertile soil will have perhaps only 10 per cent exchangeable hydrogen and 90 per cent total exchangeable bases, such as calcium (75 per cent), magnesium (10 per cent), and potassium (5 per cent). In arid regions, sodium usually is in greater quantities than hydrogen.

The exchangeable cations are the primary source of calcium, magnesium, and potassium in plant nutrition. Since the exchangeable cations do not leach from the soil, they represent nature's mechanism for assuring a continuous storehouse of cations that are readily available for plant growth.

ADSORPTION AND EXCHANGE OF ANIONS

Under certain conditions, soils have the ability to hold a *small* amount of anions in the exchangeable form. Nutrient anions, such as NO_3^-, SO_4^-, and $H_2PO_4^-$, are sometimes held in a exchangeable form for use by plants. Nitrates are capable of slight anionic exchange under acid conditions; but almost no exchange is exhibited when the soil is nearly neutral. This means that anionic exchange with nitrates is negligible because they do not form readily under acid conditions. Sulfates are also present in greater amounts in the exchangeable form when the soil is acid.

Phosphates are held in the soil in fairly large amounts, and the more acid the soil, the more the phosphates are retained. But most of the adsorption is not exchangeable. In acid soils, iron and aluminum readily form relatively insoluble compounds with phosphates.

In general, the relative order of anionic exchange is

$$H_2PO_4^- > SO_4^= > NO_3^-$$

Organic matter, kaolinite, and amorphous iron increase the amount of anionic exchange.

SOIL pH

Freshly distilled water contains the same number of H^+ and OH^- ions; it is therefore neutral in reaction. When the soil solution contains the same number of H^+ and OH^- ions, it also is neutral. Add $Ca(OH)_2$ to the neutral soil and there will be more OH^- than H^+; then the soil will be alkaline. Conversely, when HCl is added to the neutral soil, the soil will contain more H^+ than OH^-, and the soil will become acid (Figure 3.6).

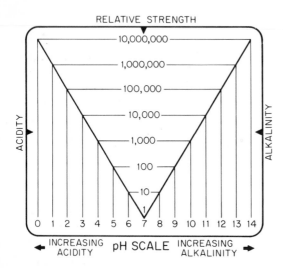

FIG. 3.6. The chart shows the relationship of pH values to relative acidities and alkalinities. Note that pH 7 is the neutral point —neither acid nor alkaline. The *acid* region is below pH 7, the pH values *decreasing* with increasing acidity. The *alkaline* region is above pH 7, the pH values *increasing* with increasing alkalinity. Note, also, that the relative strength changes 10-*fold* for each unit change in pH. *Source:* Beckman Instruments.

The most convenient method of expressing the relationship between H^+ and OH^- is pH. By pH is meant the logarithm of the reciprocal of the active hydrogen ions in grams per liter, usually written

$$pH = \log \frac{1}{(H^+)}$$

At neutrality, the hydrogen-ion concentration has been determined to be 0.000 000 1 or 1×10^{-7} gram of hydrogen per liter of solution. Substituting this concentration into the formula, we get

$$pH = \log \frac{1}{0.000\ 000\ 1}$$
$$= \log 10,000,000$$
$$= 7$$

At a pH of 6, there would be 0.000 001 gram of active hydrogen, or 10 times the concentration of H^+. At each smaller pH unit, the H^+ increases by 10 in concentration. It therefore follows that a pH of 5 is 10 times more acid than a pH of 6 and 100 times more acid than a pH of 7; and so on.

The entire pH range is from 0 to 14, but the most common range of *soil* pH is between 4 and 8. Acid forest humus layers in the Northeast have been known to test as low as pH 3.5, and some high-sodium soils in arid regions may test pH 9 (Figure 3.7).

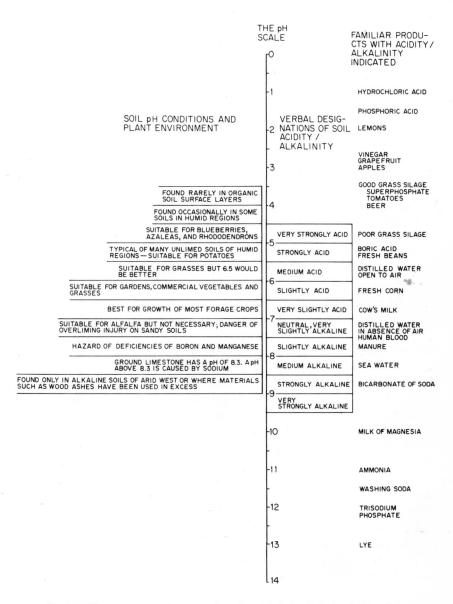

FIG. 3.7. The entire pH scale ranges from 0 to 14, but soils under field conditions vary between pH 3.5 and 9.0. Outside this range of 3.5 to 9.0 the pH scale is used primarily in chemical laboratories. In general, most plants are best suited to a pH of 5.5 on organic soils and a pH of 6.5 on mineral soils. *Adapted from:* Winston A. Way, *The Whys and Hows of Liming,* U. of Vt. Brieflet 997, 1968.

BUFFER CAPACITY OF SOILS

The buffer capacity of a soil refers to its ability to resist a change in pH. The principal factors influencing buffer capacity are

1. The amount and kind of clay
2. The amount of organic matter

Soils containing more clay and organic matter have a greater buffer capacity. Clay with an expanding lattice, such as montmorillonite, has a higher buffer capacity than one with a nonexpanding lattice, such as kaolinite.

FIXATION OF AMMONIUM

In recent years, it has been demonstrated that approximately 5 per cent of the total nitrogen in surface soils and as much as 60 per cent of the total nitrogen in subsoils is held as nonexchangeable (fixed) ammonium. The mechanism of fixation appears to be the same as that of potassium fixation —that is, from a replacement by the ammonium ion for interlayer cations such as calcium, magnesium, sodium, and hydrogen, in the expanded lattice of clay minerals (crystals). When ammonium ions replace other cations, the lattice of the clay crystal contracts, entrapping the ammonium ions in a nonexchangeable (fixed) form. The fixed ammonium can be slowly released by cations that expand the lattice—that is, calcium, magnesium, sodium, and hydrogen. Potassium contracts the lattice and therefore does not replace the fixed ammonium. The total amount of fixed ammonium in a soil is directly related to the amount and kind of clay present: the more the clay, the more the fixed ammonium; and the more the percentage of clay present that is of the expanding-lattice type (montmorillonite), the more the fixed ammonium. Soils containing kaolinite, a clay with a nonexpanding lattice, exhibit less fixed ammonium.

FIXATION OF POTASSIUM

Long-time experiments have shown that there is some mechanism in the soil that fixes available potassium in a form which is not available to plants. The factors influencing the amount of potassium fixation are as follows:

1. The kind of clay minerals present. Kaolinite does not appear to fix potassium, whereas large quantities are fixed by illite, vermiculite, and montmorillonite.
2. The relative amount of exchangeable potassium. The greater is the percentage of exchangeable potassium in relation to the total exchange capacity, the greater is the potassium fixation.
3. Wetting and drying of the soil. Soils that are wetted and dried fix large amounts of exchangeable potassium. One explanation for this mechanism is that potassium ions move inside the clay-crystal lattice when it is wet and expanded, and, upon drying, the ions are trapped inside. Any soil treatment that would keep the soil more uniform in moisture content, such as shading or the use of a mulch, would therefore tend to reduce potassium fixation.
4. The presence of organic matter. Humus particles exist in the soil in all sizes. Some particles are small enough to enter the clay-crystal lattice and to reduce the amount of contraction upon drying. This mechanism tends to lessen the amount of entrapped potassium.

PHOSPHORUS FIXATION AND CHELATION

When phosphorus fertilizers are added to a soil, the plant may recover only from 2 to 25 per cent of the phosphorus. Most of the phosphorus is tied up in relatively insoluble compounds of iron, aluminum, orthocalcium phosphates, and apatite. If sugar, starch, manure, cover crops, or any readily decomposable organic materials are then applied to such soils, a larger percentage of the unavailable phosphorus soon becomes available (Figure 3.8).

FIG. 3.8. An example of chelation. Both pots contained the same kind of soil and received the same amounts of manure and superphosphate. In the pot on the left, the superphosphate was mixed with the soil and the manure was added later. In the pot on the right, the superphosphate was added to the manure, then mixed with the soil. The manure contains chelating compounds that reduced the fixation of phosphorus by the soil. (Vermont.) *Source:* A. R. Midgley, Vt. Agr. Exp. Sta.

Research scientists have examined the reason for this and have found that decomposing organic materials form citrates, tartrates, acetates, oxalates, malates, malonates, and other organic anions which release fixed phosphorus. These anions form highly stable complex ions with Ca^{++}, Fe^{+++}, and Al^{+++}, and, in fact, have a greater affinity for these cations than does phosphorus. That is to say, in the relatively insoluble iron, aluminum, and tricalcium phosphates, the organic anions such as the citrates readily combine with the iron, aluminum, and calcium and thereby release the phosphorus for plant growth.

Inorganic anions also act as chelating agents, but they are much less effective than organic anions. Fluorides, arsenates, borates, sulfates, chlorides, and nitrates to some extent solubilize phosphorus from iron, aluminum, and tricalcium phosphates.

SUMMARY

The chemically active part of the soil is on the surfaces of clay crystals and humus particles. Clays are secondary minerals which exist as a two-layer unit, such as kaolinite, or a three-layer unit, such as montmorillonite or illite. Organic soil material in the colloidal state is called *humus*. Humus is amorphous (noncrystalline), dark brown to black, and contains approximately 5 per cent nitrogen and 60 per cent carbon.

Cations in the soil are freely exchanged among the surfaces of plant roots, clay crystals, and humus, as well as between these and the soil solution. Cationic exchange capacity may vary from 1 milliequivalent for sandy soils to 100 milliequivalents per 100 grams of soil for a muck. The exchange capacity varies directly with the amount and kind of clay or humus in a soil. The exchange capacity for humus, however, may be as much as 10 times that of clay.

The pH is the logarithm of the reciprocal of the hydrogen ion concentration in grams per liter. Most soils have a pH between 4.0 and 8.0, but the extreme range is from 3.5 to 10.0.

Certain soils are capable of fixing large amounts of ammonium and potassium. Montmorillonite fixes more ammonium and potassium than does kaolinite. Wetting and drying tends to fix potassium in all soils. Humus reduces the fixation of potassium.

Citrates, tartrates, acetates, oxalates, malates, malonates, and other products of organic decay are capable of releasing, through the process of chelation, fixed phosphates and of rendering them available for plant growth. Fluorides, arsenates, borates, sulfates, chlorides, and nitrates, to a much lesser extent, are capable of chelation.

QUESTIONS

1. Draw and label a kalolinite, a montmorillonite, and an illite mineral.
2. Illustrate cationic exchange in a sample of soil by the use of a salt solution (NaCl).
3. What is the pH of a soil with an H^+ concentration of 0.000 001 gram per liter?
4. What can be done to reduce K fixation?
5. Explain chelation.

REFERENCES

Bear, Firman E. (ed.), *Chemistry of the Soil* (2nd ed.). New York: Reinhold Publishing Corp., 1964.

Bollard, E. G., and G. W. Butler, "Mineral Nutrition of Plants," *Annual Review of Plant Physiology*. Vol. 17 (1966).

Buckman, Harry O., and Nyle C. Brady, *The Nature and Properties of Soils* (7th ed.). New York: The Macmillan Company, 1969, 653 pp.

Coleman, H. T., and A. Mehlich, "The Chemistry of Soil pH," in *Soil: The Yearbook of Agriculture* (1957). United States Department of Agriculture, pp. 72–79.

Jackson, M. L., *Soil Chemical Analysis*. Englewood Cliffs, N. J.: Prentice-Hall, Inc., 1958.

Rich, C. I., and G. W. Kunze (eds.), *Soil Clay Mineralogy: A Symposium*. Chapel Hill, N. C.: The University of North Carolina Press, 1964, 330 pp.

Science for Better Living: The Yearbook of Agriculture (1968). United States Department of Agriculture.

CHAPTER 4

PARENT MATERIALS OF SOILS

The geological deposit, the glacial till, the sandy deposits of the sand plains, the lake-laid or marine clays, sands, silts, and gravels, the residual earth resulting from rock decay constitute soil materials or parent material of soils.—C. F. MARBUT

Soils have developed from minerals and rocks that were weathered by climatic forces until sufficient nutrients became available to support plants. Productive soils develop only from minerals and rocks plentifully supplied with all essential elements. Soils rich in available phosphorus are found only on parent materials that are well supplied with this element. The same is true for soils containing large amounts of available calcium, magnesium, potassium, and other mineral nutrients.

But the converse of this statement is not always true. Intense weathering and leaching may produce soils low in available phosphorus, calcium, magnesium, potassium, and other minerals, even though these nutrients are plentiful in the minerals and rocks comprising the parent materials. The Maury silt loam in central Tennessee is an example of a soil that responds to applications of phosphorus and calcium, even though the soil was developed on parent materials very rich in these elements. Warm temperatures cause year-round weathering, and a high rainfall leaches downward some weathered phosphorus and calcium (Figure 4.1).

Certain soils in the humid Tropics (Oxisols) have been so intensely weathered that the surface 5 to 10 feet may not have any detectable weatherable minerals remaining.

Of the 16 elements essential for plants, carbon, hydrogen, oxygen, and

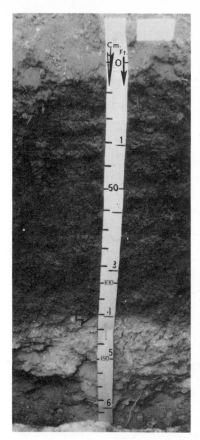

FIG. 4.1. Soils develop from rock by climatic weathering, modified by plants, animals, and slopes, over decades. *Right:* A basalt rock is being turned into soil material by climate (mostly rainfall and temperature), modified by the influence of higher (seed-bearing) plants (at "A"), and lower plants (lichens, a symbiotic union of fungi and algae, at "B"), and by slopes, in time. *Left:* After sufficient time the surface of the basalt rock weathers to become soil, as shown here. The basalt rock, instead of being more than 4 feet *above* ground level (as in photo at right) is now more than 4 feet *below* ground level (120 cm) at arrow. (Ethiopia.) (Scale is in centimeters and feet.) *Photo by Roy L. Donahue.*

nitrogen came originally from the atmosphere and the other 12 came from rocks and minerals to parent materials and from parent materials to soils.

CLASSIFICATION OF PARENT MATERIALS

Parent materials from which soils are derived may be classified as *residual, transported,* or *cumulose.* Residual materials are those that have remained in place long enough for a soil to develop from them. Transported materials are mineral and rock fragments that have been moved into place by one or more of these agents: water, wind, ice, and gravity. Cumulose materials are peats and mucks that have developed in place from plant residues and have been preserved by a high water table.

An outline may help the reader to visualize the relationships among the parent materials of soils.

I. Residual material
    A. Igneous—granite, basalt, and lava
    B. Sedimentary—limestone, sandstone, and shale
    C. Metamorphic—marble, quartzite, and gneiss
II. Transported material
    A. Water
        1. Alluvial—running water
        2. Lacustrine—lakes
        3. Marine—ocean
    B. Wind
        1. Eolian
        2. Loess
    C. Ice
        1. Moraine
        2. Till plain
        3. Outwash plain

FIG. 4.2. A soil in the process of developing from granite, an *igneous* rock, in Yosemite National Park, California. *Source:* H. W. Turner, U.S. Geological Survey.

D. Gravity
    1. Colluvial
III. Cumulose material (organic)
    A. Fibric (peat)
    B. Folic (leaf "mold")
    C. Hemic (peat or muck)
    D. Sapric (muck)

RESIDUAL MATERIALS (MINERALS AND ROCKS)

A mineral is a substance that occurs in nature, has distinct physical properties, and is of a chemical composition which can be written as a formula.

Rocks are mixtures of minerals; for that reason, their physical and chemical composition varies with the characteristics of the minerals comprising them. But a single mineral is seldom the parent material from which a soil develops; a soil is formed mainly from rocks.

Rocks are classified into *igneous, sedimentary,* and *metamorphic.* Igneous rocks have formed from molten material that has solidified. Sedimentary rocks have formed from sediments accumulated at the surface of the earth. These sediments may have been derived from either minerals, rocks, or organisms, or from precipitates of sea water. Metamorphic rocks have come from either igneous or sedimentary rocks that have later been transformed beneath the surface of the earth by heat, pressure, and chemically active liquids.

It is estimated that 3 per cent of the soils in the United States are residual and have developed in place from the underlying igneous, sedimentary, or metamorphic bedrock. The principal examples are soils from

1. Igneous rocks such as granites in the Piedmont Region and in the Rocky Mountains, and lava in Arizona (Figures 4.2 and 4.3)
2. Sedimentary rocks, such as limestone in central Tennessee, central

FIG. 4.3. Lava is an igneous rock that has cooled rapidly; therefore, there was not sufficient time for minerals to crystallize, as in the formation of granite. This is a lava flow on Sunset Mountain, Coconino National Forest, Arizona, where the 15 to 20 inches of annual precipitation have not been sufficient to develop a soil. The lone western yellow pine with stunted growth is typical in such plant environments. (Elevation 6,500 feet.) *Source:* J. O. Veatch.

FIG. 4.4. This residual soil has developed from limestone, a *sedimentary* bedrock. *Source:* W. T. Carter, Tex. Agr. Exp. Sta.

FIG. 4.5. Granite gneiss, a *metamorphic* rock, is in the foreground. Note its layer-like structure. *Source:* T. C. Richardson, *The Farmer-Stockman.*

Kentucky, along the Grand Canyon in Arizona, and in southern Missouri (Figure 4.4); and sandstone and shale in the Appalachian Mountains and in Oklahoma, northwest Arkansas, and north central Texas

3. Metamorphic rocks. There are only a few areas where the soils have developed from metamorphic rocks, such as marble, quartzite, or gneiss (Figure 4.5).

MOON "SOIL"

On July 20, 1969, Neil A. Armstrong and Edwin E. Aldrin, Jr., first walked on the surface of the moon and collected samples of "soils" and rocks for analysis (Figure 4.6).

Some soil scientists say that the surface of the moon is not soil because it is not on the surface of the earth and because it contains no organic matter. Perhaps the material on the moon can best fit into this chapter, Parent Materials of Soils, but the moon "soil" certainly looks and acts like soil. The moon materials are apparently of igneous origin, and are subjected to daily surface temperature variations of 530°F, from 250°F above zero during the day to 280°F below zero at night. Below a depth of 40 inches (1 meter), however, no temperature variations occur. Someday, soil scientists may agree to call it *soil*.[1]

---

[1] A mixture of moon soil and earth soil is even more productive than earth soil alone. Furthermore, moon soil has the ability to kill certain earth bacteria.

FIG. 4.6. Man's first footprints on the moon. On July 20, 1969, this footprint was made on the moon by Neil A. Armstrong or Edwin E. Aldrin, Jr. Source: National Aeronautics and Space Administration, Houston, Texas 77058.

The moon "soil" samples continue to be analyzed, but some information is already fairly well agreed upon, as follows:

1. The moon resembles a dehydrated cinder.
2. The moon is between 3.5 and 4.5 billion years old, as compared with 3.3 billion years for the age of this earth.
3. No evidence of life has yet been found, either living or in fossils.
4. The moon rocks resemble basalt, a basic, fine-grained, igneous rock, as we know it on earth.
5. The dark-colored moon soil is between 2 and 8 inches deep, with a texture of fine sand and a bulk density slightly greater than earth soils.

A comparison of the chemical analyses of moon soil and the earth's crust (earth soil) is given in Table 4.1. There is very little difference in

TABLE 4.1 A COMPARISON OF THE ABUNDANCE OF 13 ELEMENTS ESSENTIAL FOR PLANT GROWTH IN MOON "SOIL" AND ON EARTH *

Element	Moon "Soil" (%)	Earth's Crust (%)
Silicon	20.20	27.72
Aluminum	7.30	8.13
Iron	12.50	5.00
Magnesium	4.60	2.09
Calcium	9.60	3.63
Potassium	0.11	2.59
Manganese	0.16	0.10
Phosphorus	0.14	0.10

Element	Parts per Million	Parts per Million
Boron	2.0	10.0
Nitrogen	30.0	20.0
Chlorine	350.0	130.0
Cobalt	40.0	25.0
Copper	9.9	55.0
Zinc	22.0	70.0
Molybdenum	0.7	1.5

* "Space Science," *Chemical and Engineering News* (Jan. 19, 1970), pp. 20–21.

Note: In addition to the preceding 13 elements that are essential for plant growth, a total of 53 elements were found in moon "soil" which were comparable in abundance with the same elements on earth.

FIG. 4.7. Horizontal stratification of soil materials (alluvial materials) that have been deposited by successive stages of river flood waters in Morton County, North Dakota and that are shown on a soil map as Havre silt loam (soil type) in the family of Ustic Torrifluvents. (See Chapters 6 and 7 for a more detailed explanation.) (Scale is in feet.) *Source:* William M. Johnson, Soil Conservation Service, U.S. Dept. of Agr.

composition of silicon, aluminum, manganese, phosphorus, nitrogen, and cobalt. Moon soil is approximately three times higher in percentage of calcium and chlorine as earth soil, twice as concentrated in iron and magnesium, one-half the composition of molybdenum, one-third as rich in zinc, one-fifth as concentrated in boron and copper, and one-eleventh as high in percentage of potassium.

MATERIALS TRANSPORTED BY WATER

Materials that have been transported by water are classified as *alluvial, lacustrine,* or *marine.*

Alluvial materials are sediments that have been deposited by flowing water, such as small streams or large rivers. If the material that lies along the river is subject to periodic flooding, the deposit is known as a *flood plain.* Older deposits that were laid down by the river but are not now subject to flooding are called *terraces* (Figure 4.7).

Lacustrine materials were deposited in fresh-water lakes during glacial times. Depressions that were filled with water from melting glaciers gradually became filled with sediment. When the glacier retreated, the level of the lake receded, exposing sediments that had been deposited in the bottom of the lake. These are called *lacustrine* materials. The principal lacustrine deposits in the United States are found bordering the Great Lakes, in northeastern North Dakota and northwestern Minnesota, northwestern

FIG. 4.8. These horizontal *lacustrine* beds of sands, silts, and clays were deposited during glacial times over a period of several thousand years in a fresh-water lake in Northeast Washington State. *Source:* F. O. Jones, U.S. Geological Survey.

FIG. 4.9. *Marine* sediments in West-Central Georgia consist of layers of laminated dark gray fossiliferous clay and fine sand that have been in place long enough to support vegetation. *Source:* D. H. Eargle, U.S. Geological Survey.

Nevada, northwestern Utah, and northeastern Washington State (Figure 4.8).

Marine materials were formed by the deposition of sediments carried into the ocean by rivers. As the land surface was uplifted or the ocean receded, the sediments became weathered into soil. The principal marine sediments in the United States occur adjacent to the Gulf Coast and the Atlantic Coast (Figure 4.9).

MATERIALS TRANSPORTED BY WIND

Sand dunes and some deposition from present-day dust storms are called *eolian* deposits. Fine-textured soil materials that were deposited following the last glacial period are known as *loess*.

Loessial soil materials are mostly silt loam in texture and occur mainly in the Mississippi Valley. There are large areas of loess in Kansas, Nebraska, Iowa, Missouri, Illinois, Indiana, Kentucky, Tennessee, and Mississippi. Extensive deposits also occur in Washington and Idaho (Figure 4.10).

MATERIALS TRANSPORTED BY ICE

From perhaps 1 million years ago to as recently as 10,000 years ago, continental ice intermittently occupied the land which is now the northern border of the United States. Some geologists claim that we are now simply in another interglacial period. Parts of Alaska, Greenland, and Iceland and the mountains of northern Europe, Switzerland, and Antartica are now occupied by a mass of ice similar to what once covered parts of the northern United States.

As snow continued to accumulate, pressure from its weight changed the snow to ice. After centuries of such buildup, the ice, under tremendous pressure, moved outward. In moving across rocks, sand, silt, and clay, the

FIG. 4.10. *Loess* deposit in Iowa. *Source:* U.S. Geological Survey.

FIG. 4.11. A summer view of a glacier that is active during the winter in the Province of Alberta, Canada, showing a fresh deposit of *lateral moraine* in the left foreground as well as directly in front of the glacier on both sides of the U-shaped outlet. *Source:* F. O. Jones, U.S. Geological Survey.

FIG. 4.12. Glacial *till plain* materials are usually stony on the surface (*above*) as well as throughout the entire deposit (*below*). *Sources: Above,* Julian P. Donahue; *below,* U.S. Geological Survey.

FIG. 4.13. Fresh deposits of coarse sand and gravel from water gushing forth from the Malaspina glacier in the Gulf Region of Alaska are known as *outwash plains.* Source: J. H. Hartshorn, U.S. Geological Survey.

FIG. 4.14. *Colluvial* material, known as a *talus cone,* has been formed by rocks moving in response to gravity in Glacier National Park, Montana. *Source:* H. E. Malde, U.S. Geological Survey.

glacial ice picked them up, making a mass of dirty and stony ice. Apparently there was a greater movement outward during the winter buildup of snow; during the summer, the ice front moved outward and some of the ice melted. Water flowing from the melting ice carried sediment, and the larger rocks were dropped in place.

When the ice front melted about as fast as it advanced, deposits of sediment were built up, resulting in a series of stony hills at the ice front known as *terminal* moraines. Stony ridges deposited along the margins of the ice front are known as *lateral* moraines (Figure 4.11). When the ice front melted faster than it advanced, a smoother deposition resulted, known as *till plains* (Figure 4.12). Water gushing forth from a rapidly melting ice front carried fairly coarse sand and gravel particles and deposited them in a level plain. These are *outwash plains* (Figure 4.13).

MATERIALS TRANSPORTED BY GRAVITY

Soil debris at the foot of a slope that moved there in response to gravity is called *colluvial* material. Colluvial material exists to some extent at the base of all slopes, but it is especially noticeable in mountainous topography, where rock slides, slips, and avalanches are common (Figure 4.14).

FIG. 4.15. *Cumulose* materials originate in shallow water where over decades generations of plants grow, die, and fall in the water where they are preserved because of the low oxygen content. Here is a mangrove swamp growing in shallow water near the Dade County and Monroe County lines in Southern Florida. Under the water is an accumulation of 5 to 8 feet of dark brown fibrous (Fibric) peat over limestone bedrock. *Source:* J. O. Veatch.

FIG. 4.16. Fibric organic materials (peat) interbedded with silt in the Yukon Region of Alaska. *Source:* S. Taber, U.S. Geological Survey.

FIG. 4.17. Sapric organic soil is comprised of organic materials decomposed beyond recognition, as shown here in Michigan. This cumulose soil is classified and mapped as Lupton muck. (Scale is in feet.) *Source:* Soil Conservation Service, U.S. Dept. of Agr.

CUMULOSE MATERIALS

Cumulose materials are primarily partly decomposed plants that have accumulated in shallow water. Many generations of plants, growing for decades, have fallen in the water in which they were growing and have been preserved because of the low oxygen content (Figure 4.15).

Organic soils are classified as follows: [2]

1. Fibric (Latin, *fibra,* fiber). The *least decomposed* of the organic soils, mostly mosses (formerly known as peat) (Figure 4.16).
2. Folic (Latin, *folium,* leaf). Partially decomposed leaves and twigs (leaf "mold").
3. Hemic (Latin, *hemi,* half). Intermediate in degree of decomposition (formerly known sometimes as peat and at other times as muck).
4. Sapric (Greek, *sapros,* rotten). The most highly decomposed of the organic materials (formerly known as muck) (Figure 4.17).

PHYSICAL WEATHERING (DISINTEGRATION)

Minerals exposed at the surface of the earth in rocks, sands, silts, and clays are constantly being broken down by chemical and physical weathering. Physical weathering processes include the following:

---

[2] *Soil Taxonomy of the National Cooperative Soil Survey.* Soil Conservation Service, U.S.D.A. (Dec., 1970), pp. 11-3 to 11-20.

FIG. 4.18. Heating and cooling of this granite boulder have caused expansion and contraction of the surface and resulted in scaling off of the rock. This is known as *exfoliation*. (India.) *Photo by Roy L. Donahue.*

1. Freezing and thawing. The expanding force of freezing water is sufficient to split any mineral or rock.
2. Heating and cooling. This gives rise to differential expansion and contraction of minerals in rocks, tending to tear them apart. Temperature changes also bring about *exfoliation,* a loosening of an entire "cap" on a mineral or rock (Figure 4.18).
3. Wetting and drying. Wetting and drying disrupt the soil by causing it to swell and contract, and abrasion within the soil makes the particles finer.
4. Erosion. Water carrying sediments in suspension or rolling them along the bottom of streams exerts a strong scouring action that grinds particles finer. Wind erosion acts in a similar manner.
5. Action of plants, animals, and man. Plants grow between rocks, splitting them apart. Animals are constantly scratching rocks, and this action on soft rocks aids in their disintegration. Man helps in physical weathering by plowing and cultivating, an action that results in breaking minerals and rocks into finer fragments (Figure 4.19).

CHEMICAL WEATHERING (DECOMPOSITION)
AND CHEMICAL TRANSFORMATION

Physical disintegration is accompanied by chemical decomposition in weathering processes. Chemical weathering includes the following:

1. Oxidation. Oxygen readily combines with nearly all minerals to aid in chemical decomposition. An example of oxidation follows.

$$4FeO \quad + \quad O_2 \quad \rightarrow \quad 2Fe_2O_3$$

ferrous oxide oxygen ferric oxide (hematite)

FIG. 4.19. Lichens (scale-like, whitish plants), shrubs, and trees are acting on these granite rocks to make them into soil. *Above:* Elizabeth Island, Massachusetts. *Right:* Central India, showing a tree root forcing a granite rock apart. *Sources: Above,* J. B. Woodworth, U.S. Geological Survey; *right,* Roy L. Donahue.

2. Reduction. Under conditions of excess water, such as is encountered in a flooded soil, oxygen is less plentiful and, as a consequence, reduction takes place. Reduction may be illustrated in this manner.

$$2Fe_2O_3 \qquad - \qquad O_2 \qquad \rightarrow \qquad 4FeO$$

ferric oxide   **oxygen**   **ferrous**
  (hematite)   **oxide**

3. Solution. Soluble minerals, such as halite (rock salt), dissolve readily in this manner.

$$NaCl \qquad \rightarrow \qquad Na^+ \qquad + \qquad Cl^-$$

halite   **sodium ion**   **chlorine ion**

4. Hydrolysis. This process consists of a reaction with water and the formation of an hydroxide. An example of hydrolysis of a mineral is shown for orthoclase.

$$K\,Al\,Si_3O_8 \quad + \quad HOH \quad \rightarrow \quad H\,Al\,Si_3O_8 \quad + \quad KOH$$

orthoclase   **water**   **acid**   **potassium**
                                                silicate   **hydroxide**
                                                clay

5. Hydration. A common example of hydration is the formation of limonite from hematite.

$$2Fe_2O_3 \quad + \quad 3H_2O \quad \rightarrow \quad 2Fe_2O_3 \cdot 3H_2O$$

hematite   **water**   **limonite**

(red)   **(yellow)**

6. Carbonation. The atmosphere contains 0.03 per cent carbon dioxide, but decomposing plants liberate it in large amounts. Carbon dioxide aids in chemical weathering because it makes minerals more soluble. For example (Figure 4.20).

$$Ca\,CO_3 \quad + \quad CO_2 \quad + \quad H_2O \quad \rightarrow \quad Ca(HCO_3)_2$$

calcite   **carbon**   **water**   **calcium**

(slightly   **dioxide**   **bicarbonate**

soluble)   **(readily**

                                                      soluble)

SUMMARY

Soils have developed from minerals and rocks. All elements essential for plant growth, except nitrogen, oxygen, carbon, and hydrogen, originally came from the parent minerals and rocks.

Parent materials of soils are classified as *residual, transported,* and *cumulose.*

Residual materials are grouped into *igneous, sedimentary,* and *metamorphic* rocks. The accepted classification of transported soil materials is based upon the agents of transport: *water, wind, ice,* and *gravity.* Cumulose materials have formed in place by the preservation of organic remains in shallow lakes.

Rocks and minerals are weathered into parent materials of soil by physical disintegration and chemical decomposition. Physical processes are freezing and thawing, heating and cooling, wetting and drying, erosion, and the action of plants, animals, and man. Chemical weathering processes include oxidation, reduction, solution, hydrolysis, hydration, and carbonation.

QUESTIONS

1. Why is it true that rich soils come only from minerals and rocks that contain an abundance of elements essential for plant growth? Why is the converse not always true?
2. Give an example of an igneous rock. Is the moon of igneous origin?

FIG. 4.20. *Solution and carbonation,* as well as *oxidation, reduction, hydrolysis,* and *hydration,* are all at work in this cave in Carlsbad, New Mexico. It is a good example of chemical weathering (decomposition) and transformation. *Source:* R. V. Davis, U.S. Geological Survey.

3. Compare marine and lacustrine sediments as to origin.
4. How does a moraine differ from a till plain?
5. Give an example of oxidation.

REFERENCES

Buckman, Harry O., and Nyle C. Brady, *The Nature and Properties of Soils* (7th ed.). New York: The Macmillan Company, 1969, 653 pp.

Soil: The Yearbook of Agriculture (1957). United States Department of Agriculture.

Soils and Men: The Yearbook of Agriculture (1938). United States Department of Agriculture.

Soil Survey Manual, United States Department of Agriculture Handbook No. 18. United States Department of Agriculture, 1962.

Soil Taxonomy of the National Cooperative Soil Survey. Soil Conservation Service, United States Department of Agriculture, Wash. D.C., Dec., 1970.

United States Geological Survey, *Abstracts of North American Geology.* United States Department of Interior (Jan., 1968), 163 pp.

CHAPTER 5

SOIL FORMATION
(SOIL GENESIS)

. . . Soils in the Russian concept were conceived to be independent natural bodies, each with a unique morphology and resulting from a unique combination of climate, living matter, parent rock materials, relief, and time. The morphology of each soil, as expressed in its profile, reflects the combined effects of the particular set of genetic factors responsible for its development.*

During the early part of this century, Russian soil scientists proposed that soil formation is the result of climate and living matter acting upon soil material on a given relief, over a period of time. This is the accepted concept today.

Climate is an active factor of soil formation that functions primarily through precipitation and temperature. Living matter (biosphere) embraces all plants and animals, including man, that stir the soil, assist in organic matter decomposition, and indirectly influence the release of nutrients from soil minerals. Parent materials, conditioned by their relief, are acted upon by climate and biosphere over a period of time to produce soil.

The five factors of soil formation and their category of activity in the the process of soil formation are as follows:

1. Climate (active)
2. Biosphere (active)

---

* *Soil Survey Manual,* U.S.D.A. Handbook No. 18, Soil Survey Staff (Aug. 1962), p. 3.

FIG. 5.1. A soil profile is in the process of being formed (at arrow) from rocks that were piled three feet thick on top of this Hindu temple in Rajasthan (Northwestern India). Mean annual rainfall is 20 inches. *Photo by Roy L. Donahue.*

3. Parent material (passive)
4. Relief (passive)
5. Time (neutral) (Figure 5.1)

CLIMATE AND SOIL FORMATION

Climate influences soil formation largely through precipitation and temperature. The average annual precipitation in the United States is presented as a hyetograph in Figure 5.2. Average annual temperature is portrayed by isotherms in Figure 5.3.

East of a line drawn from Chicago to New Orleans, the precipitation averages mostly 40 to 50 inches a year. West of this line, the precipitation decreases fairly regularly to approximately 10 inches, from which point again annual precipitation increases to more than 100 inches on some of the western slopes in Washington and Oregon.

Average annual temperatures in the West are primarily a result of differences in elevation; in the East, temperatures are influenced mostly by latitude. The 40-degree isotherm is dominant across the United States–Canadian border. From New York to Chicago to Denver, the 50-degree isotherm is present. High elevations in the West are also represented by this line. Sixty degrees average annual temperature is reported for parts of North Carolina, Tennessee, Arkansas, Oklahoma, northwestern Texas, southern New Mexico, southern Arizona, and California. Central Florida and southern Texas lie within the 70-degree isotherm.

Over the face of the earth, climate is the dominant factor in soil formation. Any soil profile is both the direct as well as the indirect result of the action of centuries of climatic forces. Some direct effects of climate on soil formation include:

1. A shallow accumulation of lime in areas of low rainfall (Figure 5.4)
2. Acid soils in humid areas due to intense weathering and leaching

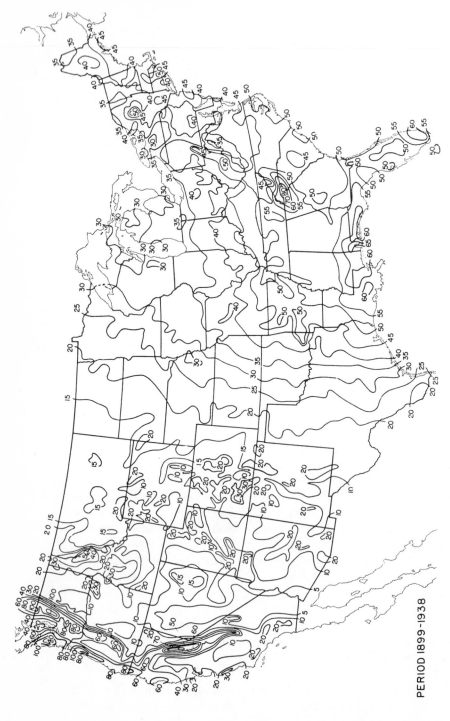

PERIOD 1899-1938

FIG. 5.2. Average annual precipitation in the United States in inches. Source: Climate and Man: The Yearbook of Agriculture (1941), U.S. Dept. of Agr.

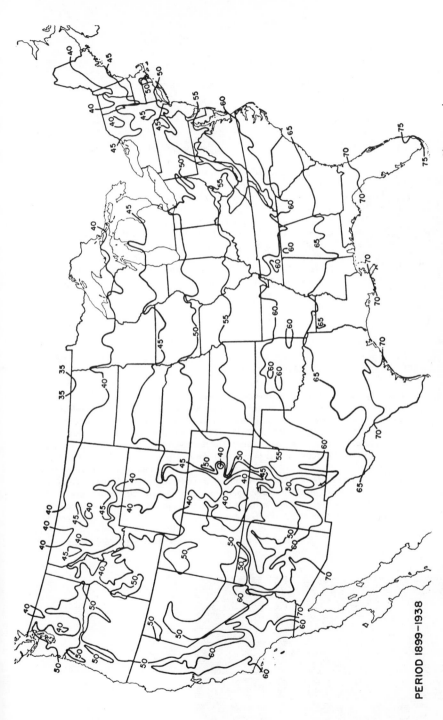

PERIOD 1899–1938

FIG. 5.3. Average annual temperature in the United States in degrees Fahrenheit. Source: *Climate and Man: The Yearbook of Agriculture* (1941), U.S. Dept. of Agr.

89

FIG. 5.4. Climate is the dominant factor in soil formation, and rainfall is the paramount factor in climate. With approximately 5 inches or less of annual rainfall, soluble salts (including lime) accumulate on the surface as shown in (1). With approximately 12 inches of annual rainfall, lime may leach downward and concentrate at a depth of 6 to 20 inches as shown by arrows in (2). With 24 inches of annual rainfall, most of the lime may have leached below 4 feet except for pockets, shown by arrows as white streaks in (3). (Scale is in feet.) *Sources:* (1), C. C. Nikiforoff, U.S. Dept. of Agr.; (2), Roy Larsen, Soil Conservation Service, U.S. Dept. of Agr.; (3), M. L. Horsch, Soil Conservation Service, U.S. Dept. of Agr.

FIG. 5.5. Sharp contrasts in climate in Alaska at 65° north latitude and in Puerto Rico at 18° 15′ north latitude are recorded in these contrasting soil profiles. *Left:* Soil that has developed on a north-facing slope in Alaska where the mean annual air temperatures range from 20° to 26°F and the mean annual precipitation ranges from 10 to 14 inches. The arrow points to the top of the permafrost (permanently frozen layer) at a depth of less than 2 feet. Overlying this is a light colored A horizon, and above this and extending to the surface is 18 inches of sphagnum moss that has accumulated. (Note profile description of Ester series that follows.) *Right:* Soil that has developed in Puerto Rico where the mean annual air temperature is 77°F (about 50°F warmer) and the mean annual precipitation is 76 inches (approximately 6 times as much). The soil profile features are fairly uniform to a depth of more than 6 feet. (See profile description of Nipe clay that follows.) (Both scales are in feet.) *Sources: Left,* J. C. F. Tedrow, Rutgers U. Agr. Exp. Sta.; *right,* Soil Conservation Service, U.S. Dept. of Agr.

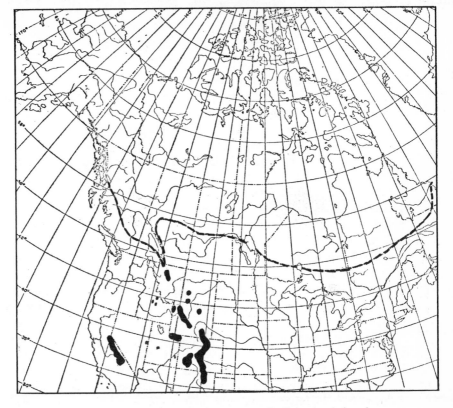

FIG. 5.6. Perpetual ice in the soil (permafrost) occurs in soils shown by darkened areas and in soils north of the dashed line. *Sources: (1) Darkened areas,* John L. Retzer, "Present Soil-Forming Factors and Processes in Artic and Alpine Regions," *Soil Science,* Vol. 99, No. 1, Jan., 1965, pp. 38–44; *(2) dashed line,* I. C. Brown and M. Drosdoff, "Chemical and Physical Properties of Soils and Their Colloids Developed from Granitic Materials in the Mojave Desert," *J. of Agr. Research,* Vol. 61, 1940, pp. 335–52.

3. Erosion of soils on steep hillsides
4. Deposition of soil materials downstream
5. More intense weathering, leaching, and erosion in warm regions where the soil does not freeze

Climate influences soil formation indirectly largely through its action on vegetation. Forests are the dominant vegetation in humid climates. The soil profile that develops in a forest has many more horizons than one that develops under grass. Semiarid climates encourage only prairie grasses, and a deep, dark, uniform surface soil results. Arid climates supply only enough moisture for sparse, short, plains grasses, which inadequately protect the soil against wind and water erosion.

Because of sharp contrasts, a brief comparison will be made of climate and soil formation in coastal, humid, tropical Puerto Rico and arid, Arctic Alaska (Figures 5.5 and 5.6). In humid, tropical Puerto Rico, the mean annual rainfall is more than 70 inches (with no dry season), and the July and January mean monthly temperatures are 80 and 75°F respectively. By strong contrast, the mean annual rainfall in Arctic Alaska is only 7 inches (0.1 as much), the July mean is 50°F (0.625 as much), and the January mean is minus 27°F (102°F lower).[1]

In continuously humid and hot Puerto Rico, the mean monthly temperatures of 75 to 80°F (24 to 27°C), and the annual variation at low elevations of the mean of less than 9°F (5°C), indicate that the weathering of rocks to make parent materials and the weathering of parent materials to make soils is continuous throughout each day in the year. The type of weathering under these conditions is mostly chemical decomposition, such as oxidation, reduction, solution, hydrolysis, hydration, and carbonation.

With low rainfall, dominantly cold weather, and a range of monthly temperature means of 77°F (50°F and minus 27°F), weathering in Arctic Alaska is mostly physical disintegration, such as exfoliation and differential expansion.

Soils resulting from hot and humid weathering in Puerto Rico are typically deep (often more than 10 feet), reddish in color, low in organic matter, and low in essential elements because of leaching. Soils in Arctic Alaska are usually very shallow (1 to 3 feet) to permafrost, very dark in color, high in organic matter, and high in essential elements because of scant leaching. Descriptions of representative soils in Alaska and Puerto Rico follow.

ESTER SERIES (Alaska)

The Ester series is a member of the loamy, mixed, acid family of Histic Lithic Pergelic Cryaquepts. These soils consist of a thick mossy O horizon, a thin A horizon, and mottled silty loess that is fairly high in mica over schist bedrock. They are perennially frozen at shallow depths.

Typifying Pedon: Ester silt loam—forest (colors are for moist conditions).

---

[1] J. C. F. Tedrow, J. V. Drew, D. E. Hill, and L. A. Douglas, "Major Genetic Soils of the Arctic Slope of Alaska," in *Selected Papers in Soil Formation and Classification,* Soil Science Society of America Special Publication Series No. 1, 1967, pp. 164–76.

Note: Although Arctic Alaska, receiving 5 to 10 inches of precipitation a year, is climatically a desert, ecologically it is a humid region because of the constantly high relative humidity that condenses as dew at night and thereby adds to the total moisture supply without being recorded as precipitation.

O11	12–6″	Raw sphagnum moss; clear smooth boundary.
O12	6–0″	Yellowish-brown (10YR 5/4), partially decomposed sphagnum moss; many twigs, leaves, and roots; extremely acid; abrupt, smooth boundary.
A1	0–2″	Very dark grayish-brown (10YR 3/2) silt loam; black (10YR 2/1) irregular streaks; weak, thin, platy structure; roots common; very strongly acid; clear, wavy boundary; 2 to 5 inches thick.
C1gf	2–12″	Gray (5Y 4/1) silt loam; many, very dark grayish-brown (2.5Y 3/2) streaks; weak, thin, platy structure; frozen with clear ice lenses; few angular rock fragments; strongly acid; clear wavy boundary; 7 to 15 inches thick.
IIC2gf	12–16″	Very dark grayish-brown (2.5Y 3/2), very gravelly silt loam; more than 60 per cent by volume of weathered schist fragments; frozen; medium acid; 0 to 10 inches thick.
IIIC	16″+	Shattered schist bedrock.

NIPE SERIES (Puerto Rico)

The Nipe series is a member of the clayey, oxidic, isohyperthermic family of Typic Acrorthox. These soils have fine-textured, highly weathered profiles with granular, dark reddish-brown A horizons and thick, dark red and dusky red, friable B horizons that extend to more than 80 inches.

Typifying Pedon:		Nipe clay—native pasture (colors are for moist soil).
Ap	0–11″	Dark reddish-brown (2.5YR 2/4) clay; strong, fine, granular structure; friable, nonsticky, slightly plastic; many fine roots; strongly acid; clear smooth boundary; 6–12 in. thick.
B1	11–18″	Dark reddish-brown (2.5YR 3/4) clay; weak, fine, angular blocky structure; very friable, nonsticky, slightly plastic; many fine pores; common fine roots; very strongly acid; clear smooth boundary; 6–10 in. thick.
B21	18–28″	Dark red (7.5R 3/8) clay; weak, fine, angular blocky structure; very friable, nonsticky, slightly plastic; many fine pores; few fine roots; very strongly acid; diffuse smooth boundary; 8–12 in. thick.
B22	28–38″	Dark red (7.5R 3/6) clay; massive; firm, nonsticky, slightly plastic; many fine pores; few fine iron concretions; strongly acid; diffuse smooth boundary; 8–12 in. thick.

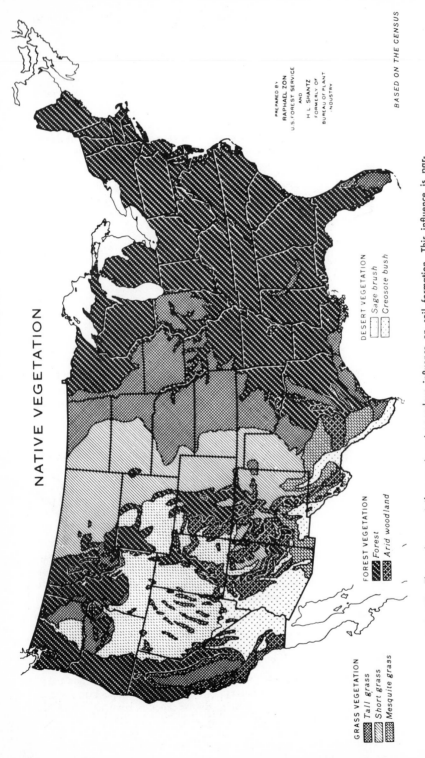

NATIVE VEGETATION

PREPARED BY
RAPHAEL ZON
U.S. FOREST SERVICE
AND
H. L. SHANTZ
FORMERLY OF
BUREAU OF PLANT
INDUSTRY

BASED ON THE CENSUS

DESERT VEGETATION
Sage brush
Creosote bush

FOREST VEGETATION
Forest
Arid woodland

GRASS VEGETATION
Tall grass
Short grass
Mesquite grass

FIG. 5.7. The native vegetation exerts a tremendous influence on soil formation. This influence is particularly noticeable in the tension zone between forests and grasslands (see Figs. 5.8 and 5.9). Source: U.S. Dept. of Agr.

FIG. 5.8. Under a forest vegetation in humid regions, the A horizon is more highly leached and there is a well-developed B horizon. (Minnesota.) (Compare with Fig. 5.9.) *Source:* U.S. Dept. of Agr.

FIG. 5.9. Under a grass vegetation in humid regions, the A horizon is dark and deep, and a B horizon is thin or absent. (Minnesota.) (Compare with Fig. 5.8.) *Source:* U.S. Dept. of Agr.

B23	38–48″	Dusky red (7.5R 3/4) clay; massive; friable, nonsticky, slightly plastic; many fine pores; few fine iron concretions; strongly acid; gradual smooth boundary; 8–12 in. thick.
B24	48–62″	Dark red (7.5R 3/6) clay; massive; friable, nonsticky, slightly plastic; many fine pores; medium acid; diffuse smooth boundary; 10–20 in. thick.
B25	62–80″+	Dusky red (7.5R 3/4) clay; massive; friable, nonsticky and nonplastic; many fine pores; medium acid.

BIOSPHERE AND SOIL FORMATION

Plant and animal life greatly influence the processes of soil formation and the character of the soil produced. Organic matter, acidity, and bulk density are the soil characteristics most quickly influenced by the kinds of plants and animals present. Differences in soils that have resulted primarily from differences in vegetation are especially noticeable in the tension zone where trees and grasses meet. Minnesota, Illinois, Missouri, Oklahoma, and Texas are some of the states in which these differences can be observed readily (Figure 5.7).

Soils developed under forest vegetation have more horizons, a more highly leached A horizon, and less humified organic matter than do soils that have developed under grass vegetation. Grassland soils near the tension zone are rich in humified organic matter for a foot or more and contain a weakly developed B horizon (Figures 5.8 and 5.9).

When they exist in large numbers, burrowing animals, such as moles, gophers, prairie dogs, earthworms, and ants, are highly important in soil

FIG. 5.10. Anthropic influences assist in making small rocks from big rocks and thus hasten soil formation. *Photo by Roy L. Donahue.*

formation. Animals in the soil tend to retard the translocation of weathered products that lead to distinct horizon differentiation. Soils with many burrowing animals will have fewer horizons because of constant mixing within the profile (see Chapter 8, Soil Ecology).

Men (or rather women and children here) also contribute toward making smaller soil material from larger rocks and thus hasten the time required for soil formation (Figure 5.10).

PARENT MATERIAL AND SOIL FORMATION

Rocks on the surface of the earth are weathered until all essential elements become available to support lichens and other lower forms of plant life. As continuing generations of lichens grow, die, and decay, they leave increasing amounts of organic matter. Organic acids further hasten decay of the rock. With an increasing buildup of organic matter and fine rock fragments, more rainwater is available for use by larger numbers of plants and animals.

In time, mobile materials near the surface will be leached downward and some of them deposited a few inches below the surface. The zone of deposition constitutes the beginnings of a B horizon. In a few hundred years, the leached surface soil (A horizon) and the concentrated subsurface (B horizon) will be well developed and contrasting in nearly all characteristics. Surface erosion removes the top of the A horizon as fast as the

FIG. 5.11. This Vertisol (soil) is very rich in calcium because it developed from parent material high in calcium. (Calcium is the whitish parent material below four feet.) (Houston Black Clay in Central Texas.) *Source:* U.S. Dept. of Agr.

A and B horizons slowly settle into the parent rock from whence they came. The soil is then in equilibrium with its environment.

The kind of soil that develops depends in part upon the kind of rock present. Granite is slow to weather, and soils developing from it are usually not very productive. From limestone as the parent rock there develops a dark-colored soil of greater productivity (Figure 5.11). Sandy soils of low fertility develop from sandstone, and shale results in silt loam soils of low productive potential.

But most parent materials of soils in the United States are not from the rocks directly beneath them. Approximately 97 per cent of the parent materials are deposits that were moved in place by water, wind, ice, or gravity, or a combination of these transporting agents. The kind of soil that develops from these unconsolidated deposits depends to a large extent on the texture, structure, nutrient content, and topographic position of the parent materials.

RELIEF AND SOIL FORMATION

Relief influences soil formation primarily through its associated water relations. With the same kind of climate and parent material, soils that have

FIG. 5.12. Relief and soil formation: The Ajanta caves in central India where Buddhists chiseled temples in solid basalt bedrock about the time of Christ are shown in panorama. On the hillsides above the caves where the relief is less steep (A), trees and grasses are growing and the soil is several feet deep. On the very steep relief where the temples have been carved (B), no soil has been formed because the rock fragments erode as fast as they are detached from the bedrock. Vegetation and soil are again present in the valley floor (C). *Photo by Roy L. Donahue.*

developed on steep hillsides have thin A and B horizons. First, the surface erodes quite rapidly and, second, less water moves down through the profile (Figure 5.12).

Soil materials on gently sloping hillsides have more water passing through them. The profile generally is deep, the vegetation more luxuriant, and the organic-matter level higher than in soils on steep topography.

Materials lying in land-locked depressions receive runoff waters from above. Such conditions favor a greater production of vegetation but a slower decomposition of the dead remains; the result is the existence of soils with large amounts of organic accumulations. If the area is wet at the surface for many months of the year, a Fibrists, Folists, Hemists, or Saprists (peat or muck) organic soil develops.

TIME AND SOIL FORMATION

The length of time required for a soil to develop horizons depends upon many interrelated factors, such as climate, nature of the parent material, burrowing animals, and relief. Horizons tend to develop faster under cool, humid, forested conditions. Acid sandy loams lying on gently rolling topography appear to be most conducive to rapid soil-profile development.

FIG. 5.13. Time and soil formation. (1): This basalt rock has not had enough time to produce a soil profile, even though the rainfall is adequate. (Ethiopia.) (A not-soil.) (2): There is not sufficient rainfall here to develop a distinct soil profile. (Colorado.) (Entisol.) (3): The soil is so high in percentage of clay that soil profile formation has been retarded, even though there has been sufficient time and ample rainfall. Note darkened area near arrow where darker colored surface soil has fallen in the deep and wide cracks. (California.) (Scale is in feet.) (Vertisol.) Sources: (1), Roy L. Donahue; (2) and (3), Soil Conservation Service, U.S. Dept. of Agr.

Under ideal conditions, a recognizable soil profile may develop within 200 years; under less favorable circumstances, the time may be extended to several thousand years (Figure 5.13).

Factors that retard soil-profile development are as follows:

1. Low rainfall
2. Low relative humidity
3. High lime or sodium carbonate content of parent material
4. Excessive sandiness, with very little silt and clay
5. A high percentage of clay
6. Resistant parent material, such as granite
7. Very steep slopes
8. High water tables
9. Constant accumulations of soil material by deposition
10. Severe wind or water erosion of soil material
11. Large numbers of burrowing animals
12. All of man's activities, such as plowing, liming, and fertilizing.

SUMMARY

The five factors that influence soil formation are climate, biosphere, parent material, relief, and time. Rainfall and temperature are the principal active ingredients in climate that affect soil formation. The depth of accumulation of lime is determined by the amount of rainfall available for leaching. Soils in northern Alaska are shallower than those in Puerto Rico largely as a result of rainfall and temperature.

QUESTIONS

1. Name the five factors of soil formation.
2. What factor is the principal one that determines the depth of a lime layer?
3. Contrast a soil profile in northern Alaska to one in Puerto Rico.
4. How does steepness of slope affect soil formation?
5. Can biosphere retard soil profile development?

REFERENCES

Chinzei, T., K. Oya, and Z. Koja, in collaboration with Roy L. Donahue and J. C. Shickluna, *Soils and Land Use in the Ryukyu Islands*. Naha, Okinawa: University of the Ryukyus, 1967, 187 pp.

Donahue, Roy L., *Soils of Equatorial Africa and Their Relevance to Rational Agricultural Development*. Research Report No. 7, Institute of International Agriculture, Michigan State University, 1970, 52 pp.

Soil Taxonomy of the National Cooperative Soil Survey. Soil Conservation Service, United States Department of Agriculture, Washington, D.C., Dec., 1970.

The following papers on soil formation are especially valuable for a greater understanding of the subject. (All of these papers have been reproduced in *Selected Papers in Soil Formation and Classification*, Soil Science Society of America, Special Publication Series No. 1, 1967, 428 pages.)

Cady, John G., "Mineral Occurrence in Relation to Soil Profile Differentiation," pp. 336–41.

Crocker, R. L., "The Plant Factor in Soil Formation," pp. 179–90.

Crompton, E., "Soil Formation," pp. 3–15.

Ruhe, Robert V., "Geomorphic Surfaces and the Nature of Soils," pp. 270–85.

Thorp, James, "Effects of Certain Animals That Live in Soil," pp. 191–208.

Note: Captions for colorplates are on pages 102, 103.

MOHAVE LOAM

A11 BROWN LOAM
A12 REDDISH BROWN SANDY LOAM
B1T BROWN SANDY CLAY LOAM

1' BROWN CLAY LOAM
B2T

PRISMATIC–
STRONG BLOCKY

2'

B3CA BROWN CLAY LOAM FIRM

3' NOTE CaCO3 DEPOSITS

MANITOBA SITE 2

Ft.

1

2

3

4

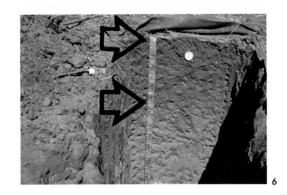

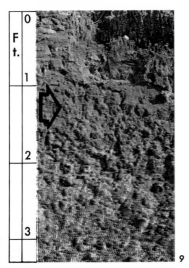

CHAPTER 6

SOIL CLASSIFICATION

(SOIL TAXONOMY)

Since 1951 we have been developing a system of soil classification that incorporates the knowledge gained since 1938. . . . In 1960 we published a draft of the classification to permit wide review and criticism. . . . In 1965 we began to use the classification. . . . No useful classification can remain static while knowledge increases and demands for interpretation change.—CHARLES E. KELLOGG

THE U. S. SYSTEM OF SOIL TAXONOMY— GUIDELINES FOR ESTABLISHMENT

Since 1951, the Soil Survey Staff of the United States Department of Agriculture has been catalyzing and blending ideas from soil scientists throughout the world in response to seven definite proposals ("Approximations") for the development of a system of soil classification within the following guidelines: [1]

---

[1] *Soil Classification: A Comprehensive System, Seventh Approximation,* Soil Survey Staff, U.S.D.A. (Aug., 1960), p. 11; and *Soil Taxonomy of the National Cooperative Soil Survey.* Soil Conservation Service, U.S.D.A., Wash., D.C., Dec., 1970.

Note: Before the adoption of the "Seventh Approximation" in 1965, the official system of soil classification used in the United States was first published in complete form in 1938 and modified in 1949; references to these two systems are: Mark Baldwin, Charles E. Kellogg, and James Thorp, "Soil Classification," in *Soils and Men: The Yearbook of Agriculture* (1938), pp. 979–1001; James Thorp and Guy D. Smith, "Higher Categories of Soil Classification: Order, Suborder, and Great Soil Groups," *Soil Science,* Vol. 67 (1949), pp. 117–26.

TABLE 6.A SOIL PROFILES IN COLOR REPRESENTING ALL TEN

Figure No.	Order	Suborder	Great Group	Subgroup	Family	Series
1	Alfisols	Boralfs	Cryoboralfs	Typic Cryoboralfs	Loamy, mixed, frigid	Waitville
2	Aridisols	Argids	Haplargids	Typic Haplargids	Fine loamy, mixed, thermic	Mohave
3	Entisols	Psamments	Ustipsamments	Typic Ustipsamments	Mixed, mesic	Valentine
4	Histosols	Hemists	Cryohemists	Terric Cryohemists	Sphagnic, borustic, euic	Grindstone
5	Inceptisols	Ochrepts	Eutrochrepts	Typic Eutrochrepts	Coarse silty, mixed, thermic	Natchez
6	Mollisols	Ustolls	Argiustolls	Typic Argiustolls	Fine, montmorillonitic, mesic	St. Paul
7	Oxisols	Torrox	Torrox	Typic Torrox	Clayey, halloysitic, isohyperthermic	Makaweli
8	Spodosols	Orthods	Haplorthods	Typic Haplorthods	Sandy, mixed, frigid	Kalkaska
9	Ultisols	Udults	Paleudults	Aquic Paleudults	Fine silty over clayey mixed, thermic	Sawyer
10	Vertisols	Usterts	Pellusterts	Udic Pellusterts	Fine, montmorillonitic, thermic	Houston black

SOIL ORDERS: THEIR TAXONOMY AND DIAGNOSTIC HORIZONS

Phases of Series	General Diagnostic Horizons or Differentiating Features		Specific Diagnostic Horizon Indicated by Arrows on Photos	Location of Soil Profile
	Epipedon	Endopedon		
Waitville sandy loam *	*High* base saturation	Argillic, ochric, and calcic horizons	Argillic endopedon	Manitoba, Canada
Mohave loam, 1–5 per cent slopes	Ochric horizon (light colored)	Calcic, natric, or gypsic horizon	Calcic endopedon	Arizona
Valentine fine sand *	—	Sandy to a depth of one meter	Sandy to a depth of one meter	Nebraska
Grindstone peat, deep phase *	Histic horizon	—	Histic epipedon	Manitoba, Canada
Natchez, silt loam 7–40 per cent slopes	Ochric horizon (light colored)	Cambic horizon	Cambic endopedon	Mississippi
St. Paul silt loam, 3–7 per cent slopes	Mollic horizon (dark colored)	Argillic (clay) horizon	Mollic epipedon	Dewey County, Oklahoma
Makaweli silty clay loam, 2–12 per cent slopes	—	Oxic horizon	Oxic endopedon	Island of Kauai, Hawaii
Kalkaska sand, 0–40 per cent slopes	—	Spodic horizon	Spodic endopedon	Osceola County, Michigan
Sawyer loamy sand, 2–8 per cent slopes	Low base saturation	Argillic (clay) horizon	Argillic endopedon	Dooly County, Georgia
Houston black clay, 1–3 per cent slopes	30 per cent or more of clay in all horizons, wide and deep cracks, moist chromas of less than 1.5 in upper foot.		50–60 per cent clay, moist chroma of 1 in upper 38 in.	Travis County, Texas

* When no slope is indicated it is 0–1 per cent.

Courtesy credits: Figures 1, 4, Canada Department of Agriculture; Figures 3, 5, 6, 7, 10, Soil Conservation Service, U.S.D.A.; Figure 8, Eugene P. Whiteside and Figure 9, David A. Lietzke, both at Michigan State University.

1. "A *natural* classification of the soils of the United States is needed. If it will accommodate soils of other continents, it will be more useful in the United States."
2. "A *natural* classification should be based on the properties of the objects classified."
3. "The properties selected should be observable or measurable. . . ." [This represents the *scientific* viewpoint; any measurable property should yield the same or similar result each time it is measured.]
4. "The properties selected should be those that either *affect* soil genesis or *result* from soil genesis."
5. "If an arbitrary selection must be made between two properties of apparent equal genetic significance, but with different significance to plant growth, the property with the greater significance to plant growth should be selected. . . ."
6. "Subdivisions of all classes in a given category need not be made according to a common property. . . ."
7. "Many soils of the world are unknown [to soil scientists], and their placement in the system should not be predetermined. . . ."
8. "A classification system should be flexible enough that it may be modified to incorporate new knowledge without . . . confusion. . . ."
9. "It is normally undesirable to have a change in the classification of a soil as the result of a fire, or a single plowing. . . ."

THE UNITED STATES SYSTEM OF SOIL CLASSIFICATION

The classification system of soils that was adopted by the United States in 1965 is natural and scientific. Some of its categories contain elements of conceptual inference and some are based upon factual characteristics.

A simplified explanation of the 1965 soil classification system follows.[2]

[2] *Soil Taxonomy of the National Cooperative Soil Survey.* Soil Conservation Service, U.S.D.A., Wash., D.C., Dec., 1970.

Note: As Guy D. Smith explained in the Preface to *Soil Classification: A Comprehensive System, Seventh Approximation*: "This is not a book for beginning students of soil classification. It is written to introduce the new system to people who are familiar with the present system. It assumes knowledge of the *Soil Survey Manual* terminology for describing soils." [*Soil Survey Manual*, U.S.D.A. Handbook No. 18, Soil Survey Staff (1951 and 1962), 503 pp.]

The authors believe, however, that the *Seventh Approximation* is sufficiently important to present here, even in abbreviated form, because they have faith in the next generation's desire and motivation to learn. Readers who are not satisfied with this "short form" are encouraged to consult the references listed here.

There are six categories in the system:

Order ⎫
Suborder ⎬ Based on conceptual inference

Great Group ⎫
Subgroup ⎪
Family ⎬ Based on factual characteristics
Series ⎭

Order. Eight of the ten soil orders are differentiated by the presence or absence of diagnostic horizons or features that are characteristic of the kinds and intensities of soil-forming processes and contrasting climates. Entisols and Histosols may occur under any climate.

Suborder. Suborders within a soil order are differentiated largely on the basis of soil properties resulting from differences in soil moisture and soil temperature. Forty-seven suborders have been recognized at present.

Great Group. Soil great groups are subdivisions of suborders. The 225 identified great groups have been established largely on the basis of differentiating soil horizons and soil features. The differentiating soil horizons include those which have accumulated clay, iron, and/or humus and those which have pans that interfere with water movement or root penetration. The differentiating soil features are the self-mulching properties of clays, soil temperature, and major differences in content of calcium, magnesium, sodium, potassium, salts, and gypsum.

Subgroup. Each soil great group is generally divided into three subgroups: one representing the central (typic) segment of the soil group; a second segment that has properties that tend toward (intergrade) other orders, suborders, or other great groups; and a third segment the properties of which are such that prevent its classification as typic or an intergrade.

Family. Soil families are separated within a subgroup primarily on the basis of soil properties important to the growth of plants or behavior of soils when used for engineering purposes. The soil properties include texture, mineralogy, reaction (pH), soil temperature, permeability, thickness of horizons, structure, and consistence. No estimates are available on the total number of soil families.

Series. Each family contains several soil series. The 9000 or more soil series in the United States have narrower ranges of characteristics than the soil family. The name of the soil series represents a prominent geographic name of a river, town, or area where the series was first recognized. Soil series are differentiated on the basis of observable and mappable soil characteristics, such as color, texture, structure, consistence, thickness, and number and arrangement of horizons in the soil profile.

Soil Type. Although not officially recognized in the United States system of soil taxonomy, soil types are subdivisions of soil series and may be unofficially considered the lowest category in the system. Soil type names combined with slope and erosion classes, represented by letters and numbers, are used as mapping units on all standard soil maps. (In 1971 soil mapping units were officially designated *Phases of Series.*)

SOIL PROFILES REPRESENTING THE SOIL ORDERS

Photographs of each of the 10 soil orders from 9 states and 1 province of Canada have been selected and their taxonomy identified in the new U. S. system. There follows under the respective soil order, all of the names of the 47 suborders and the derivation of each name.

A brief summary of the characterization of each soil order is given, including the diagnostic horizons and differentiating features, followed by the official description of the typifying pedon (Table 6.1).[3]

At the end of this chapter, additional and relevant information is given: Table 6.2, Derivation of Names of Soil Great Groups and Subgroups Portrayed in this Chapter; Master Soil Horizons and Layers; Diagnostic Soil Horizons; and Key to Soil Orders.

Soil Order: *Alfisols* (origin: coined from Ped*alfer*)

Suborders: *Aqualfs* (Latin, *aqua,* water)
Boralfs (Greek, *boreas,* northern)[4]
Udalfs (Latin, *udus,* humid)
Ustalfs (Latin, *ustus,* burnt)
Xeralfs (Greek, *xeros,* dry)

Alfisols were formerly classified as Gray-Brown Podzolic, Noncalcic Brown, Gray Wooded, Degraded Chernozem, and associated Half-Bog and Planosol.

According to the 1967 soil map entitled "Patterns of Soil Orders and Suborders of the United States," Alfisols occur in large bodies in the East North Central States and in the Mountain States.

On the map "Soils of the World," Alfisols occupy 14.7 per cent of the land surface of the world (ranking third in area) and occur on all con-

---

[3] *Note:* The authors realize that this chapter on soil classification is too brief for the strongly motivated scholar to do more than whet his desire for further information on this historic subject. To such a person, the authors suggest further readings from the many references listed.

[4] Suborder illustrated by photograph.

FIG. 6.1. On soil maps this soil is shown as Nebish loam and is classified in the fine-loamy, mixed, frigid family of Typic Eutroboralfs. The complete taxonomy follows: Order—Alfisols; Suborder—Boralfs; Great Group—Eutroboralfs; Subgroup—Typic Eutroboralfs; Family—Typic Eutroboralf, fine-loamy, mixed, frigid; Series—Nebish; Type—Nebish loam. (Minnesota.) (Scale is in feet.) Source: W. M. Johnson, Soil Conservation Service, U.S. Dept. of Agr.

tinents. Humid and subhumid climates, and tall grasses, savanna, and oak-hickory forests characterize the climate and native vegetation where Alfisols occur.

Alfisols have marks of pedogenic processes characterized by the following:

1. Movement of clay into a Bt (clay accumulation) horizon, shown here between 7 and 17 inches in depth (between arrows)
2. A medium-to-high supply of basic cations in the soil, such as calcium and magnesium, which is evidence of only mild leaching (This is in contrast to Ultisols that have had severe leaching.)
3. Water available for plant growth for three or more warm-season months (Figure 6.1).

NEBISH SERIES

The Nebish series is a member of the fine-loamy, mixed, frigid family of the Typic Eutroboralfs. Typically, these soils have thin very dark gray, loam A1 horizons, moderately thick grayish brown loam or sandy loam

TABLE 6.1   SOIL PROFILES REPRESENTING THE TEN SOIL ORDERS:

Figure No.	Order	Suborder	Great Group	Subgroup	Family	Series
6.1	Alfisols	Boralfs	Eutroboralfs	Typic Eutroboralfs	Fine loamy, mixed, frigid	Nebish
6.2	Aridisols	Argids	Haplargids	Xerollic Haplargids	Fine loamy over sandy skeletal, mixed, mesic	Dixie
6.3	Entisols	Orthents	Torriorthents	Ustic Torriorthents	Fine silty, mixed, calcareous, mesic	Colby
6.4	Histosols	Hemists	Cryohemists	Terric Cryohemists	Spagnic, borustic, euic	Grindstone
6.5	Inceptisols	Ochrepts	Eutrochrepts	Aquic Dystric Eutrochrepts	Coarse loamy, mixed, mesic	Minoa
6.6	Mollisols	Udolls	Argiudolls	Typic Argiudolls	Fine, montmorillonitic, mesic	Sharpsburg
6.7	Oxisols	Ustox	Haplustox	Tropeptic Haplustox	Fine silty, halloysitic, isohyperthermic	Molokai
6.8	Spodosols	Orthods	Haplorthods	Typic Haplorthods	Sandy, mixed, frigid	Adams
6.9	Ultisols	Udults	Paleudults	Typic Paleudults	Fine loamy, siliceous, thermic	Ruston
6.10	Vertisols	Usterts	Pellusterts	Udic Pellusterts	Fine, montmorillonitic, thermic	Houston black

* Refer to Key to Soil Orders at the end of this chapter.

THEIR TAXONOMY AND DIAGNOSTIC HORIZONS *

Phases of Series	General Diagnostic Horizons or Differentiating Features		Specific Diagnostic Horizon Indicated by Arrows on Photos	Location of Soil Profile
	Epipedon	Endopedon		
Nebish loam, 2–6 per cent slopes	High base saturation	Argillic (clay) horizon	Argillic (clay) endopedon	Minnesota
Dixie gravelly loam, 1–6 per cent slopes	Ochric horizon (light colored)	Calcic, natric, or gypsic horizon	Calcic and gypsic endopedon	Utah
Colby silt loam, 1–30 per cent slopes	Ochric horizon (light colored)	—	Ochric epipedon	Colorado
Grindstone peat, deep phase	Histic (organic) horizon	—	Histic epipedon	Manitoba, Canada
Minoa very fine sandy loam, 1–3 per cent slopes	Ochric horizon (light colored)	Cambic horizon	Cambic endopedon	New York
Sharpsburg silty clay loam, 2–14 per cent slopes	Mollic horizon (dark colored)	Argillic (clay) horizon	Mollic epipedon	Nebraska
Molokai silty clay loam, 0–30 per cent slopes	—	Oxic horizon	Oxic endopedon	Hawaii
Adams sandy loam, 0–25 per cent slopes	—	Spodic horizon	Spodic endopedon	Vermont
Ruston fine sandy loam, 1–30 per cent slopes	Low base saturation	Argillic (clay) horizon	Argillic (clay) endopedon	Louisiana
Houston black clay, 0–8 per cent slopes	Dark colored, deep-cracking clays; may have slickensides and/or gilgai micro relief; compression and separation planes.		Compression and separation planes in soil mass	Texas

A2 horizons, brownish clay loam B horizons with clay accumulation, and light olive brown calcareous loam C horizons.

Typifying Pedon:		Nebish loam—forested (colors are for moist conditions unless otherwise noted).
A1	0–3″	Very dark gray (10YR 3/1) loam; weak, very fine granular structure; very friable; neutral; abrupt smooth boundary; 1–4 in. thick.
A2	3–9″	Grayish brown (10YR 5/2) sandy loam; weak, very thin, platy structure; very friable; neutral; abrupt wavy boundary; 4–10 in. thick.
B21t	9–17″	Dark yellowish brown (10YR 4/4) clay loam; strong, fine, and medium subangular blocky structure; firm; thin, nearly continuous porous grainy light gray (10YR 7/1, dry) coats on ped faces; medium patchy, very dark grayish brown (10YR 3/2) clay films on ped faces; slightly acid; gradual wavy boundary; 5–12 in. thick.
B22t	17–29″	Dark yellowish brown (10YR 4/4) clay loam; moderate, medium, and coarse subangular blocky structure; firm; medium, nearly continuous, very dark grayish brown (10YR 3/2) clay films on ped faces; slightly acid; gradual wavy boundary; 5–14 in. thick.
B3t	29–33″	Light olive brown (2.5Y 5/4) light clay loam; weak, medium, and coarse subangular blocky structure; firm; thin and medium patchy very dark grayish brown (2.5Y 3/2) clay films on ped faces; slightly acid; clear wavy boundary; 3–8 in. thick.
C	33–48″	Light olive brown (2.5Y 5/4) loam; weak, thin, platy structure parting to weak, fine, subangular blocky structure; friable; few white (10YR 8/2) lime threads; strongly effervescent; moderately alkaline.

Soil Order: *Aridisols* (origin: Latin, *aridus,* dry)

Suborders: [5] *Argids* (Latin, *arare,* to plow)
                 Orthids (Greek, *orthos,* true)

Aridisols include the former groups: Desert, Reddish Desert, Sierozem, Solonchak, a few Brown and Reddish Brown, and associated Solonetz.

Aridisols are located in the United States primarily in the western Mountain States and Pacific States in areas of low rainfall where scattered

---

[5] Suborder illustrated by photograph.

FIG. 6.2. This Aridisol is shown on a soil map as Dixie gravelly loam and would be classified in the fine-loamy over sandy skeletal, mixed, mesic family of Xerollic Haplargids. The complete taxonomy follows: *Order*—Aridisols; *Suborder*—Argids; *Great Group*—Haplargids; *Subgroup*—Xerollic Haplargids; *Family*—Xerollic Haplargid, fine-loamy over sandy skeletal, mixed, mesic; *Series*—Dixie; *Type*—Dixie gravelly loam. (Utah.) (Scale is in feet.) *Source:* Soil Conservation Service, U.S. Dept. of Agr.

grasses and desert shrubs dominate the vegetation. Worldwide, Aridisols rank first in area of all 10 soil orders with 19.2 per cent. The Sahara Desert is shown on the map as Aridisols, as are large areas of central and southern Asia, Australia, southern Africa, and southern South America.

Aridisols develop in areas of low rainfall and from almost any kind of parent material; their characteristics are therefore extremely variable. Little leaching has occurred in Aridisols; for this reason, in one or more horizons there is a concentration of calcium carbonate, gypsum, and/or soluble salts. This Aridisol has a conspicuous horizon of enrichment of secondary carbonates that extends from 15 to 36 inches in depth, marked by arrows (Figure 6.2).

DIXIE SERIES

The Dixie series is a member of the fine-loamy, mixed, mesic family of Xerollic Haplargids. Typically, Dixie soils have brown, neutral, gravelly loam, A horizons; reddish brown, mildly alkaline, gravelly sandy clay loam B2t horizons and white, weakly to strongly cemented gravelly loam Cca horizons, underlain by very gravelly heavy sandy loam.

Typifying Pedon: Dixie gravelly loam—rangeland (colors are for dry soils unless otherwise noted).

A1   0–6"   Brown (10YR 5/3) gravelly loam, dark brown (10YR 4/3) moist; strong, fine and very fine, granular structure; soft, friable; the immediate surface layer, to a depth of about 1 in., is lighter colored and has weak, thin, platy structure and very fine vesicular porosity; few fine and medium roots; about 20 per cent gravel; neutral (pH 7.2); clear smooth boundary; 4–10 in. thick.

B2t	6–15″	Reddish brown (5YR 5/4); gravelly sandy clay loam; reddish brown (5YR 4/4) when moist; moderate, fine, subangular, blocky structure; hard, friable; few fine and medium roots; few thin, clay films on some ped faces; about 15 per cent gravel; mildly alkaline (pH 7.4); abrupt wavy boundary; 5–12 in. thick.
C1ca	15–36″	White (10YR 8/2); weakly to strongly cemented caliche; (gravelly loam) very pale brown (10YR 7/3) when moist; massive and breaks into chips and fragments; roots penetrate only where it is cracked; moderately alkaline, (pH 8.1); abrupt wavy boundary; 12–25 in. thick.
C2	36–72″	Very pale brown (10 YR 7/4); very gravelly, heavy, loamy, skeletal; yellowish brown (10YR 5/4) when moist; becomes increasingly gravelly and cobbly with depth; massive; soft, very friable; about 60 per cent gravel; strongly calcareous; moderately alkaline (pH 7.9).

Soil Order: *Entisols* (origin: from rec̲e̲n̲t̲ soils)

Suborders: *Aquents* (Latin, *aqua,* water)
Arents (Latin, *arare,* to plow)
Fluvents (Latin, *fluvius,* river)
Orthents (Greek, *orthos,* true)[6]
Psamments (Greek, *psammos,* sand)

Entisols are a soil order that includes the former Azonal soils and a few Low Humic Gley.

Entisols occur widely distributed in the United States, including river flood plains, rocky soils of mountainous areas, and beach sands. World distribution of Entisols is on all continents, with a total of 12.5 per cent of the land surface and a rank of fourth among the 10 soil orders. Rainfall may be high or low, and vegetation may be forest, savanna, or grass. Precipitation and vegetation are therefore not diagnostic of the soil order.

There is no genetic horizon in any Entisol, but possibly a plow layer exists. Entisols are unique among the soil orders by:

1. The dominance of mineral soil materials (in contrast to organic soil materials which are classified as Histosols)
2. The absence of distinct pedogenic horizons
3. The presence of an Ochric horizon (between arrows in Figure 6.3).

The absence of distinct pedogenic horizons in Entisols may be due to:

---

[6] Suborder illustrated by photograph.

1. Presence of a parent material too inert to develop soil horizons, such as quartz sand
2. Formation of the soil from a parent material that dissolves almost completely with very little residue, such as limestone
3. Insufficient time to develop horizons, as in recently deposited volcanic ash or river alluvium
4. Ecological conditions not conducive to horizon formation, as is true of soil on the moon
5. Occurrence of steep slopes where the rate of surface erosion equals or exceeds the rate of soil profile formation.

COLBY SERIES

The Colby series is a member of the fine-silty, mixed, calcareous, mesic family of Ustic Torriorthents. These soils have grayish-brown, calcareous, silt loam A1 horizons, and calcareous silt loam C horizons that are pale brown in the upper part and very pale brown in the lower part.

Typifying Pedon:	Colby silt loam—grassland (colors are for dry soil unless otherwise noted).
A1 0–4″	Grayish brown (10YR 5/2) silt loam; dark grayish brown (10YR 4/2) moist; weak, fine, platy structure to a depth of 2 in., weak, fine, granular structure below; friable; calcareous; moderately alkaline; gradual smooth boundary; 3–6 in. thick.

FIG. 6.3. This soil is mapped as Colby silt loam and is classified in the fine-silty, mixed, calcareous, mesic family of Ustic Torriorthents. The complete taxonomy follows: Order—Entisols; Suborder—Orthents; Great Group— Torriorthents; Subgroup—Ustic Torriorthent; Family— Ustic Torriorthent, fine-silty, mixed, calcareous, mesic; Series—Colby; Type—Colby silt loam. (Colorado.) (Scale is in feet.) Source: Arvad Cline, Soil Conser. Serv.

AC	4–8″	Light brownish gray (10YR 6/2) silt loam; grayish brown (10YR 5/2) moist; weak, fine, granular structure; friable; calcareous, moderately alkaline; gradual smooth boundary; 3–6 in. thick.
Cca	8–20″	Pale brown (10YR 6/3); silt loam, brown (10YR 5/3) moist; structureless, massive; friable; few fine roots and root channels; porous; few soft lime accumulations; calcareous, moderately alkaline; gradual wavy boundary; 10–30 in. thick.
C	20–60″	Very pale brown (10YR 7/3) silt loam; light yellowish brown (10YR 6/4) moist; structureless, massive; friable; porous; strongly calcareous, moderately alkaline.

Soil Order: *Histosols* (origin: Greek, *histos,* tissue)

Suborders: *Fibrists* (Latin, *fibra,* fiber)
Folists (Latin, *folia,* leaf)
Hemists (Greek, *hemi,* half)[7]
Saprists (Greek, *sapros,* rotten)

Histosols were formerly known as Bog soils. They occur in large bodies in Florida and Georgia and in numerous areas in the Northeast and Lake States. Outside the United States, the largest continuous body of Histosols is in Canada south of Hudson Bay, and another large area is in northwestern Canada. Histosols do occur on all continents but in areas too small to be shown on the soil map of the world. The world-wide extent of Histosols is only 0.8 per cent of all soils, and they rank last in area of all 10 soil orders.

Precipitation and humidity are high, and vegetation may be marsh grasses and sedges. In the northern United States, ericaceous shrubs, black spruce, tamarack, white spruce, black ash, and red maple are common; in the southeastern United States, cypress, black gum, and mangrove may predominate.

The classification of Histosols is based upon their physical properties (stage of decomposition, thickness, bulk density, water-holding capacity, temperature, and permeability) and chemical properties (carbonates, bog iron, sulfates, sulfides, pH, base saturation, and carbon:nitrogen ratio).

Based upon their degree of decomposition, Histosols are classified into four suborders, as follows:

1. Fibrists (Latin, *fibra,* fiber) soil materials. These are the least decomposed, and the original plant remains can readily be identified

---

[7] Suborder illustrated by photograph.

FIG. 6.4. This organic soil (Histosol) is mapped as Grindstone peat, deep phase, and is classified in the sphagnic, borustic, euic family of Terric Cryohemists. The complete taxonomy follows: *Order*—Histosols; *Suborder*—Hemists; *Great Group*—Cryohemists; *Subgroup*—Terric Cryohemists; *Family*—Terric Cryohemists, sphagnic, borustic, euic; *Series*—Grindstone; *Type*—Grindstone peat, deep phase. (Manitoba, Canada.) (Scale is in six inch intervals.) *Source:* Canadian Dept. of Agr.

as to botanical origin.

2. Folists (Latin, *folium,* leaf). These materials have originated from the accumulation of very slowly decomposing leaves and twigs.

3. Hemists (Greek, *hemi,* half; implying half-decomposed) soil materials intermediate in degree of decomposition.

4. Saprists (Greek, *sapros,* rotten) soil materials. These organic materials are the most highly decomposed of the four suborders. The original plants comprising the organic materials cannot be identified because of the advanced stages of decomposition (Figure 6.4).

GRINDSTONE SERIES [8]

Classification: Terric Mesisol

Family: Sphagnic, borustic, euic

Series: Grindstone—deep phase

Classification, U.S.D.A.: Terric Cryohemist

---

[8] *Note:* The horizon designation is that used by the Canadian Department of Agriculture.

Typifying Pedon:

Of	0–12"	Yellowish brown (10YR 5/6 moist) to very pale brown (10YR 7/4 moist), nonwoody fibrous, spongy; compacted or layered; sphagnum moss with woody intrusions (tamarack roots and stems), medium acid, unrubbed fiber content 76 per cent.
Oml	12–24"	Very dark brown to dark reddish brown (10YR 2/2 to 5YR 3/3 moist) moderately decomposed, mixed woody and nonwoody fibrous material composed of mosses, shrubby remains, and herbaceous remains, becoming more herbaceous near bottom of layer; medium acid; unrubbed fiber content 61 per cent.
Om2	24–48"	Dark brown to very dark brown (10YR 3/3 to 2/2 moist) moderately decomposed; medium fibered herbaceous material, matted or feltlike; medium acid; unrubbed fiber content 54 per cent near top to 45 per cent near bottom of layer.
IIAhg	48–51"	Black (5Y 3/0 wet) clay; strong, fine, granular; sticky, very plastic; neutral; abrupt, wavy boundary.
IICg	51"+	Light gray (5Y 5/1 wet) clay; amorphous; sticky; neutral to mildly alkaline.

Soil Order: *Inceptisols* (origin: Latin, *inceptum,* beginning, inception)

Suborders: *Andepts* (Japanese, *ando: an,* black; *do,* soil)
Aquepts (Latin, *aqua,* water)
Ochrepts (Greek base of *ochros,* pale)[9]
Plaggepts (modified from German, *Plaggen,* sod)
Tropepts (modified from Greek, *tropikos,* of the solstice, tropical)
Umbrepts (Latin, *umbra,* shade)

Inceptisols before 1965 were officially known in the United States as Brown Forest, Low Humic Gley, Humic Gley, Ando, and Sol Brun Acide.

Inceptisols occur mostly in the Middle Atlantic States and the Pacific States. They develop in humid climates from the Arctic to the Tropics mostly under trees, but sometimes under grasses. On a global basis, Inceptisols occupy 15.8 per cent of the land surface, second in area of the 10 soil orders. They occur on all continents, but the largest area is in mainland China.

---

[9] Suborder illustrated by photograph.

FIG. 6.5. This soil is mapped as Minoa very fine sandy loam and is classified in the coarse-loamy, mixed, mesic family of Aquic Dystric Eutrochrepts. The complete taxonomy follows: *Order*—Inceptisols; *Suborder*—Ochrepts; *Great Group*—Eutrochrepts; *Subgroup*—Aquic Dystric Eutrochrepts; *Family*—Aquic Dystric Eutrochrept, coarse-loamy, mixed, mesic; *Series*—Minoa; *Type*—Minoa very fine sandy loam. (New York State.) (Scale is in feet.) *Source:* Soil Conservation Service, U.S. Dept. of Agr.

There is a uniformity in texture of the various horizons in Inceptisols, and soil textures are finer than loamy fine sand, some accumulation of organic matter on the surface of virgin soils, and some slight evidence of weathering. A Cambic endopedon lies between 10 and 32 inches on the photograph and is indicated by an arrow (Figure 6.5).[10]

MINOA SERIES

The Minoa series is a member of the coarse-loamy, mixed, mesic family of Aquic Dystric Eutrochrepts. Minoa soils have very dark grayish-brown loamy A horizons, mottled brown and grayish-brown loamy B horizons, and mottled brownish-gray C horizons, with the entire soil having a high content of very fine sand and few or no coarse fragments.

Typifying Pedon: Minoa very fine sandy loam—plowed field (colors are for moist soil unless otherwise stated).

Ap   0–10″   Very dark grayish-brown (10YR 3/2) very fine sandy loam; light brownish-gray (10YR 6/2) dry; moderate fine granular structure; very friable; many fine roots; medium acid; abrupt smooth boundary; 7–10 in. thick.

---

[10] *Note:* Further characterizations of Inceptisols are in: *Soil Taxonomy of the National Cooperative Soil Survey.* Soil Conservation Service, U.S.D.A., Wash., D.C., Dec., 1970.

B2	10–22″	Brown (10YR 5/3) loamy very fine sand; common, medium and fine, faint and distinct, yellowish brown and pale brown mottles; very weak, very fine granular structure; very friable; few fine roots; many fine and medium pores; medium acid; clear wavy boundary; 10–15 in. thick.
B3	22–32″	Grayish brown (10YR 5/2) loamy very fine sand; many coarse and medium, distinct and prominent, yellowish-brown and strong brown mottles; massive; friable; few fine roots; few fine pores; few, firm, dark yellowish-brown (10YR 4/4), very fine sandy loam lamellae ¼ in. thick; slightly acid; clear wavy boundary; 8–16 in. thick.
C	32–46″	Light, brownish-gray (10YR 6/2), loamy very fine sand; many medium and coarse, distinct and prominent, brown, dark yellowish-brown and strong brown mottles; massive; firm; neutral.

Soil Order: *Mollisols* (origin: Latin, *mollis,* soft)

Suborders: *Albolls* (Latin, *albus,* white)
Aquolls (Latin, *aqua,* water)
Borolls (Greek, *boreas,* northern)
Rendolls (From *Rendzina,* Polish word for noise in plowing dark, limey, clay soils)
Udolls (Latin, *udus,* humid)[11]
Ustolls (Latin, *ustus,* burnt)
Xerolls (Greek, *xeros,* dry)

Mollisols were known formerly as Rendzina, Prairie, Chernozem, Chestnut, Brunizem, and associated Solonetz and Humic Gley.

The largest contiguous body of Mollisols in North America is in the Great Plains, extending north into Canada and south almost to the Gulf of Mexico. Other areas occur in the Intermountain Region in the West and Northwest. The largest body of Mollisols in the world is in central Europe, central Asia, and northern mainland China. A second large area is in Argentina and adjoining countries in South America. Mollisols rank sixth and total 9 per cent of all soils of the world. Precipitation is subhumid to semiarid in regions varying from North Temperate to alpine (mountainous) to Tropical.

Mollisols are characterized by the presence of a surface horizon (mollic

---

[11] Suborder illustrated by photograph.

FIG. 6.6. On a soil map, this soil is shown as Sharpsburg silty clay loam and is classified in the fine, montmorillonitic, mesic family of Typic Argiudolls. The complete taxonomy follows: *Order*—Mollisols; *Suborder*—Udolls; *Great Group*—Argiudolls; *Subgroup*—Typic Argiudolls; *Family*—Typic Argiudoll, fine, montmorillonitic, mesic; *Series*—Sharpsburg; *Type*—Sharpsburg silty clay loam. (Lancaster County, Nebraska.) (Scale is in feet.) *Source:* Soil Conservation Service, U.S. Dept. of Agr.

epipedon) that is thick, dark in color, strong in structure, and more than 50 per cent saturated with bases (mostly calcium). In this profile, the mollic epipedon extends from the surface to a depth of more than 2 feet, at arrows (Figure 6.6).

SHARPSBURG SERIES

The Sharpsburg series is a member of the fine, montmorillonitic, mesic family of Typic Argiudolls. They typically have very dark brown, friable, silty clay loam A horizons, brown and dark yellowish-brown, firm, heavy, silty clay loam B2 horizons that are mottled in the lower part, and mottled yellowish-brown and grayish-brown, friable, light, silty clay loam B3 and C horizons.

Typifying Pedon:		Sharpsburg silty clay loam—cultivated (colors are for moist soil unless otherwise stated).
Ap	0–8″	Very dark brown (10YR 2/2), silty clay loam, dark gray (10YR 4/1) dry; weak, fine, granular structure; friable; slightly acid; abrupt smooth boundary; 6–8 in. thick.
A12	8–11″	Very dark brown (10YR 2/2), medium, silty clay loam; weak, very fine, subangular blocky structure parting to moderate fine granular structure; friable; medium acid; gradual smooth boundary; 0–6 in. thick.

A3   11–17″   Very dark grayish-brown (10YR 3/2), medium, silty clay loam, some brown (10YR 4/3) peds; moderate, very fine, subangular blocky structure; friable; medium acid; gradual smooth boundary; 4–8 in. thick.

B21t   17–24″   Brown (10YR 4/3), heavy, silty clay loam; weak, medium, prismatic structure parting to moderate, fine, and very fine, subangular blocky structure; firm; common, very dark gray (10YR 3/1) coats; thin discontinuous clay films; strongly acid; gradual smooth boundary; 6–10 in. thick.

B22t   24–32″   Brown (10YR 4/3) and dark yellowish-brown (10YR 4/4), heavy, silty clay loam; few, fine, faint, distinct, strong brown (7.5YR 5/6) and fine, faint, grayish-brown (2.5Y 5/2) mottles; moderate, medium, prismatic structure parting to moderate, fine, subangular, blocky structure; firm; thick discontinuous clay films; few fine dark bodies; strongly acid; gradual smooth boundary; 6–10 in. thick.

B23t   32–44″   Dark yellowish-brown (10YR 4/4), medium, silty clay loam; common fine and medium, faint grayish-brown (2.5Y 5/2) mottles; few fine distinct yellowish-brown (10YR 5/6) mottles; weak, medium, prismatic structure parting to weak, medium, subangular blocky structure;

FIG. 6.7. This Oxisols is mapped as Molokai silty clay loam and is classified in the fine, halloysitic, isohyperthermic family of Tropeptic Haplustox. The complete taxonomy follows: *Order*—Oxisols; *Suborder*—Ustox; *Great Group* —Haplustox; *Subgroup*—Tropeptic Haplustox; *Family*—Tropeptic Haplustox, fine-silty, halloysitic, isohyperthermic; *Series*—Molokai; *Type* —Molokai silty clay loam. (Island of Molokai, Hawaii.) (Scale is in feet.) *Source:* W. M. Johnson, Soil Conservation Service, U.S. Dept. of Agr.

		friable; few, thin, discontinuous clay films; many very fine dark bodies; medium acid; gradual smooth boundary; 8–12 in. thick.
B3t	44–49″	Yellowish-brown (10YR 5/4), light, silty clay loam; common fine and medium faint grayish-brown (2.5Y 5/2) and yellowish-brown (10YR 5/6) mottles; massive, some vertical cleavage; few thin discontinuous clay films; many fine dark bodies; medium acid; gradual smooth boundary; 4–12 in. thick.
C	49–70″	Grayish-brown (2.5Y 5/2) and yellowish-brown (10YR 5/4), light silty clay loam; common fine distinct brown (7.5YR 4/4) mottles; massive; friable; many fine dark bodies; slightly acid.

Soil Order: *Oxisols* (origin: French, *oxide,* oxide)

Suborders: *Aquox* (Latin, *aqua,* water)
      Humox (Latin, *humus,* earth)
      Orthox (Greek, *orthos,* true)
      Torrox (Latin, *torridus,* hot and dry)
      Ustox (Latin, *ustus,* burnt)[12]

Oxisols, although not officially known to occur in the conterminous United States, were formerly known in Hawaii, Puerto Rico, and other parts of the Tropics as Laterites and Latosols. On the map, Patterns of Soil Orders and Suborders of the United States, Oxisols are not in the legend, since their occurrences in Hawaii are too small in area to be shown. Oxisols are very extensive, however, in tropical South America and Africa. On a world basis, they rank fifth and total 9.2 per cent of all soil orders.

The typical climate under which Oxisols develop is continuously hot and humid, and the vegetation is usually a forest.

Oxisols have developed on old upland, medium-to-fine textured parent materials that have weathered into crystalline kaolinitic-type clays with a net negative charge, and amorphous iron and aluminum oxides with a net positive charge. The oxic horizon often has a net positive charge.

Weatherable minerals in Oxisols are either absent or present only in trace amounts. There are therefore no reserves of bases in these soils except those on the exchange complex and in plant tissue. This fact explains the principal cause of the usual rapid decline in crop yields when Oxisols are cleared and cultivated (Figure 6.7).

---

[12] Suborder illustrated by photograph.

There is almost no translocation of clay in Oxisols; Bt horizons are therefore usually faint, diffuse, or absent. Clays in Oxisols are usually not dispersable by shaking in distilled water. High permeability and low erodibility are further characteristics of Oxisols.

The oxic horizon in Figure 6.7 is the B horizon, between 15 and 72 inches, between arrows.

MOLOKAI SERIES

The Molokai series is a member of the fine-silty, halloysitic, isohyperthermic family of Tropeptic Haplustoxs. Typically, these soils have dark reddish-brown, friable A horizons that have weak granular structure, dark reddish-brown B horizons that have weak, coarse, prismatic structure in the upper part and moderate fine and very fine, subangular blocky structure in the lower part, and fine black manganese concretions throughout the pedon that effervesce with hydrogen peroxide.

Typifying Pedon:		Molokai silty clay loam—pineapple plantation (colors are for moist soil unless otherwise noted).
Apl	0–7″	Dark reddish-brown (2.5YR 3/4), silty clay loam, dark red (2.5YR 3/6) dry; weak, very fine, fine and medium granular structure; slightly hard, friable, slightly sticky, plastic; many roots; many interstitial pores; many very fine black concretions that effervesce strongly with hydrogen peroxide; extremely acid (pH 4.4); clear wavy boundary; 6–7 in. thick.
Ap2	7–15″	Dark reddish-brown (2.5YR 3/4), silty clay loam, dark red (2.5YR 3/6) dry; weak, medium and coarse, subangular blocky structure breaking to moderate, fine and very fine, granular structure; slightly hard, friable, sticky, plastic; roots common; common very fine tubular and interstitial pores; common fine black concretions that effervesce strongly with hydrogen peroxide; very strongly acid (pH 4.6); clear smooth boundary; 8–9 in. thick.
B21	15–35″	Dark reddish-brown (2.5YR 3/4), silty clay loam, red (2.5YR 4/6) dry; weak, coarse, prismatic structure breaking to weak, coarse, subangular blocky structure; slightly hard, friable, sticky, plastic; many very fine and fine tubular pores; few shiny patchy faces on prisms; common fine black concretions that effervesce strongly with hydrogen peroxide; slightly acid (pH 6.5); gradual wavy boundary; 14–22 in. thick.

| B22 | 35–64" | Dark reddish-brown (2.5YR 3/4), silty clay loam, red (2.5YR 4/6) dry; weak, coarse, prismatic structure breaking to strong, very fine and fine, subangular blocky structure; moderately compact in place, slightly hard, firm, sticky, plastic; many very fine and common tubular pores; common patchy pressure faces on peds; common patchy clay films on ped faces; few very fine black concretions that effervesce moderately with hydrogen peroxide; neutral reaction (pH 6.6); gradual wavy boundary; 27–30 in. thick. |
| B3 | 64–72" | Dark reddish-brown (5YR 3/3) clay loam; dark reddish-brown (5YR 3/4) dry; moderate, fine and very fine, subangular blocky, and angular blocky structure; slightly hard, friable, slightly sticky, plastic; common, very fine and fine, tubular pores; thin patchy clay films on peds; red (2.5YR 4/6) clay films on the walls of larger pores; common hard earthy lumps; few very fine black concretions that effervesce slightly with hydrogen peroxide; neutral reaction (pH 6.6). |

Soil Order: *Spodosols* (origin: Greek, *spodos,* wood ash)

Suborders: *Aquods* (Latin, *aqua,* water)
Ferrods (Latin, *ferrum,* iron)
Humods (Latin, *humus,* earth)
Orthods (Greek, *orthos,* true)[13]

Spodosols were formerly known in soil classification as Podzol, Ground-Water Podzol, and Brown Podzolic. They exist in large tracts in New England, at high elevations in the Middle Atlantic States, and/or in the northern part of the Great Lakes States. Between latitude 42°N and 60°N, Spodosols are common in Canada, northern Europe, and northern Asia. They rank eighth and occupy 5.4 per cent of the world's land surface.

Humid climate with cool temperatures and a forest vegetation are the conditions under which most Spodosols have formed.

Spodosols exhibit evidence of translocation of humus, aluminum, and/or iron. A small amount of silicate clay also may be translocated into the B horizon as amorphous materials. The most unique characteristic of strongly developed Spodosols is a sandy B horizon indurated (cemented) with humus and noncrystalline aluminum and/or iron, which is black, brown, or reddish in color (spodic horizon, shown between arrows in photograph). The percentage base saturation in spodic horizons is usually

[13] Suborder illustrated by photograph.

Ft.

FIG. 6.8. This soil of northern Temperate and Sub-arctic latitudes is mapped as Adams sandy loam and is classified in the sandy, mixed, frigid family of Typic Haplorthods. The complete taxonomy follows: *Order*—Spodosols; *Suborder*—Orthods; *Great Group* —Haplorthods; *Subgroup*—Typic Haplorthods; *Family*—Typic Haplorthod, sandy, mixed, frigid; *Series* —Adams; *Type*—Adams sandy loam. (East Middle-bury, Vermont.) (Scale is in feet.) *Source:* A. R. Midgley, Vt. Agr. Exp. Sta.

less than 2.5 per cent. In virgin soils, an albic horizon (light colored and coarse-to-medium textured) usually overlies the B horizon. No Spodosol series are listed in clayey families in the United States (Figure 6.8).

ADAMS SERIES

The Adams series is a member of the sandy, mixed, frigid family of Typic Haplorthods. These acid, typically gravel-free, sand soils have plowed layers rich to poor in organic matter, depending upon use, and subsoils enriched in humus and iron/aluminum over loose sands or gravel and sand.

Typifying Pedon:		Adams sandy loam—forested (colors refer to moist, broken soil, unless specified otherwise).
02	4–0″	Black (10YR 2/1) humus in a mat of roots and fungal hyphae; very strongly acid; abrupt smooth boundary; 2–6 in. thick, if undisturbed.
A2	0–4″	Pinkish-gray (7.5YR 7/2) loamy sand; single grain; loose; plentiful medium roots; few pores; no coarse fragments; very strongly acid; abrupt wavy boundary; 2–6 in. thick.
B21h	4–6″	Dark reddish-brown (5YR 2/2) loamy sand; weak, fine granular; very friable, except few firm sections ½ to 2 in. across; plentiful medium roots; porous; no coarse fragments; very strongly acid; clear wavy boundary; 1–3 in. thick.

B22ir	6–10″	Brown to dark brown (7.5YR 4/4) loamy sand; very weak, very fine granular; very friable; plentiful roots; porous; no coarse fragments; very strongly acid; gradual wavy boundary; 3–7 in. thick.
B23ir	10–16″	Brown (7.5YR 5/4) sand; single grain; loose; large roots present; few pores; no coarse fragments; very strongly acid; diffuse wavy boundary; 4–8 in. thick.
B3	16–26″	Yellowish-brown (10YR 5/4) sand; single grain; loose; few large roots; few pores; no coarse fragments; very strongly acid; diffuse wavy boundary; 6–12 in. thick.
C	26–40″+	Grayish-brown (10YR 5/2) sand; single grain; loose; few large roots; few pores; no coarse fragments; strongly acid.

Soil Order: *Ultisols* (origin: Latin, *ultimus,* last)

Suborders: *Aquults* (Latin, *aqua,* water)
Humults (Latin, *humus,* earth)
Udults (Latin, *udus,* humid)[14]
Ustults (Latin, *ustus,* burnt)
Xerults (Greek, *xeros,* dry)

Ultisols were formerly classified as Red-Yellow Podzolic, Reddish Brown Lateritic, and associated Planosol and Half-Bog. Ultisols exist mostly in the South Atlantic States, the eastern South Central States, and in the Pacific States, mostly in subtropical climates. Central America, South America, western Africa, southeastern Asia, and Australia all have large areas of Ultisols. Ranking seventh in area among the 10 soil orders, Ultisols occupy 8.5 per cent of the world's land area.

Most Ultisols have had their genesis under humid climates, subtropical to tropical temperatures, and a forest or forest-plus-grass (savanna) vegetation.

Characteristic of Ultisols is the presence of an horizon in which clay has accumulated. Intensive weathering of the clay (argillic) horizon is responsible for the low (less than 35 per cent) base saturation. Some weatherable minerals (micas or feldspars) must be present, as well as some illitic or montmorillonitic clay. The mean annual soil temperature is more than 47°F (8°C).

This profile of an Ultisol, Ruston sandy loam, has developed from marine or alluvial sediments of Pleistocene age or older, in the Gulf Coastal Plain in the southern United States. The soil is well drained,

---

[14] Suborder illustrated by photograph.

FIG. 6.9. This Ultisol is mapped as Ruston fine sandy loam and is classified in the fine-loamy, siliceous, thermic family of Typic Paleudults. The complete taxonomy follows: *Order*—Ultisols; *Suborder*—Udults; *Great Group*—Paleudults; *Subgroup*—Typic Paleudults; *Family*—Typic Paleudult, fine-loamy, siliceous, thermic; *Series*—Ruston; *Type*—Ruston fine sandy loam. (Louisiana.) (Scale is in feet.) *Source:* Soil Conservation Service, U.S. Dept. of Agr.

moderately permeable, and acid. The argillic endopedon occurs at a depth of 16 to 41 inches, marked by arrows (Figure 6.9).

RUSTON SERIES

The Ruston series is a member of a fine-loamy, siliceous, thermic family of Typic Paleudults. These soils have fine sandy loam A horizons that are less than 20 in. thick, sandy clay loam Bt horizons with appreciable silt and hues of 5YR or redder, and sola more than 60 in. thick.

Typifying Pedon:		Ruston fine sandy loam—forest, formerly culti-vated (colors are for moist soil).
Ap	0–4″	Brown (10YR 5/3), fine sandy loam; weak, medium granular structure; friable; medium acid; clear smooth boundary; 3–6 in. thick.
A2	4–16″	Pale brown (10YR 6/3), fine sandy loam; a few small pockets of yellowish-red (5YR 5/6) in lower part; weak, medium, subangular blocky structure; very friable; many fine pores; small streaks of bleached sand grains common; medium acid; clear smooth boundary; 6–15 in. thick.

B2t 16–27" Yellowish-red (5YR 4/6), sandy clay loam; moderate, fine, subangular blocky structure; friable; almost complete (2.5YR 3/6) clay films on ped surfaces; sand grains coated and bridged with clay; strongly acid; clear wavy boundary; 9–20 in. thick.

B3t 27–41" Yellowish-red (5YR 5/6), fine sandy loam; weak, medium, subangular blocky structure; friable; almost complete (2.5YR 3/6) clay films on ped surfaces; sand grains coated and bridged with clay; strongly acid; clear wavy boundary; 10–40 in. thick.

A'2 & B 41–47" Yellowish-red (5YR 5/6), fine sandy loam; weak, medium, subangular blocky structure; firm; common fine pores; ½ to 2 in. diameter pockets of somewhat brittle, light yellowish-brown (10YR 6/4), fine sandy loam that makes up approximately one-half the horizon; few patchy clay films; strongly acid; clear wavy boundary; 0–10 in. thick.

B'2't 47–67" Coarsely mottled, yellowish-red (5YR 5/6), yellowish-brown (10YR 5/4), red (2.5YR 4/6), and light gray (10YR 7/2), sandy clay loam; moderate, medium, subangular blocky structure; firm, somewhat brittle; patchy clay films in upper part and continuing in lower; strongly acid; clear wavy boundary; 0–28 in. thick.

B'22t 67–92" Coarsely mottled, red (2.5YR 4/6), yellowish-brown (10YR 5/4), and strong brown (7.5YR 5/6), fine sandy loam; moderate, medium, subangular blocky structure; firm and brittle; continuous clay films on ped surfaces in upper part grading to patchy in lower part; very strongly acid; gradual boundary; 0–38 in. thick.

C 92–96" Coarsely mottled, red (2.5YR 4/8) and light gray (10YR 7/1) clay loam; massive; friable; very strongly acid.

Soil Order: *Vertisols* (origin: Latin, *verto,* turn)

Suborders: *Torrerts* (Latin, *torridus,* hot and dry)
     Uderts (Latin, *udus,* humid)
     Usterts (Latin, *ustus,* burnt)[15]
     Xererts (Greek, *xeros,* dry)

Vertisols at one time were known as Rendzina and Grumusol. They exist primarily in central and southeastern Texas and to a lesser extent in Alabama, Mississippi, and California. Outside the United States, large

---

[15] Suborder illustrated by photograph.

areas of Vertisols occur in India, Australia, and eastern Africa. Ranking ninth in area, Vertisols comprise 2.1 per cent of the total land surface of the world.

Vertisols have developed under subhumid to semiarid climates with vegetation dominated by tall grasses and scattered trees and shrubs.

Soils in the order of Vertisols have developed from parent materials high in limestones, marls, or basic rocks, such as basalt. They are high (more than 35 per cent) in montmorillonitic clay, high (more than 30 milliequivalents) in exchange capacity, and when dry have wide, deep cracks (more than 1 centimeter wide at a depth of 50 centimeters).

Vertisols have profiles that exhibit evidence of pedogenic processes of seasonal turning of the soil surface by peds falling into dry-weather cracks. Vertisols also show evidence of vertical and horizontal mass movement by the presence of uneven soil surfaces (gilgai micro relief), slickensides, and wedge-shaped (parallelepiped) compound subsoil aggregates tilted at an angle from the horizontal. Vertical movement throughout the year of a soil similar to the one in this profile was measured at a 3-foot depth and was reported to be 1.5 inches.[16] (See also Vertisol profile in Ethiopia, page 250).

The light-colored horizon (below 3 feet) in this Vertisol contains more than 50 per cent calcium carbonate.

Evidence of vertical and horizontal movement of soil mass may be seen as diagonal compression and separation planes at arrow (Figure 6.10).

HOUSTON BLACK SERIES

The Houston Black series is a member of the fine, montmorillonitic, thermic family of Udic Pellusterts. These cyclic clayey soils have very dark gray A horizons, thick grayish-brown upper AC horizons, and mottled olive brown and gray lower AC horizons.

Typifying Pedon:		From center of microdepression—pasture (colors are for dry soil unless otherwise noted).
A11	0–8″	Very dark gray (10YR 3/1) clay; black (10YR 2/1) moist; moderate, fine, subangular blocky, and moderate, medium, granular structure; extremely hard, very firm, very sticky and plastic; many fine roots; common worm casts; few snail shell fragments; shiny ped faces; few, fine,

---

[16] R. M. Smith, R. C. Henderson, and O. J. Tippit, *Summary of Soil and Water Conservation Research from the Blackland Experiment Station, Temple, Texas, 1942–53*, Texas Agricultural Experiment Station Bulletin 781 (1954), 54 pp.

FIG. 6.10. This soil is mapped as Houston Black clay and is classified in the fine, montmorillonitic, thermic family of Udic Pellusterts. The complete taxonomy follows: Order—Vertisols; Suborder—Usterts; Great Group—Pellusterts; Subgroup—Udic Pellusterts; Family—Udic Pellustert, fine, montmorillonitic, thermic; Series—Houston Black; Type—Houston Black clay. (San Marcos, Hays County, Texas.) (Scale is in feet.) Source: Alan Anderson, Soil Conservation Service, U.S. Dept. of Agr.

		black, weakly cemented iron–manganese concretions; few, fine, strongly cemented $CaCO_3$ concretions; calcareous in matrix; moderately alkaline; clear wavy boundary; 6–12 in. thick.
A12	8–24″	Very dark gray (10YR 3/1) clay, black (10 YR 2/1) moist; moderate fine and very fine, angular blocky structure; remaining characteristics same as preceding horizon; gradual wavy boundary; 0–38 in. thick.
A13	24–38″	Dark gray (10YR 4/1) clay, very dark gray (10 YR 3/1) moist; coarse, grooved, intersecting slickensides that form parallelepipeds; common fine roots; remaining characteristics same as preceding horizon; 0–20 in. thick.
AC1	38–80″	Grayish-brown (10YR 5/2) clay; dark grayish-brown (10YR 4/2) moist; with few medium, distinct mottles of olive brown (2.5YR 4/4), and many coarse, faint mottles of gray (10YR 5/1); structure and consistence same as preceding horizon; few fine roots; few worm casts; shiny ped faces; few streaks of dark gray from above; few fine black weakly cemented concretions and fine brown masses of iron–manganese; few fine and medium, strongly and weakly cemented $CaCO_3$ concretions; few medium, powdery masses of $CaCO_3$; calcareous in matrix; moderately alkaline; gradual wavy boundary; 10–50 in. thick.
AC2	80–104″+	Distinctly and coarsely mottled, light olive brown (2.5Y 5/4) and gray (10YR 6/1) clay; common fine mottles of olive and brown; weak medium and coarse angular blocky structure; few intersecting slickensides that form parallelepipeds; very hard; very firm; very sticky and plastic; few fine roots; few fine brown masses of iron and manganese; few powdery masses of $CaCO_3$; calcareous in matrix; moderately alkaline.

THE U. S. SYSTEM OF SOIL TAXONOMY—ADDENDA

Since the United States system of soil taxonomy is so new, important, and complex, these addenda are placed here to serve as a supplement for greater in–depth understanding by the more highly motivated student.

Addendum A. Derivation of Names of Soil Great Groups and Subgroups Portrayed in this Chapter.

Addendum B. Master Soil Horizons and Layers.

Addendum C. Diagnostic Soil Horizons.

Addendum D. Key to Soil Orders.

ADDENDUM A

TABLE 6.2 DERIVATION OF NAMES OF SOIL GREAT GROUPS AND SUBGROUPS PORTRAYED IN THIS CHAPTER

Formative Element	Derivation of Formative Element	Mnemonicon (Reminder)	Connotation of Formative Element
Aqu	Latin, *aqua*, water	Aquarium	Aquic moisture regime
Arg	Modified from argillic horizon; Latin, *argilla*, white clay	Argillite	An argillic horizon
Eutr, eu	Modified from Greek, *eu*, good; *eutrophic*, fertile	Eutrophic	High base saturation
Frag	Modified from Latin, *fragilis*, brittle	Fragile	Presence of fragipan
Hapl	Greek, *haplous*, simple	Haploid	Minimum horizon
Med	Latin, *media*, middle	Medium	Of temperate climates
Pale	Greek, *paleos*, old	Paleosol	Old development
Pell	Greek, *pellos*, dusky	*Pellaea*, genus of ferns with dark stripes	Low chroma
Torr	Latin, *torridus*, hot and dry	Torrid	Torric moisture regime
Trop	Modified from Greek, *tropikos*, of the solstice	Tropical	Humid and continually warm
Ud	Latin, *udus*, humid	Udometer	Udic moisture regime
Ust	Latin, *ustus*, burnt	Combustion	Ustic moisture regime
Xer	Greek, *xeros*, dry	Xerophyte	Xeric moisture regime

Note: Typic, modified from "typical," is used as a part of the subgroup name when the soil is thought to typify the great group.

ADDENDUM B
MASTER SOIL HORIZONS AND LAYERS [17]

ORGANIC HORIZONS

O—Organic horizons of mineral soils. Horizons (1) formed or forming in the upper part of mineral soils above the mineral part; (2) dominated by fresh or partly decomposed organic material; and (3) containing more than 30 per cent organic matter if the mineral fraction is more than 50 per cent clay, or more than 20 per cent organic matter if the mineral fraction has no clay. Intermediate clay content requires proportional organic matter content.

O1—Organic horizons in which essentially the original form of most vegetative matter is visible to the naked eye.

O2—Organic horizons in which the original form of most plant or animal matter cannot be recognized with the naked eye.

MINERAL HORIZONS AND LAYERS

Mineral horizons contain less than 30 per cent organic matter if the mineral fraction contains more than 50 per cent clay or less than 20 per cent organic matter if the mineral fraction has no clay. Intermediate clay content requires proportional content of organic matter.

A—Mineral horizons consisting of: (1) horizons or organic matter accumulation formed or forming at or adjacent to the surface; (2) horizons that have lost clay, iron, or aluminum with resultant concentration of quartz or other resistant minerals of sand or silt size; or (3) horizons dominated by (1) or (2) but transitional to an underlying B or C.

A1—Mineral horizons, formed or forming at or adjacent to the surface, in which the feature emphasized is an accumulation of humified organic matter intimately associated with the mineral fraction.

A2—Mineral horizons in which the feature emphasized is loss of clay, iron or aluminum, with resultant concentration of quartz or other resistant minerals in sand and silt sizes.

---

[17] *Supplement to Agricultural Handbook No. 18,* Soil Conservation Service, U.S.D.A. (May, 1962).

A3—A transitional horizon between A and B, and dominated by properties characteristic of an overlying A1 or A2 but having some subordinate properties of an underlying B.

AB—A horizon transitional between A and B, having an upper part dominated by properties of A and a lower part dominated by properties of B, and the two parts cannot conveniently be separated into A3 and B1. Such combined horizons are normally thin; they should be separated if thick enough to permit separation.

B—Horizons in which the dominant feature or features is one or more of the following: (1) an illuvial concentration of silicate clay, iron, aluminum, or humus, alone or in combination; (2) a residual concentration of sesquioxides or silicate clays, alone or mixed, that has formed by means other than solution and removal of carbonates or more soluble salts; (3) coatings of sesquioxides adequate to give conspicuously darker, stronger, or redder colors than overlying and underlying horizons in the same sequum but without apparent illuviation of iron and not genetically related to B horizons that meet requirements of (1) or (2) in the same sequum; (4) an alteration of material from its original condition in sequums lacking conditions defined in (1), (2), and (3) that obliterates original rock structure, that forms silicate clays, liberates oxides, or both, and that forms granular, blocky, or prismatic structure if textures are such that volume changes accompany changes in moisture.

B1—A transitional horizon between B and A1 or between B and A2 in which the horizon is dominated by properties of an underlying B2 but has some subordinate properties of an overlying A1 or A2.

B2—That part of the B horizon where the properties on which the B is based are without clearly expressed subordinate characteristics indicating that the horizon is transitional to an adjacent overlying A or an adjacent underlying C or R.

B3—A transitional horizon between B and C or R in which the properties diagnostic of an overlying B2 are clearly expressed but are associated with clearly expressed properties characteristic of C or R.

C—A mineral horizon or layer, excluding bedrock, that is either like or unlike the material from which the solum is presumed to have formed, relatively little affected by pedogenic processes, and lacking properties diagnostic of A or B but including materials modified by: (1) weathering outside the zone of major biological activity; (2) reversible cementation, development of brittleness, development of high bulk density, and other properties characteristic of fragipans; (3) gleying; (4) accumulation of calcium or magnesium carbonate or more soluble salts; (5) cementation by such accumulations as calcium or magnesium carbonate, or more soluble salts; or (6) cementation by alkali-soluble siliceous material or by iron and silica.

R—Underlying consolidated bedrock, such as granite, sandstone, or limestone. If presumed to be like the parent rock from which the adjacent overlying layer or horizon was formed, the symbol R is used alone. If presumed to be unlike the overlying material, the R is preceded by a Roman numeral denoting lithologic discontinuity, as explained under this heading.

LETTER SUBSCRIPTS

b—Buried soil horizon. This symbol is added to the designation of a buried genetic horizon or horizons. Horizons of another solum may or may not have formed in the overlying material, which may be similar to, or different from, the assumed parent material of the buried soil.

ca—An accumulation of carbonates of alkaline earths, commonly of calcium carbonate.

cs—An accumulation of calcium sulfate.

f—Frozen soil. The suffix f is used for soil that is thought to be permanently frozen (permafrost).

g—Strong gleying. The suffix g is used with a horizon designation to indicate intense reduction of iron during soil development, or reducing conditions due to stagnant water, as evidenced by base colors that approach neutral, with or without mottles. In aggregated material, ped faces in such horizons generally have chroma of 2 or less as a continuous phase and commonly have few or faint mottles. Interiors of peds may have prominent and many mottles but commonly have a network of threads or bands of low chroma surrounding the mottles. In soils that are not aggregated, a base chroma of 1.0 or less, with or without mottles, is indicative of strong gleying. Hues bluer than 10Y are also indicative of strong gleying in some soils. Horizons of low chroma in which the color is due to uncoated sand or silt particles are not considered strongly gleyed.

h—Illuvial humus. Accumulations of decomposed illuvial organic matter, appearing as dark coatings on sand or silt particles, or as discrete dark pellets of silt size, are indicated by h. If used, this suffix follows the letter B or a subdivision of B, as Bh or B2h.

ir—Illuvial iron. Accumulations of illuvial iron as coatings on sand or silt particles or as pellets of silt size; in some horizons the coatings have coalesced, filled pores, and cemented the horizon.

m—Strong cementation, induration. The symbol m is applied as a suffix to horizon designations to indicate irreversible cementation. The symbol is not applied to indurated bedrock. Contrary to previous usage, m is not used to indicate firmness, as in fragipans, but is confined to indurated horizons which are essentially (more than 90 per cent) continuous, although they may be fractured.

p—Plowing or other disturbance. The symbol p is used as a suffix with A to indicate disturbance by cultivation or pasturing. Even though a soil has been truncated and the plow layer is clearly in what was once B horizon, the designation Ap is used. When an Ap is subdivided, the Arabic number suffixes follow, as Ap1 and Ap2, for the Ap is considered comparable to A1, A2, or B2.

t—Illuvial clay. Accumulations of translocated silicate clay are indicated by the suffix t (German, *ton,* clay). The suffix t is used only with B, as B2t, to indicate the nature of the B.

x—Fragipan character. The symbol x is used as a suffix with horizon designations to indicate genetically developed properties of firmness, brittleness, high density, and characteristic distribution of clay that are diagnostic of fragipans. Fragipans, or parts of fragipans, may qualify as A2, B, or C. Such horizons are classified as A2, B, or C, and the symbol x is used as a suffix to indicate fragipan character. Unlike comparable use of supplementary symbols, the symbol x is applied to B without the connotative Arabic numeral normally applied to B. Arabic numerals used with C to indicate only vertical subdivision of the horizon precede the x in the symbol, as C1x, C2x.

SUBDIVISION OF HORIZONS

In a single profile it is often necessary to arbitrarily subdivide the horizons for which designations are provided—for example, to subdivide Ap, A1, A2, A3, B1, B2, B3, or C so that detailed studies of morphology can be correctly recorded. In all such cases, the subdivisions are numbered consecutively, with Arabic numbers, from the top of the horizon downward. For example, subdivisions of a B2 horizon would be labeled B21, B22, and B23.

LITHOLOGIC DISCONTINUITIES

Roman numerals are prefixed to the appropriate horizon designations when it is necessary to number a series of layers of contrasting material consecutively from the surface downward. A soil that is all in one kind of material is designated by the Roman numeral I. This numeral therefore can be omitted from the symbol, as it is understood that all the material is I. Similarly, the uppermost material in a profile having two or more contrasting materials is always designated I. Consequently, for the topmost material, the numeral I can be omitted from the symbol because it is always understood. Numbering starts with the second layer of contrasting material, which is designated II, and each contrasting material below this second layer is numbered consecutively, III, IV, and so on, downward as part of each horizon designation. Even though a layer below a layer designated by II is similar to the topmost layer, it is given the appropriate consecutive number in the sequence. Where two or more horizons developed in one of the numbered layers, the Roman number is applied to all the horizon designations in that material.

Following are two examples of horizon sequences using this convention:

A1—A2—B1—B21—IIB22—IIB3—IIC1—IIIC2
A1—A2—B1—B2—IIA′2—IIB′x—IICIx—IIIC2x—IIIC3—IVR

In the first example, the first contrasting layer is unnumbered; the second layer, starting in the B2, is indicated by Roman II, as IIB22; the third, within the C, by the symbol IIIC. In the second example, the first contrasting layer is unnumbered; the second, starting at the top of A′2, is numbered II; the third, starting in the middle of the fragipan, is numbered III, even though the fragipan is partly in C; and the fourth, starting below C, is indicated by IVR. Note that Arabic numerals are used independently of the Roman numerals, in the conventional manner, both as connotative symbols and for vertical subdivision.

ADDENDUM C
DIAGNOSTIC SOIL HORIZONS

EPIPEDONS (Surface Horizons) [18]

Anthropic epipedon. (Greek, *anthropos,* man). Same as mollic epipedon except high in phosphate due to long cultivation and fertilization. Recognized principally in old farming areas of Asia and Europe.

Histic epipedon. (Greek, *histos,* tissue). A peaty or mucky (organic) horizon less than a foot thick, normally water saturated. Present in Histic Humaquepts. (Histosols have thicker organic horizons.)

Mollic epipedon. (Latin, *mollis,* soft). High organic matter content; must have moist colors with value darker than 3.5; horizon must be at least 7 inches thick, or occupy one-third of the solum. Diagnostic of Mollisols.

Ochric epipedon. (Greek, *ochros,* pale). Lighter colored, lower in organic matter, or thinner than mollic, umbric, anthropic, or histic epipedons; present in Alfisols, Aridisols, and Ultisols.

Plaggen epipedon. (German, *Plaggen,* sod). Contains an accumulation from centuries of additions of sod, straw, manure, and farm litter; more than 50 centimeters thick. Common in old farming areas of Europe.

Umbric epipedon. (Latin, *umbra,* shade, hence dark). Same as mollic epipedon except hydrogen saturated. Present in Umbraquults.

ENDOPEDONS (Subsurface Horizons)

Albic endopedon. (Latin, *albus,* white). Clay and iron oxides have leached out, leaving light-colored sand and silt. Usually underlain by a spodic or argillic horizon. Common in Spodosols.

Argillic endopedon. (Latin, *argilla,* white clay). B horizon is significantly enriched with clay from above; must have clay skins on surfaces of peds. Present in Alfisols and Ultisols.

Cambic endopedon. (Late Latin, *cambiare,* to exchange). A slightly weathered horizon between the A and C horizons; soil structure present

---

[18] *Pedon* (origin: Greek, *pedon,* ground) may be considered a three-dimensional soil profile, and is the smallest unit volume that satisfactorily represents a soil.
  *Epi*pedon (origin: Greek, *epi,* on, upon).
  *Endo*pedon (derived from Greek, *endon,* in, within).

(or absence of rock structure) in at least half of the volume; less evidence of illuviation than in argillic or spodic endopedons; no clay skins evident; does not develop in sands. May be present in Inceptisols and Mollisols.

Natric endopedon. (New Latin, *natrium,* sodium). Same as argillic endopedon except columnar or prismatic structure and over 15 per cent exchangeable sodium. Occurs in some Aridisols and some Mollisols.

Oxic endopedon. (French, *oxide,* oxide). A concentration of hydrated oxides of iron and/or aluminum, and kaolinitic clays with few weatherable minerals remaining. No clay skins. Formed only at low elevations on old land surfaces in the humid Tropics. Diagnostic of Oxisols.

Spodic endopedon. (Greek, *spodos,* wood ash). B horizon with an accumulation of humus or sesquioxides, or both. No clay accumulation; no clay skins; may be cemented into ortstein. Diagnostic of Spodosols.

Notes:

1. Other endopedons include: agric, gypsic, petrocalcic, petrogypsic, placic, salic, duripan, fragipan, calcic, and sulfuric.
2. Soil features that are not diagnostic but are characterizing and important in soil descriptions and in man's use of the soil include: durinodes, permafrost, paralithic contact, gilgai, plinthite, and slickensides.

ADDENDUM D

KEY TO SOIL ORDERS [19]

A. Organic soils—**Histosols, p. 114.**
B. Other soils that have a spodic horizon with an upper boundary within 2 m of the surface, or that have a placic horizon that meets all the requirements of a spodic horizon except thickness, and that rests on a fragipan or on a spodic horizon, or on an albic horizon that rests on a fragipan, and that have no plaggen epipedon—**Spodosols, p. 123.**
C. Other soils that (1) have an aquic moisture regime and have plinthite that forms a continuous phase within 30 cm of the mineral surface of the soils, or (2) have an oxic horizon but have no plaggen epipedon and no argillic or natric horizon that overlies the oxic horizon
     —Oxisols, p. 121.

---

[19] In this key the diagnostic horizons and properties mentioned do not include the properties of buried soils except their organic carbon and base saturation. Properties of buried soils are considered in the categories of subgroups, families, and series, but not in the higher categories.

D. Other soils that have a mean annual soil temperature of 8°C or more and:

  1. Have no lithic or paralithic contact or petrocalcic horizon or duripan within 50 cm of the surface, and

  2. After the upper 18 cm are mixed, as by plowing, have 30 per cent or more clay in all subhorizons down to 50 cm or more, and

  3. Have, at some period in most years, unless irrigated or cultivated, cracks open at the surface [20] and the cracks are at least 1 cm wide at a depth of 50 cm, and one or more of the following is present:

    a. Gilgai

    b. At some depth between 25 cm and 1 m, slickensides close enough to intersect

    c. At some depth between 25 cm and 1 m, wedge-shaped natural structural aggregates with their long axes tilted 10 to 60 degrees from the horizontal—**Vertisols, p. 127.**

E. Other soils that have an ochric or anthropic epipedon and either:

  1. Have no argillic or natric horizon but:

    a. Have a salic horizon with its upper boundary within 75 cm of the surface and are saturated with water within 1 m of the surface for one month or more in some years, or

    b. Have one or more of the following horizons with upper boundaries within 1 m of the surface; petrocalcic, calcic, gypsic, petrogypsic, or cambic horizons, or a duripan; and either (1) have an aridic moisture regime or (2) have an ustic or xeric moisture regime and a conductivity of the saturation extract at 25°C that is 2 mmho per cm or more in some subhorizon that is shallower than the least of the following: (a) a lithic or paralithic contact, petrocalcic horizon, or duripan shallower than 25 cm; (b) 1.25 m if the weighted average particle-size class is sandy, or sandy-skeletal between 25 cm and 1 m, or between 25 cm and a lithic or paralithic contact, petrocalcic horizon or duripan if any of these is shallower than 1 m; (c) 90 cm if the weighted average particle-size class is loamy between 25 cm and 1 m or between 25 cm and a lithic or paralithic horizon or duripan if any of these is shallower than 1 m; (d) 70 cm

---

[20] An open crack is interpreted to be a separation between gross polyhedrons. If the surface horizons are strongly self-mulching, that is, a mass of loose granules, or if the soil is cultivated while the cracks are open, the cracks may be largely filled with granular materials from the surface. But they are considered to be open in the sense that the polyhedrons are separated.

if the weighted average particle-size class is clayey between 25 cm and 1 m or between 25 cm and a lithic or para- lithic contact, petrocalcic horizon or duripan if any of these is shallower than 1 m; or

2. Have an aridic moisture regime and have:
   a. Either an argillic or natric horizon and
   b. An epipedon that is not both hard and massive when dry
   —Aridisols, p. 110.

F. Other soils that have a mesic or isomesic or warmer temperature regime and either:

1. Have an argillic horizon but have no fragipan, and have base saturation (by sum of cations) that is less than 35 per cent at 1.25 m depth below the upper boundary of the argillic horizon or 1.8 m below the soil surface, or immediately above a lithic or paralithic contact, whichever is shallower; or

2. Have a fragipan that:
   a. Underlies an argillic horizon, or
   b. Meets all requirements of an argillic horizon, or
   c. Has clay skins more than 1 mm thick in some part, and
   d. Has base saturation (by sum of cations) that is less than 35 per cent at a depth of 75 cm below the upper boundary of the fragipan, or above a lithic or paralithic contact, whichever is shallower—**Ultisols, p. 125.**

G. Other soils that:

1. Have one of the following:
   a. A mollic epipedon or
   b. Have a surface horizon that, after mixing the upper 18 cm, meets all requirements of a mollic epipedon except thick- ness
   c. And, in addition, have an upper subhorizon that is more than 7.5 cm thick in an argillic or natric horizon, that meets the color, organic carbon, base saturation, and structure re- quirements of a mollic epipedon, but that is separated from the surface horizon by an albic horizon, and, in addition,

2. Have base saturation of 50 per cent or more (by NH_4OAc) as follows:
   a. If there is an argillic or natric horizon, from its upper boundary to a depth of 1.25 m below that boundary, or to 1.8 m below the soil surface, or to a lithic or paralithic contact, whichever is shallower, or
   b. If there is a cambic horizon, in all subhorizons to a depth of 1.8 m below the soil surface or to a lithic or paralithic contact, whichever is shallower, and

3. If the exchange complex is dominated by amorphous materials, have, to a depth of 35 cm or more or to a lithic or paralithic contact shallower than 35 cm, a bulk density (at $\frac{1}{3}$ bar water retention) of the fine earth fraction of 0.85 or more and have less than 60 per cent vitric [21] volcanic ash, cinders or other vitric pyroclastic material in the silt, sand, and gravel fractions, and

4. If the temperature regime is isomesic, isothermic, or isohyperthermic, either the mollic epipedon rests on a soil horizon containing 40 per cent or more $CaCO_3$ equivalent or the soil

    a. Has horizons less than 50 cm thick that have more than 35 per cent clay and COLE of 0.09 or more and cracks at some period in most years that are 1 cm or more wide at a depth of 50 cm, or

    b. If there is a lithic or paralithic contact or altered rock retaining its rock structure within 50 cm of the surface, the soil has horizons totalling less than 25 cm in thickness with 35 per cent or more clay with montmorillonitic mineralogy or with COLE of 0.09 or more—**Mollisols, p. 118.**

H. Other soils that:

1. Have an argillic or natric horizon but no fragipan, and either (a) have a frigid mean annual soil temperature, or (b) have base saturation (by sum of cations) that is 35 per cent or more at a depth of 1.25 m below the upper boundary of the argillic horizon or 1.8 m below the surface of the soil, or immediately above a lithic or paralithic contact, whichever depth is shallower, or

2. Have a fragipan that

    a. Is in or underlies an argillic horizon or

    b. Meets all requirements of an argillic horizon, or

    c. Has clay skins more than 1 mm thick in some part, and

    d. Has a mean annual soil temperature of $< 8°C$ or has base saturation (by sum of cations) of 35 per cent or more at a depth of 75 cm below the upper boundary of the fragipan or immediately above a lithic or paralithic contact, whichever is shallower—**Alfisols, p. 106.**

I. Other soils that have an umbric, mollic, or plaggen epipedon; or have a histic epipedon composed of mineral soil materials; or have a cambic horizon; or within 1 m of the surface have a calcic, petrocalcic, or placic horizon, or a duripan; or a fragipan; or any combination of these; or have a sulfuric horizon with an upper boundary within 50 cm

---

[21] Included in the meaning here are crystalline particles that are coated with glass and partially devitrified glass, as well as glass.

of the surface—**Inceptisols, p. 116.**
J. Other soils—**Entisols, p. 112.**

SUMMARY

A new system of soil classification has been developing in the United States since 1951, was officially adopted in 1965, and is presented here in simplified form. The new system of soil classification is based upon conceptual inferences and factual characteristics. The six categories in the system are: soil order, suborder, great group, subgroup, family, and series. Each of the 10 soil orders is represented by a generalized description, a soil profile in the order, and an official description of a typifying pedon.

QUESTIONS

1. Name the six categories in the new system of soil classification.
2. On what basis has the category of soil family been separated?
3. Explain what is meant by soil series.
4. What is the relationship between mapping unit and soil series?
5. Select a soil order and write a short paper on the subject.

REFERENCES

Field Manual of Soil Engineering (5th ed.). Lansing, Mich.: State Highway Commission, Michigan Department of State Highways (Jan., 1970), 474 pp., $6.00.

Soil Classification: A Comprehensive System, Seventh Approximation. Soil Survey Staff, Soil Conservation Service, United States Department of Agriculture (Aug., 1960).

Soils and Men: The Yearbook of Agriculture (1938). United States Department of Agriculture.

Soil Taxonomy of the National Cooperative Soil Survey. Soil Conservation Service, United States Department of Agriculture, Washington, D.C., Dec., 1970.

The following papers on soil classification are recommended for those persons who want more enrichment in this subject. (All appear in *Selected Papers in Soil Formation and Classification,* Soil Science Society of America, Special Publication Series No. 1, 1967, 428 pp.)

Cline, Marlin G., "Basic Principles of Soil Classification," pp. 381–92.

Manil, G., "General Considerations on the Problem of Soil Classification," pp. 393–401.

Oakes, Harvey, and James Thorp, "Dark Clay Soils of Warm Regions Variously Called Rendzina, Black Cotton Soils, Regur, and Tirs," pp. 136–49.

Simonson, Roy W., "Soil Classification in the United States," pp. 415–28.

CHAPTER 7

SOIL SURVEYS AND THEIR USE
(SOIL CARTOGRAPHY)

What is a soil survey? . . . An institution devoted to the study of the soil in its natural habitat. It is concerned primarily with the determination of soil characteristics . . . that constitute a soil individual, the fixing of these [soil] groups by a proper nomenclature, and the determination of the area and distribution [mapping] of each soil unit.

Soil surveys have created a new branch of soil science—soil anatomy.—C. F. MARBUT

For 70 years many observing farmers and ranchers over the nation have seen soil surveyors at work. In making soil maps, the soil surveyor walks briskly across the fields with an auger or spade in his hand and a map case over his shoulder. Every few minutes he stops to bore or dig a hole in the soil. Sometimes he merely looks at the surface soil; at other times he rubs the soil between his thumb and forefinger, reflects a moment, and then puts something on a map. He takes a small instrument from his belt and sights through it, looking directly up or down the slope. Occasionally he bores or digs a deep hole, studying the subsoil intently. He glances around, evidently observing farm crops, relief, and erosion; then he sketches for a few moments and moves on. He crosses farm boundaries but doesn't appear to be taking any special notice of them. He is alert and preoccupied, as if he were counting steps and making mental notes between stops. When he is questioned about his work, he explains exactly what he is doing and how this information can help the public.

The surveyor is carrying a base map which is an aerial photograph large enough to show every road, stream, woodland, field, and house. He is studying the color, texture, structure, and moisture condition of the soil—and in so doing he is identifying the soil type. His frequent borings enable him to determine the depth of the various horizons and to estimate the amount of soil that has been lost through erosion. The small instrument through which he sights is an Abney hand level, which is used to determine the per cent of slope. He sketches on the aerial photograph what he finds out about the soil, slope, and degree of erosion (Figure 7.1).

Work in soil survey started in 1899 with the mapping of four areas: one in the Pecos River Valley of New Mexico, one in the Salt Lake area of Utah, and the other two in the Connecticut River Valley of Massachusetts and Connecticut. The smallest unit of land mapped during the early surveys was 40 acres; now it is approximately 1 acre. With continuous refinements, soil survey work on an area or county basis has been carried on each year since its beginning.

In 1935, the Soil Conservation Service began a program of soil mapping with the farm or ranch as the mapping unit instead of a county. These maps were made only where the farmer requested such services, and the maps were used primarily for individual farm planning.

In 1952, all agencies of the Federal Government concerned with soil survey work were consolidated to establish the National Cooperative Soil Survey. This agency makes soil maps in response to requests by farm and ranch cooperators in the Soil Conservation District; it also makes *standard soil surveys* of areas lying between these farms and ranches and eventually publishes county soil survey maps and reports.

The cumulative total acres of soils that have been surveyed in the 50 states in the United States, including Puerto Rico and the Virgin Islands, to June 30, 1969 are as follows (Table 7.1):

TABLE 7.1 CUMULATIVE TOTAL ACREAGE OF SOILS SURVEYED IN THE UNITED STATES, BY KINDS, TO JUNE 30, 1969 *

Kind of Soil Survey	Cumulative Total Acres
Detailed Surveys	654,956,107
Conservation Surveys	26,871,208
Reconnaissance Surveys	18,458,135
Grand total	700,285,450

* *Note:* Detailed Soil Surveys and Conservation Surveys are both made on the same scale; they differ primarily in the fact that in the Detailed Surveys the mapping units have been correlated into a standard U.S. system of soil classification. Reconnaissance Surveys are more generalized, are made more quickly, and the mapping units are not correlated. (Source: U.S. Soil Conservation Service.)

FIG. 7.1. In making a soil map, the soil surveyor walks over the landscape, observing and measuring the slope of the land, its stoniness, and the vegetation [(1): Coconino National Forest, Arizona]. Periodically he stops to bore or dig a hole in the soil to determine the profile features and to describe and classify the soil for mapping. He will classify and map these three profiles as follows. (2): Mapping unit— Harney silt loam, 3 to 10 per cent slopes; classified in the fine, montmorillonitic, mesic family of Typic Argiustolls. (Kansas.) (3): Mapping unit—Hudson silty clay loam, 2 to 6 per cent slopes; classified in the fine-silty, illitic, mesic family of Glossoboric Hapludalfs. (New York). (4): The mapping unit has not been identified as to soil type and slope, but the soil has been classified in the family of Aquentic Haplorthods. (Minnesota.) (Scales are in feet.) *Sources:* (1), J. O. Veatch, Mich. State U.; (2), (3), and (4), Soil Conservation Service, U.S. Dept. of Agr.

FIG. 7.2. Soil surveys are used for predicting suitability of soils for field crops (1), forests (2), and farm ponds (3); for locating suitable sources of sand, gravel, or road fill; and for determining suitable areas for urban development, recreation sites, airfields, and highways. *Source:* All photos by Soil Conservation Service, U.S. Dept. of Agr.

USES OF SOIL SURVEYS

Soil surveys are needed to provide our nation with an inventory of the soil resources in order that public policies may be more wisely made and administered. Farmers who have a modern soil map can obtain accurate predictions of the yield they may expect of a certain crop on that soil and can determine whether or not a new variety would be more desirable. The soil surveys are used in planning management practices, including cropping systems for long-range productivity of the land. Soil surveys are basic to opening new lands to irrigation agriculture and in solving salt, alkali, drainage, and pollution problems.

One of the earliest results of the surveys, arising from the studies in the Connecticut Valley, was the introduction of Sumatra tobacco on certain soils. Some soil types produced good tobacco crops whereas others did not. Similar experiences with other crops are reported by farmers and research scientists.

Other more recent uses of soil survey data are to improve tax assessment, to aid in estimating forest site predictions, and to help in the design and location of roads, airports, and housing developments. Soil surveys have also been used as a basis for land classification and rural zoning.

The texture of the several horizons in a soil also affects the nonagricultral uses of soil. Recent county soil survey reports contain such information. The soils within a county are classified according to suitability, as sources of topsoil, sand and gravel, and for farm ponds, septic tanks, disposal fields, irrigation, and foundations for buildings (Figure 7.2).

FIG. 7.3. Information obtained by soil surveys and supported by field plot research and laboratory analyses is taught to boys and girls in land judging contests, as shown here. *Photo by Roy L. Donahue.*

Survey methods have changed greatly. The first soil scientists equipped their buggy wheels with odometers to measure distances. Few maps of any kind were then available on which to plot the soil types. The soil surveyor started with a blank sheet of paper and made his own base maps on which he put the soil boundaries. Today, aerial photographs on a scale of approximately 4 inches per mile are used as a base map for showing the soil types.

Actual inspection and interpretation are carried out on the land. To find out exactly what the soil is, scientists dig down to the parent material. Broken rock or other parent material may vary from a few inches to many feet in depth. But most agricultural soils in the United States support plant roots to depths of from 3 to 5 feet. Toward the arctic regions, the soils get shallower; toward the Equator, they get deeper, other things being equal. Laboratory analyses aid greatly in determining the final classification and in indicating the fertility and water relations of the soil (Figure 7.3).

The soil survey reports give a variety of data, such as:

1. How stony the soil is
2. How much sand or clay is in the various parts of the profile
3. Whether the soil is acid or alkaline
4. Such below-the-surface information as the height of the water table, the presence of a hardpan, and other usually unseen factors that have an effect on soil productivity and farm management decisions.

Erosion tendencies are carefully noted, and ways to control erosion are explained in the report. Crop yield predictions are made for each soil.

With this basic information, the best possible use and management of the soil can be planned. Experience of farmers and results of research at state experiment stations are used. Estimates are made on how well the soil type will produce and what is needed for its long-range management.

The published report contains a detailed map of the county. From this map, a farmer can locate his land and note the type of soils in his fields. Also, the farmer will find complete descriptions of the soils, together with crop management predictions, the relationships of soil and climate, and other pertinent data.

Soil surveys are an important element in the program for world-wide sustained food production. Soil survey experts from several countries have in recent years come to the United States to learn about our soil survey principles and practices. Also, the United States has loaned soil surveyors to several nations to help them solve their agricultural problems.

The primary use of soil surveys is to aid in finding an effective cultural balance between people and the land. In our changing world, soil surveys serve as a basis for the constant and necessary adjustments between man and his land.

NATIONAL COOPERATIVE SOIL SURVEY

Making a soil survey is like taking inventory of cattle, tractors, and buildings; neither is essential for day-to-day existence, but both are necessary for long-time efficiency of management.

Since 1899, an agency of the Federal Government, in cooperation with the land-grant colleges, has made and published soil surveys of problem areas or counties. Beginning in 1935, another agency of our government began a program of soil mapping on farms as a basis for farm planning. These maps were never published. In November of 1952, by mutual agreement, these two agencies were consolidated, placed in the Soil Conservation Service, and designated the National Cooperative Soil Survey.

Although soil surveys are made by the best of modern technicians, they cannot be made to satisfy all future demands. With an ever-advancing technology, uses of the soil will be made which cannot now be predicted. As a consequence of constant change, a soil map once thought useful for *all* future needs must now be remade on a larger scale each time there is an intended intensification of soil use.

The National Cooperative Soil Survey continues to make farm maps in cooperation with Soil Conservation Districts; it also maps the soils between the farms. As a result, the objectives of both former agencies are now incorporated in the consolidated program.

Field mapping of soils is now conducted on a scale of approximately 4 inches to the mile. Aerial photographs are used as a base. Published maps, formerly 1 inch to the mile, are now approximately 3 inches to the mile, and they are on a photographic base.

SOIL SERIES, TYPES, AND PHASES

In the process of mapping the soils of a county, a tentative legend is made describing all the soils in neighboring counties where the survey has been completed. As new soils are encountered during the progress of the survey, these new soils are described and mapped.

After completing the soil survey of a county, a soil correlator inspects the legends and the field mapping to determine if some kinds of soil called by a certain name are not the same soils mapped in another county or state under a different name. Such a system is a guarantee that soil names shall refer to the same kind of soil wherever they are mapped.

Each soil series is named after a river, town, mountain, lake, or some other geographic feature near the area where the soil was originally mapped. There is a Mohave soil in the Mohave Desert; a Dalhart soil near the town of Dalhart on the High Plains of Texas; a Fargo soil near Fargo, North Dakota; Miami soils near Miami, Ohio; and Norfolk soils near Norfolk, Virginia.

These geographic names apply to *soil series*. A soil series is a group of soils having a similar:

1. Kind, thickness, and arrangement of horizons in the profile
2. Color of horizons
3. Structure of horizons
4. Acidity or alkalinity
5. Consistence
6. Organic-matter content
7. Mineralogical composition
8. Texture.

Two particular kinds of soils may have all of these features almost identical but the surface horizon of one may be a loam and the surface horizon of the other may be a silt loam. In this instance, both soils would have the same series name but they would be classified and mapped as different *types*.

A type name consists of the series name plus the name of the texture of the surface horizon. Thus, two soils may be classified in the Miami series, but the types may be *Miami loam* and *Miami silt loam*.

A Miami silt loam may vary from the original description as to:

1. Slope
2. Degree of erosion
3. Depth to bedrock
4. Stoniness.

When such variations exist, they are called *phases*. There may be a Miami silt loam, *steep phase,* or *eroded phase,* or *shallow phase,* or *stony phase.* The modern way to designate phases is to incorporate the phase name into the type name, such as Miami *stony* silt loam.

SOIL PROFILE AND SOIL PEDON

A soil profile has traditionally been considered a vertical exposure of the soil horizons to a depth of plant root penetration, and a width of perhaps 3 to 4 feet. A more scientific approach to the study of a soil in the field is to consider the *complete* variations of horizons in a *wider* exposure. One-half the cycle of horizon shapes and variations in a profile, plus a third dimension, is now known as a *pedon,* with approximately 1 square meter as the minimum soil volume to be recognized as a pedon in soil taxonomy (Figure 7.4).

FIG. 7.4. The relationship between a soil profile and a pedon. On the left is a two-dimensional exposure of the soil from the surface downward to depth of rooting of plants, showing horizons (layers) in a soil *profile.* On the right is the same profile but with a third dimension, known as a *pedon.* (Soil is mapped as Minoa very fine sandy loam, classified in the coarse-loamy, mixed, mesic family of Aquic Dystric Eutrochrepts.) (Genesee County, New York.) *Source:* Soil Conservation Service, U.S. Dept. of Agr.

SOIL PEDON DESCRIPTION

To insure uniformity in describing soils, standard forms have been prepared by the National Cooperative Soil Survey. These forms are reproduced as Figure 7.5, and an explanation of terms is given.

Soil Type: Name the soil type, such as Cecil sandy loam, Miami silt loam, Charlton loam, Fruita very fine sandy loam, or Drummer clay loam.

Classification: Give the soil family name, such as Typic Argiudoll; fine, montmorillonitic, mesic.

Native Vegetation (or Crop): Indicate forest type, such as White Oak

FIG. 7.5. Standard form for recording the field description of a soil type. *Above:* front; *below:* reverse side, same sheet.

Soil type			Date		Stop No.		Soil type
Classification		Area					
Location							
N. veg. (or crop)		Climate					
Parent material							
Physiography							
Relief	Drainage			Salt or alkali			
Elevation	Gr. water			Stoniness			
Slope	Moisture						
Aspect	Root distrib.						
Erosion							
Permeability							File No.
Additional notes							

Hori-zon	Depth	Thick-ness	Bound-ary	Color — Check. D (ry) or M (oist)	Tex-ture	Structure	Consistence	Reac-tion	Spec. Feat.
				D M					
				D M					
				D M					
				D M					
				D M					
				D M					
				D M					
				D M					
				D M					
				D M					
				D M					
				D M					

—Hickory, Loblolly pine—Shortleaf pine—Oak, or Spruce-Fir; grass type, such as Buffalograss, Little Bluestem—Side-Oats Grama, or Muttongrass; or crop, such as apple orchard, cotton, or corn.

Climate: Indicate whether the climate is classified as humid semi-tropical, cold continental, or warm marine and soil moisture class.

Parent Material: Indicate, such as acid clay till, sandy alluvium, unconsolidated limestone, or residuum from basalt; also soil depth class.

Physiography: Designate whether first bottom (flood plain), moraine, outwash plain, coastal plain, loess, or talus slope.

Relief: Indicate if slope is convex, concave, simple, or complex; and the appropriate slope class, based on relative per cent of slope.

Elevation: Give the approximate elevation in feet above sea level.

Slope: Indicate the approximate gradient of slope.

Aspect: State the general direction of the slope, whether facing north, east, south, or west.

Erosion: Use appropriate number for the relative degree of erosion.

Permeability: Indicate the proper name, such as very slow, moderate, or very rapid.

Drainage: Use standard names, such as poorly drained, moderately well drained, or well drained.

Ground Water: Give depth to ground water.

Moisture: Show soil moisture and soil temperature classes.

Root Distribution: Indicate relative numbers of roots in each horizon.

Salt or Alkali: Describe general saltiness or alkali condition, when such conditions exist.

Stoniness: Use terms such as fragmental, sandy-skeletal, or cindery.

SOIL SURVEY REPORT

Published soil maps include a report that is written to aid in understanding and using the information on the maps. The modern report includes the following items:

1. A description of soil characteristics and their significance to the classification and use of soils
2. Actual and predicted responses of various soils to different levels of management
3. Information on the climate, geology, and physiography of the county

4. Sections on forestry practices or grazing management, depending on the relative importance of each in the county
5. Suggestions for erosion control, cropping systems, and fertilization practices as they are influenced by soil characteristics
6. Relationships of soils to irrigation practices, drainage characteristics, and salinity control
7. Predictions on the engineering properties of the soil types as described and mapped
8. Special problems of soil management in relation to the soil types shown on the soil map, such as reclamation of alkali soils, watershed management, and the management of hard-to-manage soils
9. Uses of soils for wildlife
10. Uses of soils for urban developments
11. Uses of soils for recreation development
12. Soils and their use in pollution control.

LAND-CAPABILITY CLASSIFICATION

The *standard soil survey* map shows the different kinds of soil that are significant and their location in relation to other features of the landscape. These maps are intended for users with widely different problems and, therefore, contain considerable detail to show important basic soil differences.

The information on the soil map must be explained in a way that has meaning to the user. These explanations are called interpretations. Soil maps can be interpreted by: (1) the individual kinds of soil on the map; and (2) the grouping of soils that behave similarly in responses to management and treatment. Because there are many kinds of soil, there are many individual soil interpretations. Such interpretations, however, provide the user with all the information that can be obtained from a soil map. Many users of soil maps want more general information than the individual soil-mapping unit provides. Soils are grouped in different ways according to the specific needs of the map user. The kinds of soil grouped and the variation permitted within each group differ according to the use to be made of the grouping.

The capability classification is one of a number of interpretive groupings made primarily for agricultural purposes. As with all interpretive groupings, the capability classification begins with the individual soil-mapping units, the building stones of the system. In this classification, the arable soils are grouped according to their potentialities and limitations for sustained production of the common cultivated crops that do not require specialized

LAND CAPABILITY CLASSES			
SUITABLE FOR CULTIVATION		**NO CULTIVATION-PASTURE, HAY, WOODLAND AND WILDLIFE**	
I	REQUIRES GOOD SOIL MANAGEMENT PRACTICES ONLY	V	NO RESTRICTIONS IN USE
II	MODERATE CONSERVATION PRACTICES NECESSARY	VI	MODERATE RESTRICTIONS IN USE
III	INTENSIVE CONSERVATION PRACTICES NECESSARY	VII	SEVERE RESTRICTIONS IN USE
IV	PERENNIAL VEGETATION – INFREQUENT CULTIVATION	VIII	BEST SUITED FOR WILDLIFE AND RECREATION

FIG. 7.6. Land-capability classes. *Source:* Soil Conservation Service, U.S.D.A.

site conditioning or site treatment. Nonarable soils (soils unsuitable for long-time sustained use for cultivated crops) are grouped according to their potentialities and limitations for the production of permanent vegetation and according to their risks of soil damage if mismanaged.

The individual mapping units on soil maps show the location and extent of the different kinds of soil. The greatest number of precise statements and predictions can be made about the use and management of the individual mapping units shown on the soil map. The capability grouping of soils is designed to: (1) help landowners and others use and interpret the soil maps; (2) introduce users to the detail of the soil map itself; and (3) make possible broad generalizations based on soil potentialities, limitations in use, and management problems.

The capability classification provides three major categories of soil groupings: capability unit; capability subclass; and capability class (Figure 7.6). The first category, capability unit, is a grouping of soils that have about the same responses to systems of management of common cultivated crops and pasture plants. Soils in any one capability unit are adapted to the same kinds of common cultivated and pasture plants, and require similar alternative systems of management for these crops. Under capable management, long-time estimated yields of adapted crops for individual soils within the unit do not vary more than about 25 per cent.

The second category, the subclass, is a grouping of capability units having similar kinds of limitations and hazards. Four general kinds of limitations or hazards are recognized: erosion hazard; wetness; rooting-zone limitations; and climate.

The third and broadest category in the capability classification places all the soils in eight capability classes (Figure 7.7). The risks of soil damage or limitations in use become progressively greater from Class I to Class VIII. Soils in the first four classes under good management are capable of producing adapted plants, such as forest trees or range plants, and the

FIG. 7.7. The extent of land capability classes in mainland United States. *Source: Agricultural Land Resources—Capabilities, Uses, Conservation Needs. U.S. Dept. of Agr., Agr. Inf. Bul. No. 263, 1962.*

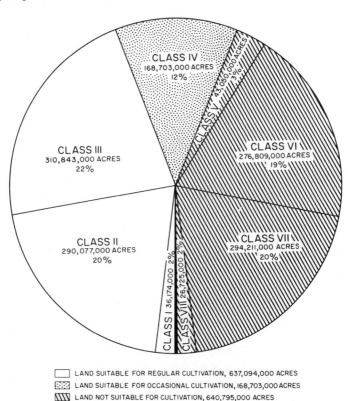

CLASS IV
168,703,000 ACRES
12%

CLASS V 43,060,000 ACRES 3%

CLASS III
310,843,000 ACRES
22%

CLASS VI
276,809,000 ACRES
19%

CLASS II
290,077,000 ACRES
20%

CLASS VII
294,211,000 ACRES
20%

CLASS I 36,174,000 2%

CLASS VIII 26,725,000 2%

☐ LAND SUITABLE FOR REGULAR CULTIVATION, 637,094,000 ACRES

▨ LAND SUITABLE FOR OCCASIONAL CULTIVATION, 168,703,000 ACRES

▩ LAND NOT SUITABLE FOR CULTIVATION, 640,795,000 ACRES

■ UNCLASSIFIED, 1,445,000 ACRES

common cultivated field crops and pasture plants. Soils in Classes V, VI, and VII are suited to the use of adapted native plants. Some soils in Classes V and VI are also capable of producing specialized crops, such as certain fruits and ornamentals, and even field and vegetable crops, under highly intensive management involving elaborate practices for soil and water conservation. Soils in Class VIII do not return on-site benefits for inputs of management for crops, grasses, or trees without major reclamation.

The grouping of soils into capability units, subclasses, and classes is done primarily on the basis of their capability to produce common cultivated crops and pasture plants without deterioration over a long period of time. To express suitability of the soils for range and woodland use, the soil-mapping units are grouped into range sites and woodland-suitability categories.

PRODUCTIVITY RATINGS

As the soil survey of a county is being made, the technician in charge obtains information on the crop adaptations and crop yield potentials of the major soils under various levels of management. The soil surveyors are constantly alert in assembling data on the present and predicted future yields of the major crops. This information may be obtained from farmers, county agricultural agents, farm planners, experimental fields, and the U.S. Census of Agriculture.

In tabular form in the soil survey report, each soil type is listed and the estimated yields are given for each of the major crops. The yields are predicted for each of two management levels; namely, yields under average management, and yields under intensive management practices. Where forestry, pasturing, and range grazing are important uses of the land, estimates are made regarding the respective yield predictions for each kind of use.

FOREST SITE CLASSIFICATION

Forest soil science has developed more slowly than agricultural soil science. Perhaps one reason for this fact has been the relatively lower value per acre of forest trees than that per acre of agricultural crops. More recently, however, the value of trees in dollars per acre, for erosion control, for wildlife food and cover, and for aesthetic value, has increased. Forest soil science has recently made accelerated progress.

The growth of each forest tree species is determined by varying combinations of soil factors, such as depth of the A horizon, texture of all

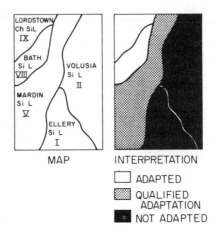

MAP   INTERPRETATION

☐ ADAPTED

▨ QUALIFIED
  ADAPTATION

■ NOT ADAPTED

FIG. 7.8. Red pine (*Pinus resinosa*) is a valuable tree species for planting; however, the species grows very slowly on poorly drained soils. From standard soil survey reports it is possible to construct a good soil suitability map for red pine. Such maps have been made in New York and an example is reproduced here. The Volusia and Ellery soils are poorly and somewhat poorly drained and are therefore not adapted to red pine. Mardin soils are moderately well drained and have qualified adaptation. Bath and Lordstown soils are well drained and well adapted to red pine. *Source: James A. De Ment and E. L. Stone, Influence of Soil and Site on Red Pine Plantations in New York. II. Soil Type and Physical Properties, Cornell U. Agr. Exp. Sta. Bul. 1020, July, 1968, 25 pp.*

horizons, depth and fluctuating characteristics of the water table, organic matter content, and nutrient supplying power of the soil.

The other significant factors of the forest site are as follows:

1. Soil moisture: whether the soil is excessively drained, well drained, moderately well drained, somewhat poorly drained, or poorly drained (Figure 7.8)
2. Effective depth to which tree roots extend
3. Relative lime content of the soil
4. Degree of stoniness
5. Steepness of slope: whether more than approximately 30 per cent or less than 30 per cent
6. Aspect: the direction that the slope faces, whether north, east, south, or west
7. Slope position: whether upper, middle, or lower, and whether the slope is convex or concave
8. Relative degree of erosion: relating primarily to an evaluation of the forest site for planting trees.

FIG. 7.9. This heap of peat on the side of a highway in southern Michigan was pushed upward as the highway collapsed and dropped 10 feet during spring thaws. A soil survey map shows areas of peat and muck which must be either avoided in locating a new highway or dug out and replaced by stones, gravel, or sand. *Source:* J. O. Veatch, Mich. State U.

Most of the information necessary for forest site evaluation can be obtained directly from the published county soil survey map and report.

SOIL MAPS AS AN AID TO ENGINEERING

The basic facts about the physical, chemical, and biological properties of soils needed to predict their behavior under farm use must be given different interpretations when the data are used for predicting their physical behavior in subgrades or foundations. Detailed soil maps are helpful in planning locations of highways and airports and in predicting answers to problems in construction and maintenance. Soil maps and reports are also useful in locating materials such as sand, gravel, clay, and stone, for use in construction. Without a scheme of classification such as that developed through soil surveys, extensive and costly testing of soils for each highway or airport project is necessary. Without a soil survey, information gained on one project cannot readily be applied to another on a contrasting soil (Figure 7.9).

NUTRITIONAL PREDICTIONS

In various places throughout the United States, certain soils contain such small amounts of essential elements that the forage is deficient, and livestock deficiency symptoms for this element are exhibited. In other places,

FIG. 7.10. This soil is classified as *Class 1* for irrigation potential because it has gentle slopes, is medium to fine textured, has granular structure, good water-holding capacity, good internal drainage, and medium in total salt content. Such information can be obtained from a standard soil survey report. *Source:* P. J. Lyerly, Tex. Agr. Exp. Sta., El Paso, Texas.

there is an excess of an element in the soil which concentrates in the forage and becomes toxic to animals.

Phosphorus is deficient in certain soils in the Northwest, the South, the Lake States, and along the Atlantic Coast. Calcium is deficient in many soils in humid regions. Cobalt deficiency occurs in the Northwest, the Lake States, the South, and the Northeast.

Toxic amounts of selenium exist in the forage grown on certain soils in Northern Plains States and the Northwest. Molybdenum toxicity exists in cattle when grazing the forage on some soils in the Southwest and in Florida.

Many correlations have been made between deficient and toxic elements in relation to the soil types as shown on the soil map. Where this information is known, it is valuable in the prediction of land use and farm management practices.[1]

LAND CLASSIFICATION FOR IRRIGATION

Areas that are considered for their irrigation possibilities must first have a modern soil survey. Upon completion of the soil survey, the soils are classified as to their suitability for irrigation into one of six classes in the following manner (Figure 7.10):

Class 1. These lands are highly suitable for irrigation farming, being capable of producing sustained yields of a wide range of climatically adapted crops at reasonable costs. They are smooth lying with gentle slopes.

---

[1] See Chapter 1 for a discussion of mineral deficiency and toxicity.

The soils are deep and of medium to fairly fine texture, with granular structure allowing for easy penetration of roots, air, and water. Yet they must have free drainage and good water-holding capacity. Land development, such as leveling and establishing a drainage system, is relatively simple to accomplish and can be attained at reasonable cost.

Class 2. These lands are measurably below Class 1 in productive capacity, adapted to a somewhat narrower range of crops. They are more costly to farm or to prepare for irrigation. These soils have one or more deficiencies in soil texture, saltiness, topography, or drainage.

Class 3. This class is inferior to Class 2 because of greater deficiencies in soils and unsuitable topography or drainage. Although approaching marginal utility for general crop production, these lands are still considered suitable for irrigation, especially if intermingled with better lands.

Class 4. These lands have an excessive deficiency or restricted utility, but special engineering and economic studies show them to be suited to irrigation under certain restricted conditions.

Class 5. These lands are temporarily held as nonarable pending completion of economic or engineering studies to determine their suitability for irrigation.

Class 6. These lands do not meet the minimum requirements of Class 5 and are small areas of suitable land lying within larger bodies of nonarable land.

LAND CLASSIFICATION FOR DRAINAGE

A modern soil map presents information pertinent to the making of accurate predictions on the feasibility of establishing drainage projects. All drainage systems need suitable outlets, and this information can be predicted from a soil map before costly engineering surveys are made. Information on soils suitable for various kinds of drainage systems can be obtained from the soil map and report. Coarse-textured soils, for example, can be tile drained successfully. Fine-textured soils usually must be drained by open ditches, although tile drainage is used in some areas.

OTHER USES OF SOIL SURVEYS

In certain places, soil surveys are used as a basis for rural zoning, watershed planning, range-site classification, land appraisal, the settlement of new lands, and tax assessment. Modern soil maps and reports are essential for the wide extension and application of research results, demonstrations, and farmer experiences from one region to another. To achieve efficient use of agricultural science, a world-wide scheme of soil classification is now being made by the U.S. Department of Agriculture in cooperation with other countries.

SOIL SURVEY REPORT—A MODEL [2]

Pertinent information from a 1969 soil survey report of Pike County, Pennsylvania, will be presented here as a representative model for all modern soil survey maps and reports made and published in the United States.

The soils in Pike County were mapped cooperatively by the Soil Conservation Service of the U.S. Department of Agriculture, the Pennsylvania State University College of Agriculture and Agricultural Experiment Station, and the Pennsylvania Department of Agriculture, State Soil and Water Conservation Commission.

The soil survey report consists of a 9 by 11-inch format with 83 pages of soil descriptions, soil classification, explanations, interpretations, references, and glossary; a colored General Soil Map; and 50 double-page, folded soil maps on an aerial base, at a scale of 1:20,000 (3.17 inches per mile). There are a total of 76 mapping units on the 50 soil maps of Pike County. These mapping units consist of soil types and soil phases (soil types divided into slope classes).

A portion of one of the 50 sections of the soil map of the county is given in Figure 7.11, together with three soil profiles representative of three of the soil type and phase mapping units: (1) Rushtown very shaly silt loam (RwE, 25–45 per cent slopes); (2) Chenango gravelly sandy loam (CmB, 0–12 per cent slopes); and (3) Tioga loamy fine sand (TgA, 0–3 per cent slopes). These three soil series are described in the soil survey report. Included also is a detailed soil pedon description of Rushtown very shaly silt loam. In the complete soil survey report, all soil series are described in similar detail.

RUSHTOWN SERIES

The Rushtown series consists of deep, well-drained soils formed in deposits of acid, gray, shale fragments at the base of steep slopes at the foot of the bluffs along the Delaware River.

---

[2] *Note:* Soil survey reports are available from each state agricultural university and from the Soil Conservation Service.

Most of the Rushtown soils are steep and for that reason have been left in woods. They contain such a large amount of coarse fragments that they have low available moisture capacity and rapid permeability. Most of the shale chips in the lower subsoil and the substratum are arranged parallel to one another.

These soils have moderate to severe limitations when used for cultivated crops because of their slope and low water-holding capacity. They are very strongly acid. The steep slopes and the unconsolidated substratum are severe limitations for building sites.

A typical Rushtown soil in woods has a dark grayish-brown silt loam surface layer that is about 75 per cent shale fragments. This layer is about 7 inches thick. The next layer (upper substratum) is yellowish-brown, very shaly silt loam. It is 95 per cent shale fragments, and the fragments have coatings of silt and clay. This layer extends to a depth of 42 inches. The lower substratum is dark gray shale fragments that are loose or are weakly cemented by silt and clay particles. Generally, this material overlies bedrock at a depth of 3 to 18 feet or more.

Description of a pedon of a Rushtown very shaly silt loam northeast of Egypt Mills, Pennsylvania-woods (colors are for moist soils):

A1	0–7″	Dark grayish-brown (10YR 4/2) very shaly silt loam; 75 per cent shale fragments; weak, fine, granular structure; loose when moist, nonsticky and nonplastic when wet; pH 4.8; gradual, wavy lower boundary; 5–10 in. thick.
C1	7–18″	Yellowish-brown (10YR 5/4), shale fragments; slightly cemented with silt and clay; single grain; friable when moist, nonsticky and nonplastic when wet; pH 4.4; diffuse lower boundary; 8–15 in. thick.
C2	18–42″	Dark yellowish-brown (10YR 4/4) shale fragments slightly cemented with silt and clay; single grain; firm when moist, nonsticky and nonplastic when wet; pH 4.8; diffuse lower boundary; 15–30 in. thick.
C3	42–72″	Dark gray (10YR 4/1) shale; single grain; loose when moist, nonsticky and nonplastic when wet; pH 5.6; 30–90 in. thick.

Depth to hard rock ranges from 3 to 18 feet or more. The amount of shale ranges from 40 to 80 per cent in the A1 horizon and from 80 to 99 per cent in the C horizon.

Rushtown soils occur between the Chenango, Tioga, and Middlebury soils of the terraces and flood plains and the Manulius, Dekalb, and other soils close to the bluffs.

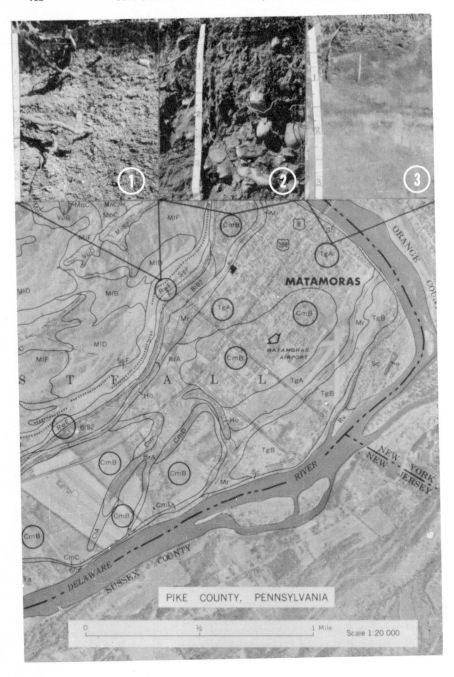

FIG. 7.11. A portion of the soil map of Pike County, Penna. at a scale of 1:20,000. (3.17 inches to the mile.) (Tapes on profiles are in feet.)

THE TAXONOMY OF THE SOILS FOR FIGURE 7.11 *

Symbol on Map	Per Cent Slopes	Order	Suborder	Great Group	Subgroup	Family	Series	Type
RwE	25–45	Entisols	Orthents	Udorthents	Typic Udorthent	Typic Udorthent; loamy fragmental, mixed, mesic	Rushtown	Rushtown very shaly silt loam
CmB	0–12	Inceptisols	Ochrepts	Dystrochrepts	Typic Dystrochrept	Typic Dystrochrept; loamy skeletal mixed, mesic	Chenango	Chenango gravelly sandy loam
TgA	0–3	Inceptisols	Ochrepts	Eutrochrepts	Dystric Fluventic Eutrochrept	Dystric Fluventic Eutrochrept; coarse loamy, mixed, mesic	Tioga	Tioga loamy fine sand

* Source: David C. Taylor. Soil Survey of Pike County, Pennsylvania. U.S. Dept. of Agriculture, Soil Conservation Service in cooperation with Pennsylvania State University and the Pennsylvania Department of Agriculture, June, 1969.

CHENANGO SERIES

The Chenango series consists of deep, well-drained, gravelly and cobbly soils that were formed in grayish glacial outwash. These soils are mainly nearly level or gently sloping soils on outwash terraces in the valleys. Small areas are on the sides of valleys where former streams dropped their loads to form deltas or kames. Most of the Chenango soils are in the major stream valleys, but some areas are in the uplands.

The Chenango soils have low to moderate available moisture capacity. They have rapid permeability and are underlain by gravel. They are generally close to a source of water for irrigation.

The nearly level Chenango soils in the valley of the Delaware River are farmed. The use of these soils for the disposal of liquid sewage or other waste is likely to cause contamination of ground water.

A typical Chenango soil in woods has a surface layer of dark-brown gravelly loam 1 to 2 inches thick, under a layer of leaf litter and a black fibrous mat. The next layer of mineral soil, which is part of the B horizon, is dark brown cobbly sandy loam that is friable and easily tilled. This layer extends 4 inches below the surface. The middle subsoil is strong brown gravelly sandy loam that is very friable and extends to 16 inches. The lower subsoil is very gravelly sandy loam that is 50 per cent or more coarse fragments. The substratum is dark grayish-brown very gravelly sandy loam that is 90 per cent coarse fragments. This material is open and very porous.

TIOGA SERIES

The Tioga series consists of deep, well-drained, loamy and sandy soils that were formed in reddish-brown to brown sediments in the larger stream valleys. These soils are nearly level or gently sloping and are on flood plains and low terraces. Most of these soils are farmed, and they make up most of the land now used for crops in the county.

A typical Tioga soil in a cropped field has a dark brown loamy fine sand surface layer that has weak, fine, granular structure. It is very easily tilled and has a thickness of about 13 inches. Below this layer there is little variation in the soil, except for slight differences in texture of the stratified deposits. The soil is mostly brown and dark brown, friable very fine sandy loam, fine sandy loam, or loamy fine sand. A sample pit was dug to a depth of 154 inches, and the entire section was composed of nearly uniform material.

Table 7.2 summarizes predicted feasible use of these same three soils for field crops, woodlands, irrigation, and homesites, and as a source of

TABLE 7.2 PREDICTED FEASIBLE USE OF THREE SOILS IN PIKE COUNTY, PENNSYLVANIA, FOR FIELD CROPS, WOODLANDS, IRRIGATION, AND HOMESITES, AND AS A SOURCE OF TOPSOIL, SAND AND GRAVEL, AND ROAD FILL *

Soil Type †	Uses of Soils for Crops (Yields from Average Soil Management)		Uses of Soils for Woodlands			Uses of Soils for Irrigation (Available Water Capacity)	Uses of Soils for Homesites with Basements	Uses of Soils as Sources of:		
	Corn (100 = 85 bu. per Acre)	Alfalfa-grass (100 = 3 tons hay per Acre)	Relative Soil Productivity for Woodlands	Native Trees to Favor in Forest Management	Trees to Favor for Planting			Topsoil	Sand and Gravel	Road Fill
Rushtown very shaly silt loam	Soil is too steep for field crops	Soil is too steep for field crops	Poor	Pitch pine, Virginia pine, black oak	Pitch pine, Virginia pine	Low	Soil too steep	Poor	Poor	Good
Chenango gravelly sandy loam	70	85	Fair	Pitch pine, Virginia pine, black oak	Pitch pine, Virginia pine	Low	Good	Poor	Good	Good
Tioga loamy fine sand	45	95	Fair	Black oak, pitch pine, Virginia pine	Pitch pine, Virginia pine	Moderate to high	Soil subject to flooding	Fair	Good	Poor

* Source: David C. Taylor, Soil Survey of Pike County, Pennsylvania, U.S. Department of Agriculture, Soil Conservation Service in Cooperation with Pennsylvania State University and the Pennsylvania Department of Agriculture (June, 1969).

† Note: Photographs of profiles of these three soil types are portrayed at the top of Figure 7.11.

Generalized from 1967 Soil Map,
Scale 1:7,500,000, U. S. Atlas

SLOPE CLASSES

Gently sloping = slopes mainly less than 10 percent
Moderately sloping = slopes mainly between 10 and 25 percent
Steep = slopes mainly steeper than 25 percent

Scale 1:17,000,000

100   0   100   200   300   400 Miles

topsoil, sand and gravel, and road fill. In addition to these data, each soil is also rated according to its recreation potential; percentage of sand, silt, and clay; bulk density; available moisture capacity; chemical composition; mineralogical composition; and classification in the U.S. (1938) system and the U.S. (1969) system of soil taxonomy (formerly known as the "Seventh Approximation").

As an example of the wise use of a soil, note on the soil map of a portion of Pike County, Pennsylvania, that the Matamoras Airport (at arrow) is located on the Chenango soil series. This soil is open, very porous, and contains many coarse fragments capable of supporting heavy loads, even during the rainy season. This information can be inferred by looking at the photograph of the soil profile labeled (2).

SOIL MAP OF THE UNITED STATES

A soil map of the United States according to patterns of soil orders and suborders, is given in Figure 7.12.

LEGEND FOR FIGURE 7.12

Only the dominant orders and suborders are shown. Each delineation has many inclusions of other kinds of soil. For complete definitions see Soil Survey Staff, *Soil Classification, A Comprehensive System, 7th Approximation,* Soil Conservation Service, U. S. Department of Agriculture, 1960, the March 1967 supplement, and *Soil Taxonomy of the National Cooperative Soil Survey,* Dec., 1970 (available from Soil Conservation Service, U. S. Department of Agriculture). Approximate equivalents in the modified 1938 soil classification system are indicated in parenthesis for each suborder.

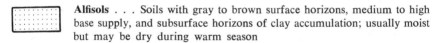

Alfisols . . . Soils with gray to brown surface horizons, medium to high base supply, and subsurface horizons of clay accumulation; usually moist but may be dry during warm season

A1   AQUALFS (seasonally saturated with water) gently sloping; general crops if drained, pasture and woodland if undrained (Some Low Humic Gley soils and Planosols)

FIG. 7.12. (Opposite.) Soil map of the United States according to the "Seventh Approximation" soil classification system, officially adopted in 1965. *Notes: 1.* Symbols on the map that are not explained include: S—Steep slopes (steeper than 25 per cent); P—Permafrost (permanent frost). 2. Of the 10 soil orders, only Oxisols are not shown on the map because they are found primarily in small areas in Hawaii. 3. Only 40 of the 47 soil suborders are delineated on the map because 7 suborders do not occupy areas large enough to be shown on a map of this scale. *Source:* Soil Survey Staff, Soil Conservation Service, U.S. Dept. of Agr.

A2   BORALFS (cool or cold) gently sloping; mostly woodland, pasture, and some small grain (Gray Wooded soils)

A2S   BORALFS steep; mostly woodland

A3   UDALFS (temperate or warm, and moist) gently or moderately sloping; mostly farmed, corn, soybeans, small grain, and pasture (Gray Brown Podzolic soils)

A4   USTALFS (warm and intermittently dry for long periods) gently or moderately sloping; range, small grain, and irrigated crops (Some Reddish Chestnut and Red Yellow Podzolic soils)

A5S   XERALFS (warm and continuously dry in summer for long periods, moist in winter) gently sloping to steep; mostly range, small grain, and irrigated crops (Noncalcic Brown soils)

 Aridisols . . . Soils with pedogenic horizons, low in organic matter, and dry more than 6 months of the year in all horizons

D1   ARGIDS (with horizon of clay accumulation) gently or moderately sloping; mostly range, some irrigated crops (Some Desert, Reddish Desert, Reddish Brown, and Brown soils and associated Solonetz soils)

D1S   ARGIDS gently sloping to steep

D2   ORTHIDS (without horizon of clay accumulation) gently or moderately sloping; mostly range and some irrigated crops (Some Desert, Reddish Desert, Sierozem, and Brown soils, and some Calcisols and Solonchak soils)

D2S   ORTHIDS gently sloping to steep

 Entisols . . . Soils without pedogenic horizons

E1   AQUENTS (seasonally saturated with water) gently sloping; some grazing

E2   ORTHENTS (loamy or clayey textures) deep to hard rock; gently to moderately sloping; range or irrigated farming (Regosols)

E3   ORTHENTS shallow to hard rock; gently to moderately sloping; mostly range (Lithosols)

E3S   ORTHENTS shallow to hard rock; steep; mostly range

E4   PSAMMENTS (sand or loamy sand textures) gently to moderately sloping; mostly range in dry climates, woodland or cropland in humid climates (Regosols)

 Histosols . . . Organic soils

H1   FIBRISTS (fibrous or woody peats, largely undecomposed) mostly wooded or idle (Peats)

H2   SAPRISTS (decomposed mucks); truck crops if drained, idle if undrained (Mucks)

Inceptisols . . . Soils that are usually moist, with pedogenic horizons of alteration of parent materials but not of accumulation

I1S   ANDEPTS (with amorphous clay or vitric volcanic ash and pumice) gently sloping to steep; mostly woodland; in Hawaii mostly sugar cane, pineapple, and range (Ando soils, some Tundra soils)

I2   AQUEPTS (seasonally saturated with water) gently sloping; if drained, mostly row crops, corn, soybeans, and cotton; if undrained, mostly woodland or pasture (some Low Humic Gley soils and Alluvial soils)

I2P   AQUEPTS (with continuous or sporadic permafrost) gently sloping to steep; woodland or idle (Tundra soils)

I3   OCHREPTS (with thin or light–colored surface horizons and little organic matter) gently to moderately sloping; mostly pasture, small grain, and hay (Sols Bruns Acides and some Alluvial soils)

I3S   OCHREPTS gently sloping to steep; woodland, pasture, small grains

I4S   UMBREPTS (with thick dark–colored surface horizons rich in organic matter) moderately sloping to steep; mostly woodland (some Regosols)

  Mollisols . . . Soils with nearly black, organic–rich surface horizons and high base supply

M1   AQUOLLS (seasonally saturated with water) gently sloping; mostly drained and farmed (Humic Gley soils)

M2   BOROLLS (cool or cold) gently or moderately sloping, some steep slopes in Utah; mostly small grain in North Central States, range and woodland in Western States (some Chernozems)

M3   UDOLLS (temperate or warm, and moist) gently or moderately sloping; mostly corn, soybeans, and small grains (some Brunizems)

M4   USTOLLS (intermittently dry for long periods during summer) gently to moderately sloping; mostly wheat and range in western part, wheat and corn or sorghum in eastern part, some irrigated crops (Chestnut soils and some Chernozems and Brown soils)

M4S   USTOLLS moderately sloping to steep; mostly range or woodland

M5   XEROLLS (continuously dry in summer for long periods, moist in winter) gently to moderately sloping; mostly wheat, range, and irrigated crops (some Brunizems, Chestnut, and Brown soils)

M5S   XEROLLS moderately sloping to steep; mostly range

  Spodosols . . . Soils with accumulations of amorphous materials in subsurface horizons

S1   AQUODS (seasonally saturated with water) gently sloping; mostly range or woodland; where drained in Florida, citrus and special crops (Ground–Water Podzols)

S2   ORTHODS (with subsurface accumulations of iron, aluminum, and organic matter) gently to moderately sloping; woodland, pasture, small grains, special crops (Podzols, Brown Podzolic soils)

S2S   ORTHODS steep; mostly woodland

Ultisols . . . Soils that are usually moist with horizon of clay accumulation and a low base supply

U1   AQUULTS (seasonally saturated with water) gently sloping; woodland and pasture if undrained, feed and truck crops if drained (Some Low Humic Gley soils)

U2S   HUMULTS (with high or very high organic matter content) moderately sloping to steep; woodland and pasture if steep, sugar cane and pineapple in Hawaii, truck and seed crops in Western States (Some Reddish Brown Lateritic soils)

U3   UDULTS (with low organic matter content; temperate or warm, and moist) gently to moderately sloping; woodland, pasture, feed crops, tobacco, and cotton (Red Yellow Podzolic soils, some Reddish Brown Lateritic soils)

U3S   UDULTS moderately sloping to steep; woodland, pasture

U4S   XERULTS (with low to moderate organic matter content, continuously dry for long periods in summer) range and woodland (some Reddish Brown Lateritic soils)

Vertisols . . . Soils with high content of swelling clays and wide deep cracks at some season

V1   UDERTS (cracks open for only short periods, less than 3 months in a year) gently sloping; cotton, corn, pasture, and some rice (some Grumusols)

V2   USTERTS (cracks open and close twice a year and remain open more than 3 months); general crops, range, and some irrigated crops (some Grumusols)

Areas with little soil . . .

X1   Salt flats

X2   Rockland, ice fields

SOIL MAP OF THE WORLD

A soil map of the world that shows the probable occurrence of soil orders, suborders, and great groups was released in January, 1971, by the Soil Conservation Service, and is exhibited as Figure 7.13 (see folded insert).

LEGEND FOR FIGURE 7.13

A   **Alfisols** . . . Soils with subsurface horizons of clay accumulation and medium to high base supply; either usually moist or moist for 90 consecutive days during a period when temperature is suitable for plant growth

A1   BORALFS cool
      A1a—with Histosols, cryic temperature regimes common
      A1b—with Spodosols, cryic temperature regimes
A2   UDALFS temperate to hot, usually moist
      A2a—with Aqualfs
      A2b—with Aquolls
      A2c—with Hapludults
      A2d—with Ochrepts
      A2e—with Troporthents
      A2f—with Udorthents
A3   USTALFS temperate to hot, dry more than 90 cumulative days during periods
      when temperature is suitable for plant growth
      A3a—with Tropepts
      A3b—with Troporthents
      A3c—with Tropustults
      A3d—with Usterts
      A3e—with Ustochrepts
      A3f—with Ustolls
      A3g—with Ustorthents
      A3h—with Ustox
      A3j—Plinthustalfs with Ustorthents
A4   XERALFS temperate or warm, moist in winter and dry more than 60 consecu-
      tive days in summer
      A4a—with Xerochrepts
      A4b—with Xerorthents
      A4c—with Xerults

D   **Aridisols** . . . Soils with pedogenic horizons, usually dry in all horizons and
      are never moist as long as 90 consecutive days during a period when tempera-
      ture is suitable for plant growth
D1   ARIDISOLS undifferentiated
      D1a—with Orthents
      D1b—with Psamments
      D1c—with Ustalfs
D2   ARGIDS with horizons of clay accumulation
      D2a—with Fluvents
      D2b—with Torriorthents

E   **Entisols** . . . Soils without pedogenic horizons; either usually wet, usually
      moist, or usually dry
E1   AQUENTS seasonally or perennially wet
      E1a—Haplaquents with Udifluvents
      E1b—Psammaquents with Haplaquents
      E1c—Tropaquents with Hydraquents
E2   ORTHENTS loamy or clayey textures, many shallow to rock
      E2a—Cryorthents
      E2b—Cryorthents with Orthods
      E2c—Torriorthents with Aridisols
      E2d—Torriorthents with Ustalfs
      E2e—Xerorthents with Xeralfs
E3   PSAMMENTS sand or loamy sand textures
      E3a—with Aridisols

E3b—with Orthox
E3c—with Torriorthents
E3d—with Ustalfs
E3e—with Ustox
E3f—with shifting sands
E3g—Ustipsamments with Ustolls

H   Histosols . . . Organic soils
H1   HISTOSOLS undifferentiated
H1a—with Aquods
H1b—with Boralfs
H1c—with Cryaquepts

I   Inceptisols . . . Soils with pedogenic horizons of alteration or concentration
but without accumulations of translocated materials other than carbonates or
silica; usually moist or moist for 90 consecutive days during a period when
temperature is suitable for plant growth
I1   ANDEPTS amorphous clay or vitric volcanic ash or pumice
I1a—Dystrandepts with Ochrepts
I2   AQUEPTS seasonally wet
I2a—Cryaquepts with Orthents
I2b—Halaquepts with Salorthids
I2c—Haplaquepts with Humaquepts
I2d—Haplaquepts with Ochraqualfs
I2e—Humaquepts with Psamments
I2f—Tropaquepts with Hydraquents
I2g—Tropaquepts with Plinthaquults
I2h—Tropaquepts with Tropaquents
I2j—Tropaquepts with Tropudults
I3   OCHREPTS thin, light-colored surface horizons and little organic matter
I3a—Dystrochrepts with Fragiochrepts
I3b—Dystrochrepts with Orthox
I3c—Xerochrepts with Xerolls
I4   TROPEPTS continuously warm or hot
I4a—with Ustalfs
I4b—with Tropudults
I4c—with Ustox
I5   UMBREPTS dark-colored surface horizons with medium to low base supply
I5a—with Aqualfs

M   Mollisols . . . Soils with nearly black, organic-rich surface horizons and high
base supply; either usually moist or usually dry
M1   ALBOLLS light gray subsurface horizon over slowly permeable horizon, season-
ally wet
M1a—with Aquepts
M2   BOROLLS cool or cold
M2a—with Aquolls
M2b—with Orthids
M2c—with Torriorthents
M3   RENDOLLS subsurface horizons have much calcium carbonate but no accumula-
tion of clay
M3a—with Usterts
M4   UDOLLS temperate or warm, usually moist

	M4a—with Aquolls
	M4b—with Eutrochrepts
	M4c—with Humaquepts
M5	USTOLLS temperate to hot, dry more than 90 cumulative days in year
	M5a—with Argialbolls
	M5b—with Ustalfs
	M5c—with Usterts
	M5d—with Ustochrepts
M6	XEROLLS cool to warm, moist in winter and dry more than 60 consecutive days in summer
	M6a—with Xerorthents

O **Oxisols . . .** Soils with pedogenic horizons that are mixtures principally of kaolin, hydrated oxides, and quartz, and are low in weatherable minerals

O1   ORTHOX hot, nearly always moist
    O1a—with Plinthaquults
    O1b—with Tropudults

O2   USTOX warm or hot, dry for long periods but moist more than 90 consecutive days in the year
    O2a—with Plinthaquults
    O2b—with Tropustults
    O2c—with Ustalfs

S **Spodosols . . .** Soils with accumulation of amorphous materials in subsurface horizons; usually moist or wet

S1   SPODOSOLS undifferentiated
    S1a—cryic temperature regimes; with Boralfs
    S1b—cryic temperature regimes; with Histosols

S2   AQUODS seasonally wet
    S2a—Haplaquods with Quartzipsamments

S3   HUMODS with accumulations of organic matter in subsurface horizons
    S3a—with Hapludalfs

S4   ORTHODS with accumulations of organic matter, iron, and aluminum in subsurface horizons
    S4a—Haplorthods with Boralfs

U **Ultisols . . .** Soils with subsurface horizons of clay accumulation and low base supply, usually moist or moist for 90 consecutive days during a period when temperature is suitable for plant growth

U1   AQUULTS seasonally wet
    U1a—Ochraquults with Udults
    U1b—Plinthaquults with Orthox
    U1c—Plinthaquults with Plinthaquox
    U1d—Plinthaquults with Tropaquepts

U2   HUMULTS temperate or warm and moist all of year, high content of organic matter
    U2a—with Umbrepts

U3   UDULTS temperate to hot, never dry more than 90 cumulative days in the year
    U3a—with Andepts
    U3b—with Dystrochrepts
    U3c—with Udalfs
    U3d—Hapludults with Dystrochrepts
    U3e—Rhodudults with Udalfs

U3f—Tropudults with Aquults
U3g—Tropudults with Hydraquents
U3h—Tropudults with Orthox
U3j—Tropudults with Tropepts
U3k—Tropudults with Tropudalfs

U4   USTULTS warm or hot, dry more than 90 cumulative days in the year
U4a—with Ustochrepts
U4b—Plinthustults with Ustorthents
U4c—Rhodustults with Ustalfs
U4d—Tropustults with Tropaquepts
U4e—Tropustults with Ustalfs

V   **Vertisols** . . . Soils with high content of swelling clays; deep, wide cracks develop during dry periods
V1   UDERTS usually moist in some part in most years, cracks open less than 90 cumulative days in the year
V1a—with Usterts
V2   USTERTS cracks open more than 90 cumulative days in the year
V2a—with Tropaquepts
V2b—with Tropofluvents
V2c—with Ustalfs

X   **Soils in areas with mountains** . . . Soils with various moisture and temperature regimes; many steep slopes, relief and total elevation vary greatly from place to place. Soils vary greatly within short distances and with changes in altitude; vertical zonation common
X1   Cryic great groups of Entisols, Inceptisols, and Spodosols
X2   Boralfs and cryic great groups of Entisols and Inceptisols
X3   Udic great groups of Alfisols, Entisols, and Ultisols; Inceptisols
X4   Ustic great groups of Alfisols, Inceptisols, Mollisols, and Ultisols
X5   Xeric great groups of Alfisols, Entisols, Inceptisols, Mollisols, and Ultisols
X6   Torric great groups of Entisols; Aridisols
X7   Ustic and cryic great groups of Alfisols, Entisols, Inceptisols, and Mollisols; ustic great groups of Ultisols; cryic great groups of Spodosols
X8   Aridisols, torric and cryic great groups of Entisols, and cryic great groups of Spodosols and Inceptisols

Z   **Miscellaneous**
Z1   Icefields
Z2   Rugged Mountains—mostly devoid of soil (includes glaciers, permanent snow fields, and, in some places, areas of soil)

SUMMARY

Soil surveys have been made since 1899, and constant refinements have kept the program both fundamental and useful. The National Cooperative Soil Survey now carries on the work in cooperation with the land-grant colleges of the United States. The primary purpose of soil surveys is to make predictions regarding national land policies and predicted responses of the soil for use in agriculture, forestry, pasture, range management, engineering, irrigation, nutrition, public health, recreation, and pollution control.

Soils are classified into series, types, and phases, the soil type being the unit of mapping. The soil pedon and its characteristics are the basis for dividing soils into series and types. A uniform system is presented for describing a pedon.

QUESTIONS

1. State five important uses of soil surveys.
2. Name the eight criteria on which a soil series is based.
3. What are productivity ratings?
4. What is the relationship between soils and trees?
5. How can soil maps be used to evaluate soils for use in urban developments?

REFERENCES

Bartelli, L. J., A. A. Klingebiel, J. V. Baird, and M. R. Heddleson (eds.), *Soil Surveys and Land Use Planning.* Madison, Wisc.: Soil Science Society of America and American Society of Agronomy, 1966, 196 pp.

Bender, William H., *Soils Suitable for Septic-Tank Filter Fields,* Agricultural Information Bulletin No. 243. Soil Conservation Service, United States Department of Agriculture, 1961, 12 pp.

Bridges, E. M., *World Soils.* Cambridge, England: Cambridge University Press, 1970, 89 pp.

Covell, R. R., and J. V. Zary, *Soil Survey Interpretations for Woodland Conservation,* Mississippi Progress Report No. 1. Jackson, Miss.: Soil Conservation Service, 1965.

Dill, Henry W., Jr., *Worldwide Use of Airphotos in Agriculture,* United States Department of Agriculture Handbook No. 344. United States Department of Agriculture, 1967, 23 pp.

Field Manual of Soil Engineering (5th ed.). Lansing, Mich.: State Highway Commission, Michigan Department of State Highways (Jan., 1970), 474 pp., $6.00.

Soils and Men: The Yearbook of Agriculture (1938). United States Department of Agriculture.

Soil Survey Manual, United States Department of Agriculture Handbook No. 18. United States Department of Agriculture, 1962.

Soil Classification: A Comprehensive System, Seventh Approximation. Soil Survey Staff, Soil Conservation Service, United States Department of Agriculture (Aug., 1960).

Soil Taxonomy of the National Cooperative Soil Survey. Soil Conservation Service, United States Department of Agriculture, Washington, D.C., Dec., 1970.

CHAPTER 8

SOIL ECOLOGY

The source of nitrogen in the soil was an early stumbling block to the balance-sheet theory of soil-plant relationships. The question was cleared up in the 1880's with the discovery that Rhizobium *organisms grow in the nodules on the roots of leguminous plants and fix nitrogen from the air into forms that plants can use.*—CHARLES E. KELLOGG

Life in the soil is concentrated in the top few inches of the soil and consists of plant roots, mammals, earthworms, arthropods, gastropods, microscopic protozoa and nematodes, and such microscopic plants as bacteria, fungi, actinomycetes, and algae. The total population of all soil life is numbered in the billions per gram of soil, and the live weight per acre may be as much as 5 to 10 tons.

Because of the production of enzymes, carbon dioxide, and organic matter, the life in the soil is responsible for making plant nutrients more readily available and in making and stabilizing desirable soil structure for better plant growth.

PLANT ROOTS

All plant roots are modified by the physical and chemical conditions of the soil. Plant roots serve as a mechanical support, absorb water and plant nutrients, act as a storehouse for plant foods, and transport water and nutrients for the growth of the aerial parts of the plant.

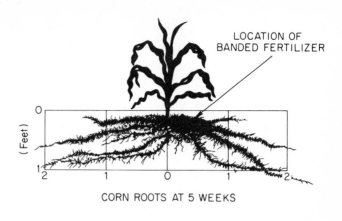

LOCATION OF
BANDED FERTILIZER

CORN ROOTS AT 5 WEEKS

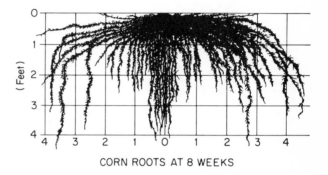

FIG. 8.1. Corn roots may grow as much as 2 inches a day and at 5 weeks of age extend one foot in depth and 4 feet laterally; at 8 weeks of age the roots may have attained a depth of 4 feet and a spread of 8 feet. Roots proliferate especially near fertilizer. *Source:* Robert Engle.

CORN ROOTS AT 8 WEEKS

The growth of plant roots is influenced by the kind of plant, the physical conditions of the soil, available nutrients, soil temperature, available water, oxygen supply, and carbon dioxide concentration.

Most grasses have a finely fibrous root system that usually extends several feet deep. Alfalfa has a tap root and may grow 20 feet deep or more, depending on soil conditions. Cotton has a combination of a tap root and strong lateral roots. Some trees, like the walnut, have a tap root, whereas most trees have lateral roots.

The rate of growth of plant roots varies widely, depending upon the kind of plant and the stage of growth of the plant, as well as environmental factors such as temperature, available moisture, and available essential elements. Plant diseases and pests may seriously retard the growth of plant roots.

Under ideal conditions, the roots of a corn plant may grow as much as 2 inches a day, and at 5 weeks of age reach a depth of 1 foot and attain a lateral spread of 4 feet. Roots proliferate especially in and near a band of fertilizer. At 8 weeks of age, corn roots may have grown to a depth of 4 feet and have attained a lateral spread of 8 feet (Figure 8.1). The tap root of a cotton plant has been reported to grow at a rate of 1 inch a day for 72

FIG. 8.2. Most tree roots grow to a depth of at least 5 feet if there are no restrictive factors in the soil such as shallow bedrock, a high water table, a compact clay, or a dry subsoil. Occasionally a toxic salt layer exists which will restrict root growth. (1): Bedrock has modified these roots of eastern white pine into ones that resemble an octopus. (Michigan.) (2): Roots of eastern white pine normally consist of many strong laterals that extend downward to depth of 5 to 10 feet when the soil conditions permit. (Michigan.) (3): The annual rainfall here is only 25 inches and therefore not enough to wet the soil below 3 feet. Roots of this western yellow pine tree grew through the moist soil to the dry soil, then stopped. (Arizona.) (4): A high water table at 2 feet has modified these roots of eastern hemlock. (Michigan.) (Scales are in feet.) *Source:* All photos by J. O. Veatch.

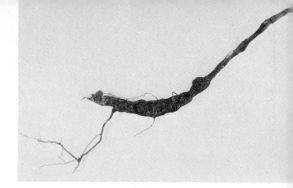

FIG. 8.3. Roots are altered when growing in soils that restrict or modify them. These apple tree roots were flattened in an irregular pattern when growing through a layer of coarse gravel at a depth of 8 feet. *Source:* J. O. Veatch.

days, whereas lateral roots may average approximately half this rate of growth.

Roots of trees are strongly modified by the physical and chemical properties of the soil in which they are grown. In the Adirondack Mountains of New York State, the senior author confirmed that yellow birch roots will extend to a depth of 4 feet on well-drained soil but to only 1 foot in depth on poorly drained soils.[1] Tree roots of eastern white pine, hemlock, and red maple in Michigan, roots of longleaf pine in Georiga and Florida, and roots of western yellow pine in Arizona are all restricted or stimulated by the physical and chemical properties of the soil and available soil moisture in which they grow (Figure 8.2).[2]

Available nutrients influence the kinds and amounts of roots produced. Roots will grow into a band of fertilizer and form a mass of fine roots. Roots will also concentrate (proliferate) in a lump of manure that has been buried in the soil.

The roots of most plants will grow at soil temperatures lower than those necessary for top growth. Apple tree roots start to grow at 35°F. Roots stop growing when soil temperatures are high. The roots of the apple tree normally are circular in cross section, but will be flattened in irregular patterns when growing in coarse, gravelly soil (Figure 8.3).

Soil water is another factor that determines the nature of plant roots. In the Great Plains, where most of the moisture comes in the winter or spring and where the summers are usually dry, most grasses and forbs have root systems to depths of at least 5 feet on well-drained soils. On soils in the Great Plains with excessive wetness, however, native meadow grasses produce more than half of their total of 7.27 tons of roots per acre in the surface 2 inches of soil.[3]

[1] Roy L. Donahue, "Forest-site Quality Studies in the Adirondacks," Part I, *Tree Growth as Related to Soil Morphology,* Cornell University Agricultural Experiment Station Memoir 229 (1940).

[2] Private communication from J. O. Veatch, Professor Emeritus of Soil Science, Michigan State University.

[3] A. W. Moore and H. F. Rhoades, "Soil Conditions and Root Distribution in Two Wet Meadows of the Nebraska Sandhills," *Agronomy Journal,* Vol. 58, No. 6 (1966).

FIG. 8.4. (a) Mole. (b) Gopher. (c) Prairie dog. (d) Woodchuck. All alter the soil by their burrowing habits. *Source:* U.S. Fish and Wildlife Service.

The roots of most crop plants cannot grow in soils when the water is held with an equilibrium tension greater than 15 atmospheres or less than $\frac{1}{3}$ atmosphere. It has been demonstrated, however, that plant roots may grow into a soil drier than at a 15-atmosphere equilibrium if a part of the roots of the same plant have adequate moisture. It appears that the root system translocates water where it is plentiful to areas where it is needed for growth.[4]

Root development of some plants is retarded when the oxygen content of the soil is below 10 per cent, and growth may cease when it is less than 5 per cent. As the soil temperature increases, the percentage of oxygen required for root growth also increases. But plants such as rice, buckwheat, and willow trees apparently do not need as much oxygen in the soil atmosphere for satisfactory root growth. The roots of tomatoes, peas, and corn may need even more oxygen than do most plants.

---

[4] See Chapter 10 for more detailed discussion of soil moisture.

Concentrations of carbon dioxide greater than 10 per cent seem to kill roots; however, 1 per cent carbon dioxide appears to be necessary for normal root growth.

SOIL-INHABITING MAMMALS

Burrowing mammals, such as prairie dogs, gophers, mice, shrews, moles, rabbits, badgers, woodchucks, armadillos, and chipmunks, all contribute to an alteration of the soil (see Table 8.1). In the long run, it appears that the burrowing activity of these mammals makes the soil more productive for plant growth.

TABLE 8.1 SOIL-INHABITING MAMMALS IN A FOREST IN THE NORTHEAST *

Species	Number per Acre	Unit Live Weight (gm)	Total Live Weight per Acre (gm)
Deer mouse	35	20	700
Red-backed mouse	12	21	252
Flying squirrel	1	60	60
Short-tailed shrew	40	18	720
Long-tailed shrew	9	6	54
Hairy-tailed shrew	3	54	162
Total	100	—	1948 gm or 4.3 lb

* W. J. Hamilton, Jr., and David B. Cook, "Small Mammals and the Forest," *Journal of Forestry*, Vol. 38 (1940), pp. 468–73.

In the short run, however, most of these animals eat vegetation and thereby damage native vegetation and many of the farmers' crops (Figure 8.4).

In the forests of the Northeast, an estimate was made of the numbers of the various species of mammals on a representative acre. Shrews dominate the soil-inhabiting mammals in a northeastern forest. The deer mouse was next in abundance, followed by the red-backed mouse and the flying squirrel. The average number of all mammals per acre was 100, and they weighted a total of 4.3 pounds.

EARTHWORMS

Earthworms are very important soil macrofauna, especially in undisturbed pasture and forest soils with a pH above 4.5. Earthworms ingest

FIG. 8.5. Earthworms are highly desirable soil macrofauna, beyond their use for fish bait. Earthworms ingest undecomposed organic matter and excrete it in their casts in a form in which the nutrients are more readily available to growing plants. Earthworms make the soil more productive for plants by "plowing" it with their burrows and by secreting body slimes that "waterproof" desirable soil peds. *Source:* Soil Conservation Service, U.S. Dept. of Agr.

undecomposed organic matter and excrete it in a form in which nutrients are more readily available to growing plants. They stir the soil in the process of making their burrows and thus improve soil aeration and hasten the infiltration of water into the soil (Figure 8.5).

It is difficult to differentiate among many of the species of earthworms. Four common species in northeastern United States are:

1. *Lumbricus terrestris,* 5–8 inches long
2. *Lumbricus rubellus,* 4–6 inches long
3. *Allolobophora caliginosa,* 4–7 inches long
4. *Octolasium lacteum,* 2½–4 inches long

Lumbricus terrestris, one of the "nightcrawlers," makes burrows as deep as 5 feet, and rolls leaves, grass, and other organic materials into the surface openings. Many species do not come to the surface but feed mostly in the first 6 inches of soil. Several earthworm species leave spherical excreta (casts) at the surface around their burrow. Almost all species feed on manure, dead grasses, legumes, and selected forest tree leaves. The leaves of elm, ash, basswood, sugar maple, and birch are readily eaten, whereas the leaves of beech and red oak are not so desirable. Pine, spruce, and hemlock needles are rejected as food by earthworms.

The soil most suitable for earthworms is one that is slightly too wet in

FIG. 8.6. Continuous tillage on the left has created an undesirable environment for earthworms. On the right, a rotation of crops has left sufficient residues to feed earthworms. Note the amount of earthworm casts. (Texas Blacklands.) *Source:* D. O. Thompson, Tex. Agr. Exp. Sta.

the spring for the best growth of most crop plants. A moderately well-drained to poorly drained fine-textured soil, which is continuously supplied with fresh palatable organic matter, is the best habitat for earthworms. Silt loam or silty clay loam soils are usually the most desirable. Clay soils are less desirable, and earthworms are seldom found in sandy soils. Continuous tillage tends to reduce the population of earthworms, although they recover quickly when perennial grasses and legumes are in the cropping system (Figure 8.6).

Near Adelaide, Australia, 78 pounds of P_2O_5 per acre per year as either rock phosphate or superphosphate on pasture increased the live weight of earthworms from 55 grams per square meter in the control plots to 90 grams per square meter in the phosphated plots.

The state law in Ohio requires that spoilbanks from coal mines be revegetated with woody plants. The soil material is so infertile, however, that many years are required to cover the ugly spoilbanks. To enhance the fertility of the soil on spoilbanks, black locust litter was added on both treatments, but on one, *live* earthworms were added; the other plot received the same number of *dead* earthworms to equalize added fertility. After 175 days, the soils in the plots were analyzed; the results are shown in Table

FIG. 8.7. Termites have aerated this soil and have altered the soil structure by tunneling and building mounds. The string across the photo represents the original soil surface. At the lower left, near the end of the ruler (marked in inches), is an "ambrosia ball," or "termite garden," made by the termites from cellulose, which serves as a substrate for a fungus that is used for termite food. (India.) *Source:* Julian P. Donahue.

8.2. The *live* earthworms increased the available nutrients from 19 per cent for potassium to 165 per cent for phosphorus.

TABLE 8.2   THE NUTRIENT CONTENT OF SOIL WITH *LIVE* EARTHWORMS VERSUS SOIL WITH *DEAD* EARTHWORMS *

Nutrient	Soil with *Dead* Earthworms	Soil with *Live* Earthworms	Increase Due to Live Earthworms (%)
Calcium (Ca) (exchangeable)	1.7	2.7	59
Potassium (K) (exchangeable)	0.16	0.19	19
Magnesium (Mg) (milliequivalents per 100 gm soil)	1.8	2.5	39
Phosphorus (P) (parts per million)	1.7	4.2	165

* J. P. Vimmerstedt, "Earthworms Speed Leaf Decay on Spoilbanks," *Ohio Report,* Vol. 54, No. 1 (Wooster, Ohio: Ohio Agricultural Research and Development Center, Jan.–Feb., 1969), pp. 3–5.

ARTHROPODS AND GASTROPODS

The common soil arthropods include mites, millipedes, centipedes, springtails, and larvae of beetles, flies, ants, and termites. These macrofauna feed mostly on decaying vegetation and help to aerate the soil with their burrows (Figure 8.7).

A light brown ant, *Formica cinera,* is credited with "plowing the prairie" during the past 3500 years. It has been estimated that the activity of this ant brings 1 inch of soil to the surface of every acre each 500 years. Its mound extends to a depth of 6 feet, is 1 foot high, and has chambers and cavities that occupy 12 per cent of the total volume of each acre of soil.[5]

Slugs and snails are important members of the gastropods that inhabit the soil. They feed mostly in decaying vegetation but will eat and damage living vegetables when other foods are scarce. Counts as high as 600,000 slugs per acre have been reported, and their weight was approximately 400 pounds. Slugs and snails typically are scavengers.

PROTOZOA AND NEMATODES

Protozoa are microscopic animals that feed mainly on bacteria; in that way, they influence the balance of microbiological populations. The three main groups of protozoa are amoebae, flagellates, and ciliates.

Nematodes are microscopic eel worms and are classified according to their activity into:

1. Omnivorous. This group is the most common of the soil nematodes. They live mainly on decaying organic matter.
2. Predaceous. This group prey on soil fauna, including other nematodes.
3. Parasitic. These nematodes infest plant roots, usually causing knots on the roots and low crop yields.

DISTRIBUTION OF SOIL MICROFLORA

The soil microflora consists of bacteria, actinomycetes, fungi, and algae. Table 8.3 gives estimates of the average number of microflora per gram of soil and the live weight per acre to plow depth.

---

[5] F. Paul Baxter and Francis D. Hole, "The Ant That Plowed the Prairie," *Crops and Soils Magazine,* Vol. 19, No. 2 (Nov., 1966), p. 11.

TABLE 8.3   THE AVERAGE NUMBER OF THE SOIL MICROFLORA AND
THEIR LIVE WEIGHT PER ACRE TO PLOW DEPTH *

Group	Average Number per Gram of Soil	Live Weight per Acre to Plow Depth (Pounds)
Bacteria	1 billion	500
Actinomycetes	10 million	750
Fungi	1 million	1000
Algae	100 thousand	150
Total	—	2400

* Francis E. Clark, A Perspective of the Soil Microflora, Soil Microbiology Conference, Purdue University (June, 1954).

Bacteria are estimated to average 1 billion per gram; actinomycetes, 10 million; fungi, 1 million; and algae, 100 thousand per gram of soil (Table 8.3). Inasmuch as the several microflora are not the same size, the live weight per acre to plow depth is not in the same order as the numbers. Fungi exceed in live weight per acre, with 1000 pounds, followed in order by actinomycetes with 750 pounds, bacteria with 500 pounds, and algae with 150 pounds. This totals 2400 pounds of live weight of microflora per acre to plow depth.

Bacteria, fungi, and actinomycetes aid in the development of desirable soil structure by their secretions of gummy substances that are not water soluble. A comparison was made of the relative amount of good soil structure that was produced by each of these three groups of plants, and the results are graphed in Figure 8.8.

From Figure 8.8, it is obvious that bacteria are responsible for the creation of the least relative amount of large soil aggregates. Whereas actinomycetes are 17 times more efficient than bacteria, fungi are the best of all for this purpose.

AUTOTROPHIC BACTERIA

The autotrophic bacteria obtain carbon from the carbon dioxide of the atmosphere and their energy from the oxidation of simple carbon compounds or of inorganic substances. Specific groups of autotrophic bacteria are capable of oxidizing ammonia, nitrite, sulfur, iron, manganese, hydrogen, carbon monoxide, or methane.

Probably the most important groups of autotrophic bacteria are those that oxidize ammonia to nitrites and nitrites to nitrates. These groups are known generally as nitrifying organisms. The environmental conditions necessary for the maximum growth of the nitrifying organisms are the following:

1. The presence of proteins to form ammonia; or the presence of ammonia salts such as ammonium sulfate
2. Adequate aeration
3. A moist but not a wet soil
4. A large amount of calcium
5. The proper temperature. The production of nitrates is greatest at 37°C (98.6°F), and process stops at 5°C (41°F) and 55°C (131°F).

The nitrification process is shown diagrammatically as follows:

Ammonification				Nitrification
NH_4	$\rightarrow$	NO_2	$\rightarrow$	NO_3
ammonium		nitrite		nitrate
(utilized		(apparently toxic		(utilized by plants)
by plants)		to plants)		

HETEROTROPHIC BACTERIA

Heterotrophic bacteria are those that depend upon organic matter for their source of energy. In this group are most of the soil bacteria. Some groups of bacteria are capable of fixing atmospheric nitrogen, whereas other groups depend upon fixed nitrogen compounds for their nutrition.

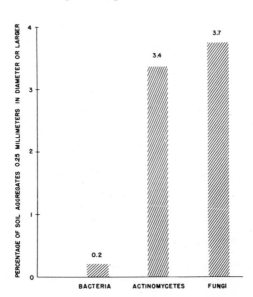

FIG. 8.8. A comparison of the aggregating influence of bacteria, actinomycetes, and fungi when incubated with sterile Gila clay at 26°C for 21 days. Source: D. S. Hubbell and Glen Staten, *Studies on Soil Structure,* N. Mex. Agr. Exp. Sta. Tech. Bul. No. 363, 1951.

FIG. 8.9. All legume seed should be inoculated before planting. *Left:* Soybeans that were not inoculated. *Right:* Soybeans that were properly inoculated. *Source:* Nitragin Co.

Members of this broad group of heterotrophic bacteria are the symbiotic [6] nitrogen-fixing legume bacteria, the nonsymbiotic nitrogen-fixing bacteria, and the nonnitrogen-fixing bacteria.

SYMBIOTIC (LEGUME) BACTERIA

The symbiotic or legume bacteria are the most important heterotrophic bacteria in a study of soils and plant growth.

It was known in ancient Greece that legumes had a beneficial effect upon the following crop. This fact is mentioned in several places in the Bible. But it was not until 1838 that Boussingault demonstrated the fact that the beneficial effect of legumes was due to their power to fix atmospheric nitrogen. The next major advance in knowledge about legume bacteria was made by Frank, in 1879, when he proved that artificial inoculation with specific bacteria resulted in nodule formation on the roots of legumes, and that bacteria in these nodules could fix atmospheric nitrogen.

Symbiotic bacteria attack the root hairs of legume plants, and the injury induces the root cells to grow around the bacteria. These cell extensions are the nodules in which the legume bacteria live. The legume roots supply to the bacteria the essential minerals and organic matter for energy;

---

[6] The word *symbiotic* has been derived from a Greek word which means "living together."

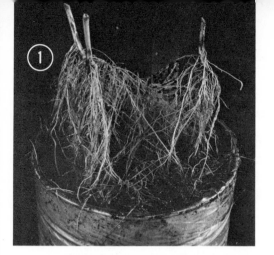
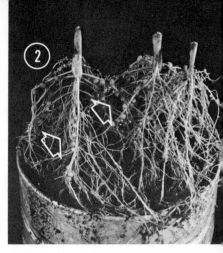

FIG. 8.10. The influence of phosphorus fertilizer on nodulation of soybeans grown on soil testing low in phosphorus. (1): No phosphorus fertilizer. (2): 800 pounds per acre of phosphorus (P) fertilizer. Note the larger number of nodules. *Source:* John Pesek, Iowa State U.

in return, the bacteria are able to use atmospheric nitrogen to build their body proteins.

Since the life span of a single bacterium is only a few hours, the bodies of some bacteria are continuously in the process of decay, releasing fixed ammonium and nitrates for use by the host legume. Likewise, entire nodules are constantly being replaced by new ones. The fixed nitrogen in the old nodules is made available by decomposition to the host legume, as well as to other plants growing in association with the legume. The nitrogen fixed by legume bacteria is therefore made available to the legume, to any associated plants, and to the crop that follows in the cropping system (Figures 8.9 and 8.10).

The symbiotic organisms belong to the genus *Rhizobium,* and there are several species named after the host legume. *Rhizobium meliloti* is the species that inoculates alfalfa and sweetclovers, and *Rhizobium trifolii* is compatible for the true clovers, such as white clover and red clover. The groups of legumes that can be inoculated by the same species of symbiotic bacteria are:

Group 1. Alfalfa, sweetclovers, and trefoils
Group 2. White clover, Ladino clover, red clover, alsike clover, and crimson clover
Group 3. Lespedezas, cowpeas, kudzu, and velvetbean
Group 4. Garden pea, sweetpea, Canada field pea, and vetches
Group 5. Soybean
Group 6. Garden bean
Group 7. Lupine

There are also many specialized groups in addition to this list.

SYMBIOTIC (NONLEGUME) BACTERIA

It has been demonstrated that the symbiotic (but nonleguminous) bacteria attached to the roots of several species of alder are capable of fixing at least 50 pounds of nitrogen (N) per acre per year. These include *Alnus incana, Alnus glutinosa, Alder tenuifolia,* and *Alnus rubra.* In the Pacific northwestern states, some species of alder are planted on mining spoils, barren glacial deposits, marine terraces, recent flood deposits, sand dunes, and other soils that are too cold, too wet, or too infertile for most other plant species. Alder is also used to interplant between other tree species on soils of low fertility to increase soil nitrogen and thereby increase tree growth.[7]

A later study in Oregon indicated that 70 pounds of nitrogen (N) per year was a more accurate value when red alder (*Alnus rubra*) was grown in pure stands and 74 pounds when red alder was grown in mixture with a stand of mixed conifers.[8]

NONSYMBIOTIC NITROGEN-FIXING BACTERIA

In 1891, Winogradsky first demonstrated that when soil was exposed to the atmosphere its nitrogen content increased. The organism responsible for the fixation of atmospheric nitrogen without the association of a legume was isolated and recognized as *Clostridium pasteurianum.* These bacteria are anerobic and belong to the same group as butyric acid bacteria.

Beijerinck in 1901 showed that there were also aerobic bacteria that could fix atmospheric nitrogen without symbiosis with legumes. The first bacteria isolated was *Azotobacter chroococcum.*[9]

Several other soil bacteria are capable of obtaining their nitrogen from the atmosphere. Among these are purple bacteria, a group known as *Granulobacter,* and several others of minor importance. As the techniques of soil bacteriology improve, there will probably be more bacteria discovered that are capable of utilizing nitrogen as a gas from the atmosphere.

---

[7] R. F. Tarrant, "Forest Soil Improvement Through Growing Red Alder (*Alnus rubra*) in Pacific Northwestern United States," *Transactions, 8th International Congress of Soil Science,* Bucharest, Roumania (1964), pp. 1029–43.

[8] R. F. Tarrant, K. C. Lu, W. B. Bollen, and J. F. Franklin, *Nitrogen Enrichment of Two Forest Ecosystems by Red Alder,* U.S.D.A. Forest Service Research Paper PNW–76 (Portland, Ore., 1968), 8 pp.

[9] There are at least five species of the genus *Azotobacter* capable of fixing atmospheric nitrogen: *Azotobacter chroococcum, Azotobacter beijerinckii, Azotobacter vinelandii, Azotobacter agilis,* and *Azotobacter indicum.*

As a general rule, *Clostridium* are more abundant in soils than *Azotobacter*. *Clostridium* develop best in poorly drained, acid soils, whereas *Azotobacter* are more abundant in well-drained, neutral soils. The amounts of atmospheric nitrogen fixed by these bacteria are variable, depending upon the soil environment. Under ideal conditions, the total nitrogen fixed by both groups varies from 25 to 50 pounds per acre per year.

NONNITROGEN-FIXING HETEROTROPHIC BACTERIA

Most of the bacteria in the soil are classified as nonnitrogen-fixing heterotrophic bacteria; these are the organisms responsible for the decomposition of organic matter and from which bacteria obtain energy. Some members of the group include the thermophilic bacteria that cause self-heating and burning of hay, myxobacteria, and certain of the genus *Bacillus,* some of which produce antibiotics.

FUNGI

Soil fungi may be parasitic, saprophytic, or symbiotic. Parasitic fungi produce plant diseases, such as cotton root rot, and many kinds of wilts, rusts, blights, and smuts. Saprophytic fungi obtain their energy from the decomposition of organic matter. Symbiotic fungi live on the roots of certain plants, and both fungus and plant are mutually benefited.

Fungi are especially useful in the soil because they break down the somewhat resistant cellulose, lignin, and gum, as well as the more readily decomposed sugars, starches, and proteins. A large part of the slowly decomposing soil humus consists of the dead remains of fungal hyphae.

Mycorrhiza, meaning "fungus root," is the name given to a symbiotic association of fungal mycelia and roots of certain trees and shrubs. It is presumed that mycorrhiza aid the host plant in the absorption of certain nutrients. Forest nurseries that have been established to raise tree seedlings not native to the area usually need an artificial inoculation of a suitable mycorrhiza. The reasoning behind this practice is that in a new region the compatible mycorrhiza are usually not present in the soil. There are two general types of mycorrhizae, based upon their manner of growth.

Ectotrophic mycorrhiza are usually formed by members of *Agaricales,* including mostly mushroom-type fruiting fungi. They grow as threadlike filaments into small roots *between* the root cells, but not *into* the cells. Their function seems to help the tree roots absorb nutrients by increasing their absorbing surfaces. Trees that have the ectotrophic type of mycorrhiza are the pines, spruces, oaks, elms, beech, hickories, chestnut, and birches.

Endotrophic mycorrhiza are usually formed from species of *Phoma* and *Pythium*. The fungal hyphae penetrate the plant root cells. Upon dying, mycorrhizal tissues are absorbed and utilized by the growing trees. Trees and other plants whose roots often have endotrophic mycorrhiza growing into them are sweet gum, poplars, maples, laurels, azaleas, rhododendrons, and orchids.

ACTINOMYCETES

Actinomycetes are taxonomically and morphologically related to both fungi and bacteria but have recently been classified as bacteria. They are characterized by branched mycelia, similar to fungi, and resemble bacteria when the mycelia break into short fragments.

In recent years, actinomycetes have attracted world-wide attention after it was discovered that they produce many antibiotics. At present, nearly 500 antibiotics have been isolated from actinomycetes. The most common antibiotics from actinomycetes are streptomycin, aureomycin, terramycin, and neomycin.

Actinomycetes are found in rather large quantities in soils where the environment is satisfactory. They thrive best where there is ample, fresh organic matter, where the soil is neutral to slightly acid, and where soil moisture is fairly abundant. They grow better than fungi, however, when the soil is fairly dry.

The primary function of actiomycetes is in decomposing organic matter, especially cellulose and other resistant forms.

Potato scab disease, an actinomycete, can be readily controlled by keeping the pH of the soil below 5.0.

ALGAE

Soil algae are microscopic chlorophyll-bearing organisms. The main groups are

1. Green
2. Blue-green
3. Yellow-green
4. Diatoms

Algae develop best in moist, fertile soils. The green color of the soil surface following the application of commercial fertilizers is due to an increase in the number of algae.

The probable effect of algae on plant growth is to:

1. Add organic matter to the soil. The organic matter is manufactured by the chlorophyll in the algae.
2. Improve soil aeration—especially of rice paddies by excreting oxygen for use by the rice plants.
3. Fix atmospheric nitrogen. Only a few groups of the blue-green algae can do this.

Certain members of the blue-green algae have been demonstrated to fix atmospheric nitrogen. The best pH range for this fixation is between 7.0 and 8.5. In flooded rice fields, this group of algae helps to maintain the nitrogen level of the soil by utilizing atmospheric nitrogen. Also, in desert soils blue-green algae are the dominant microorganisms and may be responsible for the high nitrogen content of many surface soils.

Cameron and Fuller in Arizona isolated several species of blue-green algae of genera *Nostoc, Scytonema,* and *Anabaena,* which were capable of fixing atmospheric nitrogen in solution cultures.[10]

Mayland, McIntosh, and Fuller later simulated field conditions with the use of N^{15} isotope and confirmed the fixation of nitrogen by blue-green algae that are found naturally as a part of the crust on semiarid desert surface soils in Arizona.[11] Annual rainfall was 19 inches (Figure 8.11).

[10] R. E. Cameron and W. H. Fuller, "Nitrogen Fixation by Some Algae in Arizona Soils," *Soil Science Society of America, Proceedings,* Vol. 24 (1960), pp. 353–56.

[11] H. F. Mayland, T. H. McIntosh, and W. H. Fuller, "Fixation of Isotope Nitrogen on a Semiarid Soil by Algal Crust Organisms," *Soil Science Society of America, Proceedings,* Vol. 30, No. 1 (1966), pp. 56–60.

FIG. 8.11. In the semiarid desert near Tuscon, Arizona, blue-green algae form a discontinuous surface crust that is capable of fixing (if it formed a continuous crust) as much as 9.6 pounds of N per acre per year. (1): Typical natural vegetation of shrubs, grasses, and cacti where algae crusts are common. (2): The surface crusts on the two stones (at arrows) are blue-green algae that fix atmospheric nitrogen during the approximately 60 days a year that there is enough rainfall. (Scale is in inches.) *Source:* H. F. Mayland, T. H. McIntire, and W. H. Fuller, Ariz. Agr. Exp. Sta.

SUMMARY

Life in the soil consists of both macro- and microplants and animals. Macroplants include the roots of higher plants. Microplants are the bacteria, fungi, actinomycetes, and algae. Macroanimals comprise mammals, earthworms, arthropods, and gastropods. Protozoa and nematodes are the principal soil-borne microanimals.

All plant and animal life in the soil helps to make nutrients more available and aids in the creation and stabilization of desirable soil structure. Fungi and actiomycetes are more effective than bacteria in creating good soil structure.

Plant root growth is influenced by the genetics of the plant, the physical condition of the soil, the availability of nutrients, soil temperature, and the supply of water, oxygen, and carbon dioxide. Shrews are the most numerous of the mammals in forest soils of the Northeast. Earthworm casts have a higher pH, a higher base saturation, and more available calcium, magnesium, nitrogen, phosphorus, and potassium than does the nearby soil. The greatest increase in availability of any nutrient is in potassium, where the earthworm casts are more than 10 times richer than the surrounding surface soil.

Bacteria are classified into autotrophic (such as ammonia bacteria) and heterotrophic, like the symbiotic bacteria. Autotrophic bacteria obtain carbon from the carbon dioxide of the atmosphere and their energy from the oxidation of specific inorganic or simple carbon compounds. Heterotrophic bacteria obtain their energy from the decomposition of organic matter.

Symbiotic bacteria may be those that live on the roots of legumes, as was demonstrated to be beneficial as early as 1838, and those that live on roots of a few nonlegumes, such as certain species of alder, proved first more than 100 years later.

Azotobacter are important aerobic nonsymbiotic bacteria, and *Clostridium* are well known as anaerobic nonsymbiotic bacteria.

Fungi may be parasitic (disease producing), saprophytic (aiding in decay), or symbiotic (helping certain plants to grow better). Actinomycetes are best known because one species causes Irish potato scab. More recently, many new antibiotics have been isolated from several species of actinomycetes. Algae contribute to the amount of soil organic matter and improve soil aeration in rice paddies, and several species of blue-green algae are capable of fixing atmospheric nitrogen.

QUESTIONS

1. What are the critical concentrations of oxygen and carbon dioxide for plant roots?
2. What are some food preferences of earthworms?
3. Explain the function of symbiotic bacteria on legumes.
4. Differentiate between *Azotobacter* and *Clostridium*.
5. What are mycorrhiza?

REFERENCES

Alexander, Martin, *Introduction to Soil Microbiology*. New York: John Wiley & Sons, Inc., 1961.

Eaton, Theodore H., Jr., and Robert F. Chandler, Jr., *The Fauna of Forest-Humus Layers in New York*. Cornell University Agricultural Experiment Station Memoir 247, 1942.

Gilman, Joseph C., *A Manual of Soil Fungi*. Ames, Iowa: The Iowa State College Press, 1945.

Plant Diseases: The Yearbook of Agriculture (1953). United States Department of Agriculture.

River of Life. Water· The Environmental Challenge. United States Department of the Interior, Conservation Yearbook Series, Vol. 6, 1970, 96 pp.

Russell, Sir E. John, and E. Walter Russell, *Soil Conditions and Plant Growth* (9th ed.). London: Longmans, Green & Company, Ltd., 1961.

Waksman, Selman A., *Soil Microbiology*. New York: John Wiley & Sons, Inc., 1952.

CHAPTER 9

ORGANIC MATTER

But then whatever weeds are upon the ground, being turned into the earth, enrich the soil as much as dung.

—XENOPHON

Organic matter in the soil comes from the remains of plants and animals. This includes crop residues, weeds, grasses, trees, bacteria, fungi, protozoa, earthworms, rodents, and animal manures.

Fresh organic matter helps physically by keeping the soil open and spongy, and aids chemically by releasing carbohydrates for energy and nutrients for the growth of organisms. Upon decomposition, organic matter releases carbon dioxide, which acts as a solvent on soil minerals to make them more available to plants. As organic matter breaks down, the nitrogen, phosphorus, and sulfur, and to some extent all nutrients, are released from the plant and animal tissues to become available for the growth of the next crop.

When organic matter decomposes, the slime that is formed helps to improve and stabilize desirable soil aggregates. Better air and water relations in the soil thereby result, and plant growth is enhanced.

FUNCTIONS OF ORGANIC MATTER

Organic matter serves many purposes in the soil that may be summarized as follows:

1. Coarse organic matter on the surface reduces the impact of the falling raindrop and permits clear water to seep gently into the soil. Surface runoff and erosion are thus reduced, and as a result there is more available water for plant growth.

With no straw mulch on the surface, a 2.8-inch summer rain in Texas wets the soil to a depth of 15 inches, but with a 16-ton straw mulch the soil was wet to a depth of 30 inches.

2. Decomposing organic matter produces slimes and microbial gums which help to form and to stabilize desirable soil structure.

3. Live roots decay and provide channels down through which new plants roots grow more luxuriantly. The same root channels are effective in transmitting water downward, a part of which is stored for future use by plants.

4. Fresh organic matter supplies food for such soil life as earthworms, ants, and rodents. These animals burrow in the soil and, in so doing, permit plant roots to obtain oxygen and to release carbon dioxide as they grow.

5. Trashy and coarse organic matter on the surface of soils will reduce losses of soil by wind and water erosion.

6. Surface mulches lower soil temperatures in the summer and keep the soil warmer in winter. In Kansas, the soil under a straw mulch during the growing season was *cooler* than a bare soil by 10°F at a depth of 0.5 inch and by 4°F at a depth of 25 inches. By contrast, the soil under a clear plastic mulch was 10°F *hotter* at a depth of one-half inch and 2°F *hotter* at a depth of 25 inches.[1]

7. Evaporation losses of water are reduced by organic mulches.

8. Upon decomposition, organic matter supplies some of all nutrients needed by growing plants, as well as many hormones and antibiotics. These nutrients are released in harmony with the needs of the plants. When environmental conditions are favorable for rapid plant growth, the same conditions favor a rapid release of nutrients from the organic matter. Organic matter contains a large part of the total reserves of boron and molybdenum, 5 to 60 per cent of the phosphorus reserves, up to 80 per cent of the sulfur, and practically all of the nitrogen.

The release of nitrogen during the growing season from soil organic matter is determined by the percentage of organic matter present, the soil texture, and the temperature and moisture conditions existing. The pounds of nitrogen per acre released during one growing season may vary from 15 pounds in a clay loam that is low in organic matter to 110 pounds in a silt loam that is high in organic matter (Table 9.1).

---

[1] R. J. Hank and S. A. Bowers, "Soil Temperature Is Modified by Mulch," *Crops and Soils* (March, 1960).

TABLE 9.1   NITROGEN RELEASED FROM SOIL ORGANIC MATTER
DURING THE GROWING SEASON *

Soil Organic Matter	Nitrogen Released (Pounds per Acre) †		
(%)	Sandy Loam	Silt Loam	Clay Loam
1	50	20	15
2	100	45	40
3	—	68	45
4	—	90	75
5	—	110	90

* Agricultural Ammonia News.

† Soils in southern regions will release more and soils in northern regions will release less N than that shown. When a good legume stand has been turned under, the crop that follows may have an additional 40 to 50 pounds of available nitrogen.

9. A soil high in organic matter has more available water capacity for plant growth than has the same soil with less organic matter.

10. Organic matter helps to buffer soils against rapid chemical changes when lime and fertilizers are added.

11. Organic acids released during the decomposition of organic matter help to dissolve minerals and to make them more available to growing plants.

12. Humus (decomposed organic matter) provides a storehouse for the exchangeable and available cations: potassium, calcium, and magnesium. Temporary, humus holds ammonium in an exchangeable and available form.

13. Fresh organic matter has a special function in making soil phosphorus more readily available in acid soils. Upon decomposition, organic matter releases citrates, oxalates, tartrates, and lactates, which combine with iron and aluminum more readily than does phosphorus. The result is the formation of less of the insoluble iron and aluminum phosphates and the availability of more phosphorus.

14. Cotton root rot is a soil-borne fungal disease that is reduced in severity of attack on the following crop of cotton in the presence of a large amount of freshly decomposing organic matter, such as sweet clover.

15. Under certain conditions, toxic substances may be encouraged by the presence of such kinds of organic residues as wheat straw. For example, at three locations in Nebraska over a period of 2 to 5 years, studies were made of toxins produced in soils that had been plowed as compared with soils on which trash tillage had been practiced. A plant toxin, *patulin,*

produced by *Penicillium urticae,* was detected in higher concentrations in soils on which the system of trash tillage of wheat stubble had been used. This toxic substance may be responsible for an occasional *depression* in yields of crops following the trash tillage of wheat stubble.[2]

16. In areas where *verticillium* wilt is prevalent, this fungal disease can be spread by applying cotton gin trash on the soil.

CHEMICAL PROPERTIES

Organic matter from plants is a very complex substance. It contains the following materials, but in varying percentages, depending upon the kind of plant and its state of decomposition:

1. Carbohydrates, including sugars, starches, and cellulose
2. Lignin
3. Tannin
4. Fats, oils, and waxes
5. Resins
6. Proteins
7. Pigments
8. Minerals, such as calcium, phosphorus, sulfur, iron, magnesium, and potassium.

By far the largest percentage of soil organic matter is lignin and protein, although humus may contain as much as 30 per cent polyuronides. In representative soils over the nation, lignin and protein percentages will each vary from approximately 25 to 50 per cent.

Fresh plant material varies in percentage of lignin and protein from one species to another. Alfalfa, for example, contains more protein and less lignin than a rye plant. However, in common with all plants approaching maturity, the percentage of lignin increases and the percentage of protein decreases.

BIOLOGICAL PROPERTIES

Organic matter supplies energy and nutrients for all forms of life in the soil. The fresher the organic matter, the more energy and nutrients it

---

[2] Fred A. Norstadt and T. M. McCalla, "Microbial Populations in Stubble-Mulched Soil," *Soil Science,* Vol. 107, No. 3 (1969), pp. 188–93.

contains and therefore the more valuable it is in the soil. There are certain
hazards to the use of fresh organic matter, however. Fresh straw plowed
under contains so much carbon and so little nitrogen that the yield of the
following crop is usually decreased. Table 9.2 gives the nitrogen-to-carbon
ratio of several common organic materials that are frequently incorporated
in the soil. Alfalfa has the narrowest ratio of nitrogen to carbon—a ratio
of 1:13—whereas oat straw has the widest ratio—namely, 1:80.

**TABLE 9.2   THE NITROGEN, CARBON, AND N:C RATIO OF
COMMON ORGANIC MATERIALS ***

Organic Material	Total Nitrogen (%)	Total Carbon (%)	N:C Ratio
Alfalfa	3.0	39	1:13
Green sweetclover	2.5	40	1:16
Mature sweetclover	1.7	39	1:23
Legume-grass hay	1.6	40	1:25
Oat straw	0.5	40	1:80

* Data from various sources.

What happens when alfalfa and oat straw are plowed under is seen
diagrammatically in Figure 9.1. Available nitrogen is not depressed but is
increased immediately when alfalfa is incorporated in the soil. This means

FIG. 9.1. The relationship between the N:C ratio of plant material and available
nitrogen following their incorporation in the soil.

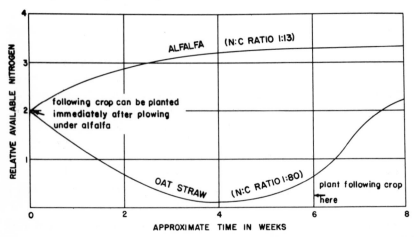

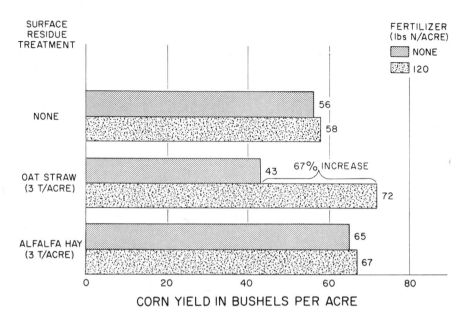

FIG. 9.2. Straw without nitrogen fertilizer depresses the yield of corn. *Source: Soil Science Society of America Proceedings,* **Vol. 21, No. 6, 1957.**

that the following crop can be planted as soon as the alfalfa has been plowed under.

By contrast, available nitrogen is depressed when oat straw is plowed under. It would be best to plant the next crop about six weeks after the straw is incorporated. However, the safe time to wait will vary with moisture, temperature, and the general fertility level of the soil.

Data to support these general relationships between the depression of available nitrogen following an application of oat straw may be seen in Figure 9.2. Without surface residue treatment, 120 pounds of nitrogen increased corn yields only from 56 to 58 bushels per acre. Following 3 tons of alfalfa hay applied on the soil surface at corn planting time, 120 pounds of nitrogen per acre increased corn yields from only 65 to 67 bushels per acre. However, when 3 tons of oat straw was applied on the surface at the time of planting, the yield of corn was depressed to 43 bushels per acre *without* nitrogen and increased to 72 bushels per acre *with* 120 pounds of nitrogen, an increase of 67 per cent in yield.

Available nitrogen in the soil, as measured by total nitrate nitrogen on July 18, where no mulch and no nitrogen fertilizer had been applied, was decreased by 24 per cent under oat straw and increased by 14 per cent under alfalfa hay.

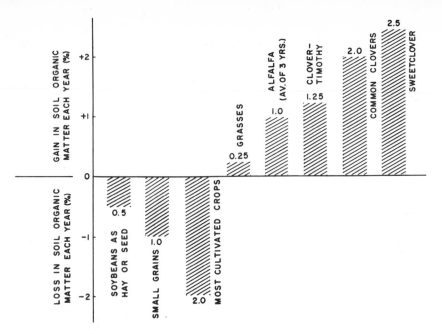

FIG. 9.3. The annual gain or loss in percentage of soil organic matter under different cropping systems in the Corn Belt. (Average annual precipitation approximately 30 to 40 inches.) Source: Robert M. Salter, R. D. Lewis, and J. A. Slipher, *Our Heritage the Soil*, Ohio Agr. Ext. Bul. No. 175, 1941.

Early European and some American plant nutritionists claimed that leachates from certain plants were toxic to other specific plants. In addition, under field conditions some crops were said to be toxic to the crops that follow them in cropping sequence. Until very recently, however, proper fertilization with nitrogen had virtually eliminated all "toxic" effects.

Now new evidence appears to confirm that the residues of *Sericea lespedeza* are toxic to the crops that follow unless a period of 4 to 6 weeks elapses between the time of turning under *Sericea lespedeza* and the planting of a following crop. This apparent toxicity is *not* reduced by the application of nitrogen fertilizer.[3]

MAINTAINING SOIL ORGANIC MATTER IN
HUMID REGIONS

Maintaining soil organic matter is difficult almost everywhere, and under continuous tillage it is nearly impossible. Figure 9.3 summarizes long-time studies in Ohio.

---

[3] Joel Giddens, George Langdale, Ned Dawson, and W. E. Adams, *"Sericea lespedeza Residues Can Be Toxic to Crops,"* *Crops and Soils Magazine* (June–July, 1969), p. 28.

In the Corn Belt, under continuous tillage of corn, potatoes, tobacco, sugar beets, or similar cultivated crops, the soil loses organic matter each year at the rate of approximately 2 per cent of the organic matter present. When the soil is supporting wheat, oats, barley, rye, or buckwheat, the annual loss of the soil organic matter is approximately 1 per cent, and for soybeans, 0.5 per cent. Mixed clover-timothy for hay or pasture results in an annual gain of 1.25 per cent; clovers alone for hay or pasture account for an annual gain of 2.0 per cent; sweetclover turned under gained 2.5 per cent a year; and 3 years of alfalfa increased 3.0 per cent, or an average of 1.0 per cent increase in soil organic matter each year for 3 years. There is no gain in soil organic matter under alfalfa after the third year. Under continuous grass, the gain in organic matter is 0.25 per cent a year.

When cultivated crops are grown, the cropping system designed to maintain organic matter must include those crops which result in increases of organic matter. The figure shows that cultivated crops such as corn result in a loss of soil organic matter equal to 2 per cent a year. Each year that a clover like red clover is grown, an increase of 2 per cent in organic matter may be expected. One year of corn and one year of red clover will therefore maintain soil organic matter. In like manner, a rotation of alfalfa–alfalfa–alfalfa–corn–wheat will just maintain the organic matter level of the soil. Grass–grass–soybeans in a rotation will accomplish the same purpose. Another rotation to maintain organic matter is sweetclover (plowed under)–corn–soybeans (Figures 9.4 and 9.5).

FIG. 9.4. Sweetclover and cotton are rotated on this field. Sweetclover helps to maintain soil organic matter in both humid and semiarid regions. (Texas.) *Source:* Tex. Ext. Serv.

FIG. 9.5. Sudangrass and cotton are rotated on this Texas Blackland farm to maintain organic matter and to reduce cotton root rot. *Source:* Dale Stockton, Enloe, Texas.

FIG. 9.6. Losses of soil organic matter during an 8 year period under different cropping systems in South Dakota (average annual precipitation, 20 inches). *Source:* Leo F. Puhr and W. W. Worzella, *Fertility Maintenance and Management of South Dakota Soils,* S.D. Agr. Exp. Sta. Cir. No. 92, 1952.

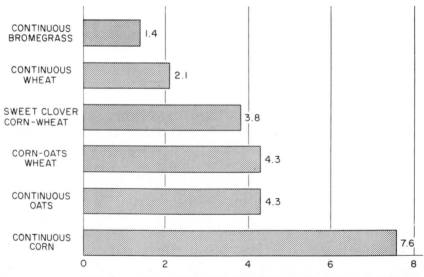

CONTINUOUS BROMEGRASS 1.4

CONTINUOUS WHEAT 2.1

SWEET CLOVER CORN-WHEAT 3.8

CORN-OATS WHEAT 4.3

CONTINUOUS OATS 4.3

CONTINUOUS CORN 7.6

0 2 4 6 8

LOSS OF ORGANIC MATTER DURING AN 8-YEAR PERIOD, IN TONS PER ACRE

Research in Illinois indicates that approximately 5 tons of straw per acre or its equivalent is necessary to maintain soil organic matter under conditions of continuous tillage.[4]

Lysimeter studies in New Jersey over a 40-year period demonstrated that unlimed soil lost 35 per cent of the original organic matter, limed soil lost 26 per cent, and limed and green-manured soil lost 14 per cent. Only when lime, green manure, and 16 tons per acre per year of barnyard manure were applied was there a gain of organic matter—in this instance, a gain of 9 per cent over the 40-year period.

MAINTAINING SOIL ORGANIC MATTER IN SEMIARID REGIONS

Losses of soil organic matter under different cropping systems are given in Figure 9.6. Continuous bluegrass is the most effective in reducing losses of soil organic matter. Next in order of desirability are continuous wheat, sweetclover–corn–wheat, corn–oats–wheat, continuous oats, and continuous corn. These losses took place during an 8-year period in South Dakota.

SUMMARY

Soil organic matter consists mainly of plant and animal residues in all stages of decomposition. Coarse organic matter serves to supply food for life in the soil, to cushion the impact of the falling raindrop, to increase available water for plant growth, and to provide a protective mulch to equalize soil temperatures and reduce evaporation losses of water from the soil. Decomposing organic matter releases some of all essential elements, provides slimes which stabilize desirable soil structure, and keeps open the channels down through which plant roots may grow. Decomposed organic matter holds more water for use by plants, maintains good structure, and serves as a storehouse for exchangeable and available nutrient cations, such as calcium, magnesium, potassium, and ammonium.

When a crop with a narrow N:C ratio, such as alfalfa, is plowed under, the following crop can be planted right away. However, if organic matter with a wide N:C ratio, like straw, is plowed under, approximately 6 weeks must be allowed for decomposition before planting the next crop.

---

[4] S. W. Melsted, *Organic Matter Management in Agronomic Practices,* Soil Microbiology Conference, Purdue University (June 23, 1954).

In humid regions, soil organic matter cannot be maintained under any system of continuous tillage. Grasses and legumes are necessary in the rotation to rebuild organic matter lost by accelerated decomposition caused by stirring the soil in the process of raising cultivated crops. Soil organic matter is even more difficult to maintain in semiarid regions.

Two *toxic* substances may increase following trash tillage in wheat production, and in growing *Sericea lespedeza.*

QUESTIONS

1. How can decomposing organic matter improve soil structure?
2. What is the principal food for most animal life in the soil?
3. Explain this statement: Nutrients are released from organic matter in harmony with the demands of the growing plant.
4. Give an example of a cropping system adapted to the humid region which will maintain soil organic matter.
5. What crops and management systems may result in the production of *toxic* organic materials in soils?

REFERENCES

Box, John, *Crop Residue to Improve Texas Soils.* Texas A & M University Agricultural Extension Service, MP–807, undated.

Cook, R. L., *Soil Management for Conservation and Production.* New York: John Wiley & Sons, Inc., 1962.

Donahue, Roy L., *Our Soils and Their Management* (3rd ed.). Danville, Ill.: Interstate Printers and Publishers, Inc., 1970.

Kellogg, Charles E., *Our Garden Soils.* New York: The Macmillan Company, 1962.

Soil: The Yearbook of Agriculture (1957). United States Department of Agriculture.

CHAPTER **10**

SOIL WATER

In 1620 Van Helmont, a noted Flemish alchemist, conducted an experiment with a willow shoot and concluded that water was the only plant nutrient.

A representative loam soil in a cultivated field contains approximately 50 per cent solid particles (sand, silt, clay, and organic matter), 25 per cent air, and 25 per cent water, only half of which is available to plants (the other half is held in thin films and in gaseous forms and plants cannot absorb it) (Figure 10.1).

FIG. 10.1. A cultivated loam soil in good physical condition for plant growth is comprised of approximately half solids and half pore spaces. The pore spaces may be occupied by half air and half water; and half of the water may be available to plants and the other half unavailable. *Source:* C. Wayne Keese, "Soil Moisture Storage," *Texas A&M University Fact Sheet,* L-754, 1968, 4 pp.

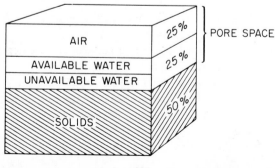

Water serves four functions in plants:

1. Water is a necessary constituent of plant protoplasm, making up 85 to 90 per cent of the fresh weight of actively growing plant parts. Even trees contain more than 50 per cent water.
2. Water is essential in the process of photosynthesis and in the conversion of starch to sugar.
3. Water is the solvent in which nutrients move into plant roots.
4. Water is necessary in maintaining the turgidity of plants, which is essential for maintaining the proper form and position of leaves and new shoots for capturing sunlight for satisfactory growth.

Although plants absorb some water from rain and dew, most of the water used by plants comes from water held by the soil. To make the maximum use of soil water, it is desirable to know how it moves into and through the soil, how it is measured and classified, and what can be done to reduce leaching losses of nutrients by percolation.

INFILTRATION

Infiltration refers to the movement of water *into* the soil. By contrast, *percolation* is the movement of water *through* a column of soil. *Permeability* permits the movement of water *within* the soil.

It is obvious that a soil should be in such physical condition as to provide channels down through which water may move as rapidly as it is received on the surface as rainfall or irrigation. Water that cannot move into the soil moves off over the surface, often carrying soil with it. The result is a reduced productivity due to the loss of fertile topsoil and less water for plant growth.

The principal factors controlling the rate of movement of water into a soil are:

1. The percentage of sand, silt, and clay in a soil. Coarse sands encourage increased infiltration.
2. The structure of a soil. Soils with large, water-stable aggregates have higher infiltration rates.
3. The amount of organic matter in the soil. The more organic matter and the coarser it is, the greater the amount of water entering the soil. Organic surface mulches are especially helpful in increasing infiltration.
4. The depth of the soil to a hardpan, bedrock, or other impervious

FIG. 10.2. The rate of infiltration of water in this Texas Vertisol is classified as very low to low, with a rate of one-tenth of an inch per hour. *Source:* Tex. Agr. Exp. Sta.

layer is a factor in infiltration. Shallow soils do not permit as much water to enter as do deep soils.

5. The amount of water in the soils. In general, wet soils do not have as high an infiltration rate as do moist or dry soils.
6. The soil temperature. Warm soils take in water faster than do cold soils. Frozen soils may or may not be capable of absorbing water, depending upon the kind of freezing that has taken place.

Infiltration rates may be classified as follows:

1. *Very Low.* Soils with infiltration rates of less than 0.1 inch per hour are classified as very low. In this group are the soils that are very high in percentage of clay (Figure 10.2).
2. *Low.* Infiltration rate of 0.1–0.5 inch per hour are considered low. This group includes soils high in clay, soils low in organic matter, or shallow soils.

FIG. 10.3. Roots in this Texas Vertisol help to make the infiltration rate high, with a value of five inches of water per hour. *Source:* Tex. Agr. Exp. Sta.

3. *Medium.* Rates of infiltration of 0.5–1.0 inch per hour are classified as medium. Most soils in this group are sandy loams and silt loams.
4. *High.* High rates include soils with greater than 1.0 inch per hour of infiltration. Deep sands, and deep, well-aggregated silt loams and some virgin black clays are in this group (Figure 10.3).

PERMEABILITY

The characteristics that determine how fast air and water move through the soil describe what is known as *permeability.* The rate of water movement through a soil is determined by the least permeable horizon. Plowpans or natural claypans reduce the permeability of a soil. Past management practices also determine permeability: continuous tillage reduces permeability, whereas the growth of deep-rooted grasses, legumes, and trees increases permeability.

Water moves in the soil as a liquid or as a vapor, mainly through the large pores. This means that the larger and more numerous the pores, the greater the permeability. Rapid permeability is essential for drainage fields below septic tanks.

FIG. 10.4. Water in the soil near the *wilting point, field capacity,* and at *saturation.* Water in the soil between the wilting point and the field capacity is considered available to plants. At saturation there is not enough air in the soil for plants to absorb the water, except in the instance of water-plants (hydrophytes) such as rice. *Note:* Soils are from the surface horizon of the Morrow Plots at Urbana, Illinois, and have a silt loam texture. The middle ped has a Munsell color notation of 10YR 2/1.

Suggested permeability classes are as follows: [1]

1. Very slow—less than 0.20 inch per hour
2. Slow—0.20–0.63 inch per hour
3. Moderate—0.63–2.0 inches per hour
4. Rapid—2.0–6.3 inches per hour
5. Very rapid—more than 6.3 inches per hour.

SOIL WATER CLASSIFIED

Water in the soil has been classified in many ways by many people. One of the most modern and meaningful classifications is based upon the energy of retention of water by the soil, usually known as *soil moisture tension.* In this way, soil water classification is more directly related to the energy that plant roots must exert in absorbing water.

When plants permanently wilt for lack of water, it means that the pull of the roots for water is not great enough to get sufficient water in time to prevent permanent wilting. This amount of water in the soil is held with a tension of approximately 15 atmospheres and is called the *wilting point.* Plants cannot grow satisfactorily in a soil wetter than the field capacity. The moisture at the field capacity is held with a tension of approximately one-third atmosphere [2] (Figure 10.4).

---

[1] As determined on undisturbed, saturated soil cores under a constant 0.5-inch head of water.

[2] The soil moisture tension at the field capacity is approximately equal to one-third atmosphere in laboratory samples. Under field conditions, tensiometer readings at the field capacity vary from 0 to 0.02 atmospheres (private communication from Sterling J. Richards).

Oven-dry weight is the basis for nearly all soil moisture calculations. The equilibrium tension of the moisture at oven-dryness is approximately 10,000 atmospheres. In its actual determination in soils, oven-dryness is determined by placing the soil in an oven at a temperature of 105°C until it loses no more water.

Air-dry weight is a somewhat variable term, mainly because the moisture and temperature of the air fluctuate. Under average conditions, moisture at air-dryness is held with a tension of approximately 1000 atmospheres. This water is not available to plants.

The *hygroscopic coefficient* is determined by placing an air-dry soil in a nearly saturated atmosphere at 25°C until it absorbs no more water. This tension is equal to approximately 31 atmospheres. Water at this tension is not available to plants but may be available to certain bacteria.

Water at the *wilting point* is held with a tension of approximately 15 atmospheres (220.5 pounds per square inch).

The *field capacity* of a soil with good drainage can readily be determined. After a soaking rain or a heavy irrigation, cover the surface of a well-drained soil to reduce evaporation losses and wait 2 or 3 days to allow free drainage. At this time, the surface soil moisture is at field capacity. In atmospheres of tension, this is approximately one-third.

Between the field capacity and saturation, water is not available to common upland crops because of too little available oxygen. It is, however, available to rice and other water-loving plants.

Between the wilting point (15 atmospheres) and the field capacity (one-third atmosphere), soil water is available to plants. This is the range of moisture in soil-plant relationships.

The common soil moisture constants are presented in Figure 10.5 in terms of atmospheres of tension.

MEASURING SOIL MOISTURE

Water in a soil may be measured in a number of ways. Three of these methods are included here.

First is the gravimetric method. This consists of obtaining a moist soil sample, drying it in an oven at 105°C until it loses no more water, and then determining the percentage of moisture. The calculation is as follows:

$$\% \text{ moisture} = \frac{\text{Loss in weight}}{\text{Oven-dry weight}} \times 100$$

As one can see, this calculation is based on the given weight of soil. The weight percentage of water can be changed to approximate volumes

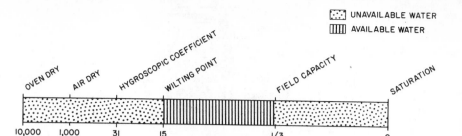

ATMOSPHERES OF TENSION

FIG. 10.5. Soil moisture constants and their approximate equivalents in atmospheres of tension as related to the relative availability of water to plants. (One atmosphere at sea level is equal to 14.7 pounds per square inch.)

of water by assuming that an average soil weighs 2 million pounds per 6-inch depth per acre. For example, if the percentage of moisture of a typical loam soil at the field capacity is 18.1, the pounds of water per acre– 6-inches of soil will be 0.181 × 2 million, or 362,000. To convert to acre-inches of water, divide by 226,584 (pounds of water per acre-inch). The answer is approximately 1.59 inches.

To carry through this calculation to its most useful conclusion, calculate the acre-inches of water at the wilting percentage and then subtract from the field capacity to obtain the inches of *available* water per 6-inch depth of soil. Assuming that the wilting percentage of the same soil is 6.8, the moisture will be 0.068 × 2 million, which equals 136,000. This number divided by the pounds per acre-inch (226,584) gives approximately 0.60 acre-inches. The acre-inches of available water per 6-inch depth is obtained by subtracting 0.60 from 1.59, which gives 0.99. Normally the figures are given on a per acre-foot basis, and in this case the answer is 2 × 0.99, or 1.98 acre-inches of *available* water per acre-foot of soil.

Second are methods that depend upon measuring the equilibrium tension of soil moisture with the use of a porous clay cup (tensiometer) filled with water. The water in the porous cup is attached to a vacuum gauge or a mercury manometer. As the soil dries out, water moves through the porous cup, setting up a negative tension or vacuum. These tension readings are then calibrated to interpret the percentage of moisture. The principal limitation of the use of tensiometers is the fact that they do not measure soil moisture as low as the wilting percentage. The actual range is from 0 to 0.85 atmosphere. The tensiometers are more useful for measuring moisture in sandy soils than in fine-textured soils.

Third are methods that are based on the changes in electrical conductivity with changes in soil moisture. In 1940 G. J. Bouyoucos at Michigan State University introduced a gypsum block inside of which were two electrodes a definite distance apart. The blocks were to be buried in the soil and the conductivity across the electrodes was to be measured

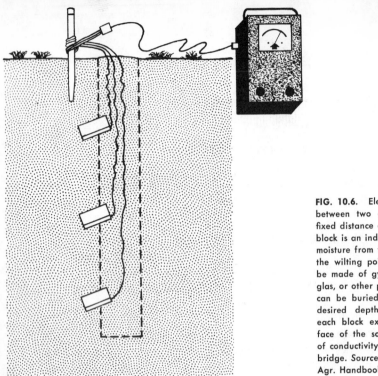

FIG. 10.6. Electrical conductivity between two electrodes set at a fixed distance apart inside a small block is an indirect measure of soil moisture from the field capacity to the wilting point. The blocks may be made of gypsum, nylon, Fiberglas, or other porous material, and can be buried in the soil at any desired depth, with wires from each block extending to the surface of the soil for easy reading of conductivity with the Bouyoucos bridge. *Source:* U.S. Dept. of Agr., Agr. Handbook No. 107, 1957.

with a modified Wheatstone bridge. With proper calibrations, the percentage of moisture from the field capacity to the wilting percentage can be readily determined (Figure 10.6).

Improved Bouyoucos blocks are made of nylon or Fiberglas that do not deteriorate in the soil as do gypsum blocks.

AVAILABLE WATER

Available water is the range of soil moisture between the wilting point and the field capacity. The amount of water to apply to a soil at the wilting point to reach the field capacity is called the available water capacity. The available water capacity also varies primarily with soil texture; for example, it is about 1.2 inches for sand soils, 2.0 inches for loams, 2.5 inches for silt loams, and 1.5 inches for clay soils, per foot of soil depth. Table 10.1 shows the generalized relationships that exist among the wilting point, field capacity, and available water capacity of eight soil textural classes. With the addition of clay in a soil, the soil's capacity to hold water at both the wilting point and the field capacity increases. The same is true of the available water capacity up to the fineness in soil texture of a silt loam. In a clay loam and in a clay, however, the available water capacity *decreases*

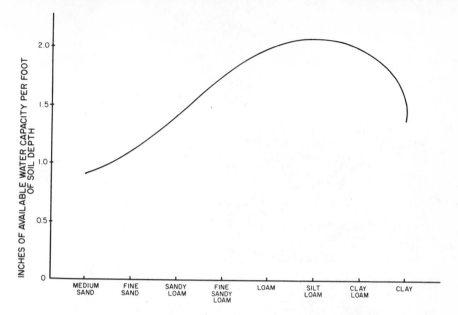

FIG. 10.7. General relationships between soil textural classes and available water capacity per foot of soil depth. Note that the greatest water capacity exists in a silt loam and that it is progressively less in a clay loam and a clay and least in coarser textural classes. *Source: Water: The Yearbook of Agriculture (1955), U.S. Dept. of Agr.*

in comparison with a silt loam. This relationship is shown graphically in Figure 10.7.

TABLE 10.1 WILTING POINT, FIELD CAPACITY, AND AVAILABLE WATER CAPACITY OF VARIOUS SOIL TEXTURES *

Soil Texture	Wilting Point		Field Capacity		Available Water Capacity	
	(%)	Water per Foot of Soil Depth (in.)	(%)	Water per Foot of Soil Depth (in.)	(%)	Water per Foot of Soil Depth (in.)
Medium sand	1.7	0.3	6.8	1.2	5.1	0.9
Fine sand	2.3	0.4	8.5	1.5	6.2	1.1
Sandy loam	3.4	0.6	11.3	2.0	7.9	1.4
Fine sandy loam	4.5	0.8	14.7	2.6	10.2	1.8
Loam	6.8	1.2	18.1	3.2	11.3	2.0
Silt loam	7.9	1.4	19.8	3.5	11.9	2.1
Clay loam	10.2	1.8	21.5	3.8	11.3	2.0
Clay	14.7	2.6	22.6	4.0	7.9	1.4

Source: Water: The Yearbook of Agriculture (1955), U.S.D.A. p. 120.
Note: It is obvious that since there is a variation in the amounts and kinds of sand, silt, and clay within any one textural group (such as within loam soils), there is also a variation in the water constants; however, for purposes of simplification, an average value is given in this table.

FIG. 10.8. Under the same amount of annual rainfall (30 inches) and approximately the same steepness of slope, the soil in (1) supports only xerophytic plants such as *Eriogonum,* a shrub, and an herb *Phacelia;* whereas, the soil in (2) supports western yellow pine. *The difference is available water.* In (1) the soil is a volcanic cinder cone which has an available water capacity less than that of coarse sand, perhaps 0.5 inch per foot of soil depth. By contrast, the clay loam soil (2) has an available water capacity of perhaps 2.0 inches per foot of soil depth. (Arizona.) *Source:* J. O. Veatch.

Soils with the same rainfall but with contrasting available soil water capacity will support contrasting vegetation (Figure 10.8).

The amount of water in the soil at the wilting point also varies slightly with the plant used for its determination. In one experiment, five plant species were used and all of them wilted premanently at different moisture levels in the soil. From higher to lower soil moisture at permanent wilting, these plants were sunflower (*Helianthus annuus*), coyote tobacco (*Nicotiana attenuata*), corn (*Zea mays*), golden cassia (*Cassia fasciculata*), and intermediate wheatgrass (*Agropyron intermedium*).[3]

---

[3] Dwane J. Sykes and Walter E. Loomis, "Plant and Soil Factors in Permanent Wilting Percentages and Field Capacity Storage," *Soil Science,* Vol. 104, No. 3 (Sept., 1967), pp. 163–73.

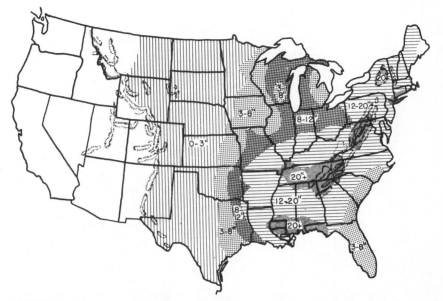

FIG. 10.9. Average annual percolation. *Source:* L. B. Nelson and R. E. Uhland, "Factors that Influence Loss of Fall Applied Fertilizers and Their Probable Importance in Different Sections of the United States," *Soil Science Society of America Proceedings,* Vol. 19, No. 4, 1955.

PERCOLATION

The movement of water through a column of soil is called *percolation.* Percolation studies are important for at least two reasons. Percolating waters are the only source of recharge of water for springs and wells. Also, percolating waters carry plant nutrients down and often out of reach of plant roots.

Not all places in the United States have sufficient precipitation to percolate through the surface soil. A glance at Figure 10.9 clearly demonstrates that the Great Plains have annual percolation rates of 0 to 3 inches per year. Coming eastward, the percolation rate varies mostly with the amount of precipitation, but with some exceptions. The notable exception is Florida, where 3 to 8 inches is the annual percolation rate. Here, the annual precipitation is in excess of 50 inches, but the temperature is high enough to evaporate a large part of the rain before it percolates through the soil.

Within these areas of generalized percolation, sandy soils permit greater percolation and clay soils will permit less water to move through them.

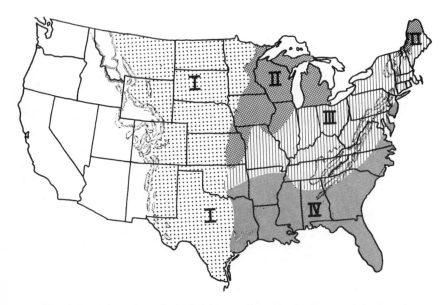

FIG. 10.10. Relative hazard of plant nutrient losses by leaching: I, slight hazard; II, moderate hazard; III, severe hazard; IV, very severe hazard. *Source:* L. B. Nelson and R. E. Uhland, "Factors that Influence Loss of Fall Applied Fertilizers and Their Probable Importance in Different Sections of the United States," *Soil Science Society of America Proceedings,* Vol. 19, No. 4, 1955.

LEACHING LOSSES OF NUTRIENTS

The relative hazard of losses of plant nutrients by leaching is represented in Figure 10.10. The relative order of the actual leaching losses of plant nutrients from a New York soil is

$$Ca > Mg > S > K > N > P$$

Only a trace of phosphorus is lost by leaching, whereas the calcium losses are the greatest of any nutrient shown.

These data compare leaching losses in Volusia silt loam, an acid soil derived from shale. No crop (continuously fallow or bare) is compared with a crop rotation with respect to the loss of plant nutrients by leaching. In every instance, there is less loss by leaching under a crop rotation than under no crop.

In the 50-inch precipitation belt in Tennessee, an experiment was conducted to determine the rate of downward movement of potassium and phosphorus when applied as fertilizers on the surface soil of an established stand of alfalfa. After 5 years, most of the phosphorus remained in the first 3 inches of soil and most of the potassium remained in the first 6 inches of soil.[4]

An experiment in the San Joaquin Valley of California was established to study the losses of nitrogen and phosphorus through tile drainage effluent on irrigated Panoche and Oxalis silty clay calcareous soils that were planted to cotton and rice. The average depth of the tile system was between 5.5 and 7.0 feet from the soil surface. The results reported are for one year, 1962.

The soil planted to cotton lost through the tile drainage effluent 9 per cent of the nitrogen fertilizer and 1 per cent of the phosphorus fertilizer applied. By contrast, 33 per cent of the nitrogen and 17 per cent of the phosphorus fertilizer were lost through the tile drainage effluent on the soil planted to rice.[5]

SUMMARY

The average acre in the United States does not absorb sufficient natural rainfall to grow corn; only range grasses can grow. Better management practices will increase infiltration and the amount of available water with no increase in rainfall. Infiltration rates of soils are influenced by texture, structure, organic matter, soil depth, wetness of the soil, and soil temperature. Infiltration rates less than 0.1 inch per hour are very low, whereas rates above 1.0 inch are high.

Permeability is a term used to express the ease of movement of air and water through a soil. Maximum permeability is limited by the permeability of the least permeable horizon. Tillage pans seriously restrict the movement of water through the soil.

Soil moisture is normally reported in terms of atmospheres of equilibrium tension. At oven-dryness, the atmosphere of soil moisture tension is 10,000; at air-dryness, 1000; at the hygroscopic coefficient, 31; at the wilting percentage, 15; and at the field capacity, $\frac{1}{3}$. Available water lies between the wilting percentage and the field capacity.

---

[4] K. L. Wells and W. L. Parks, "Vertical Distribution of Soil Phosphorus on Several Established Alfalfa Stands That Received Various Rates of Annual Fertilization," *Soil Science Society of America Proceedings,* Vol. 25, No. 2 (1961).

[5] William R. Johnson, F. Ittihadieh, Richard M. Daum, and Arthur F. Pillsbury, "Nitrogen and Phosphorus in Tile Drainage Effluent," *Soil Science Society of America Proceedings,* Vol. 29 (1965), pp. 287–89.

On a weight percentage basis, soil moisture is calculated in this way:

$$\% \text{ soil moisture} = \frac{\text{Loss in weight in oven}}{\text{Oven-dry weight of soil}} \times 100$$

Gypsum, nylon, or Fiberglas is used in making Bouyoucos blocks for measuring soil moisture by the electrical-conductivity method. This method permits the measurement of soil moisture throughout its available range, from the field capacity to the wilting percentage.

Available water capacities are low for sandy soils and for very fine-textured clay soils. Loams and silt loams have the greatest available water capacities of all mineral soils.

Water percolates downward through a deep section of soil only in areas where the precipitation is high. In the Great Plains, percolation seldom exceeds 3 inches of water a year. In the high-rainfall areas of the Appalachian Mountains and the lower Mississippi and Alabama Gulf Coast, percolation losses often reaches 20 inches a year.

Leaching losses of nutrients are proportional to the amounts of water percolated through the soil. Plant nutrients are lost in greatest amounts when no crop is growing to absorb the nutrients. Regardless of the use of the land, however, the relative rate of nutrient losses by leaching from the soil are in this order, from high to low:

$$Ca > Mg > S > K > N > P$$

QUESTIONS

1. Differentiate between infiltration, percolation, and permeability.
2. How can a soil be managed to increase permeability? How important is rapid permeability for drainage fields for septic tanks?
3. In the surface foot of soil, if there is 0.3 inch of water in the soil at the wilting point and 2.1 inches at the field capacity, what is the available water capacity of the soil?
4. How is it possible for two fine sandy loam soils to have different available water capacities?
5. Why are there almost no percolation losses in the Great Plains?

REFERENCES

The Hydrologic Cycle. Environmental Science Services Administration, Department of Commerce, 1967.

Marshall, T. J., *Relations Between Water and Soil,* Technical Communication No. 50. Harpenden, England: Commonwealth Bureau of Soils, 1959.

Pierre, W. H., D. Kirkham, J. Pesek, and R. H. Shaw (eds.), *Plant Environment and Efficient Water Use.* American Society of Agronomy, 1967.

River of Life. Water: The Environmental Challenge. United States Department of the Interior, Conservation Yearbook Series, Vol. 6, 1970, 96 pp.

Russell, M. B. (ed.), *Water and Its Relation to Soils and Crops,* Vol. 11 of *Advances in Agronomy.* New York: Academic Press Inc., 1959.

Soil: The Yearbook of Agriculture (1957). United States Department of Agriculture.

Water: The Yearbook of Agriculture (1955). United States Department of Agriculture.

CHAPTER 11

SOILS AND PLANT NUTRITION

An element is not considered essential unless a deficiency of it makes it impossible for the plant to complete its life cycle; such deficiency is specific to the element in question and can be prevented or corrected only by supplying this element; and the element is directly involved in the nutrition of the plant quite apart from possible effects in correcting some unfavorable microbial or chemical condition of the soil or other culture medium.—D. I. ARNON

There are at present 16 elements known to be essential for the growth and reproduction of higher plants. These elements are carbon, hydrogen, oxygen, phosphorus, potassium, nitrogen, sulfur, calcium, iron, magnesium, boron, manganese, copper, zinc, molybdenum, and chlorine.

Since farm livestock are mostly forage-eating animals, it is of vital concern to know whether the elements essential for plants are also essential for animals. Nineteen elements are known to be essential for animals; they are carbon, hydrogen, oxygen, phosphorus, potassium, nitrogen, sulfur, calcium, iron, magnesium, manganese, copper, zinc, sodium, iodine, chlorine, selenium, molybdenum, and cobalt. Animals require 15 of the 16 elements needed by plants (boron is not required by animals) but in addition, animals need sodium, cobalt, selenium, and iodine.

FORMS IN WHICH NUTRIENTS ARE ABSORBED
BY PLANTS

Plants take in nutrients from the soil mostly in the form of ions. The 16 essential elements move into the plant primarily in the following forms:

C	CO_2 (mostly through leaves)
H	H^+, HOH
O	O^{--}, OH^-, CO_3^{--}, SO_4^{--}, CO_2 (mostly through leaves)
P	$H_2PO_4^-$
K	K^+
N	NH_4^+, NO_3^-
S	SO_4^{--}
Ca	Ca^{++}
Fe	Fe^{++}, Fe^{+++}
Mg	Mg^{++}
B	BO_3^{---}
Mn	Mn^{++}
Cu	Cu^{++}
Zn	Zn^{++}
Mo	MoO_4^{--}
Cl	Cl^-

THE MECHANISM OF NUTRIENT UPTAKE

Plants may obtain nutrients by absorption through either the leaves or the roots. Carbon as carbon dioxide enters the plant almost entirely through the stomata of the leaves. Water also is absorbed through the stomata, but the relative amount is small in comparison with that entering the roots. Radioactive research has shown that from water, only hydrogen is utilized by the plant; oxygen in water is liberated as a gas. Many nutrients are capable of being absorbed by the leaves of plants when sprayed as nutrient solutions, but this will be discussed in the following sections.

Absorption by roots may be by the mechanism of:

1. Root interception
2. Mass-flow
3. Diffusion.

As plant roots grow by extension in the soil, they *intercept* and absorb from intimate contact such ions as nitrates and sulfates from the soil solution (Figure 11.1). They also absorb calcium, magnesium, and potassium ions that are held in an exchangeable form on the surfaces of clay and humus.

Large amounts of water are transpired by plants. This water is absorbed by plant roots, moves through roots to stems, and is lost as vapor through the stomata of the leaves. Water in the soil contains nitrates,

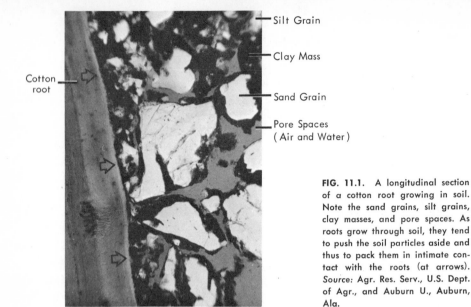

Silt Grain

Clay Mass

Cotton root

Sand Grain

Pore Spaces
(Air and Water)

FIG. 11.1. A longitudinal section of a cotton root growing in soil. Note the sand grains, silt grains, clay masses, and pore spaces. As roots grow through soil, they tend to push the soil particles aside and thus to pack them in intimate contact with the roots (at arrows). *Source:* Agr. Res. Serv., U.S. Dept. of Agr., and Auburn U., Auburn, Ala.

sulfates, calcium, magnesium, potassium, phophorus, and micronutrients. As waters of soil solution move into the plant roots, other soil water flows toward the roots, carrying in *mass* the dissolved nutrients. This is known as *mass flow*. (Absorption of nutrients by the roots, however, is independent of absorption of water.)

When plant roots absorb nutrients from the soil solution, other dissolved nutrients move toward the roots because of the laws of *diffusion* —that is, from areas of higher concentration of nutrients to areas of lower concentration.[1]

One hypothesis on the mechanism of plant absorption is that a plant root excretes a cation in exchange for another cation that it absorbs, and an anion is released by the root for each anion that it uses.[2]

The diagram in Figure 11.2 is shown to illustrate cationic exchange between a root and a clay and humus particle, and between a root and a soil solution. The source of additional cations from a limestone fragment is also given in the diagram.

As the limestone fragment slowly decomposes, it releases calcium ions (Ca^{++}) to the soil solution and to clay and humus particles. The root hair releases two hydrogen ions for each calcium ion it absorbs from either the soil solution, a clay particle, or a humus particle.

---

[1] Stanley A. Barber, "Water—Essential to Nutrient Uptake," *Plant Food Review,* National Plant Food Institute (Summer, 1964), pp. 5–7.

[2] H. Jenny and R. Overstreet, "Contact Effects Between Plant Roots and Soil Colloids," *Proceedings of the National Academy of Science,* Vol. 24 (1938), pp. 384–492.

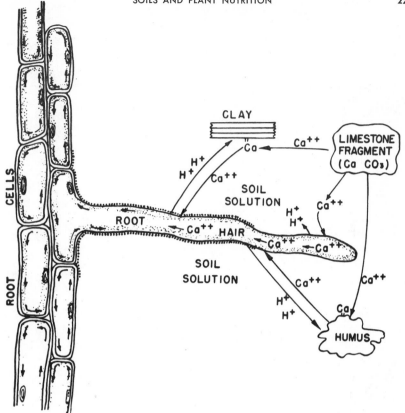

FIG. 11.2. A diagrammatic scheme for showing how a root hair takes in nutrients from the soil solution and from exchangeable ions on a clay crystal and on humus. These nutrients originally came from readily decomposable minerals such as limestone (shown here). A root hair is an extension of one of the epidermal cells of the plant root and is responsible for nearly all intake of water and nutrients.

The nutrient cations such as Ca^{++}, K^+, Mg^{++}, and NH_4^+ are held by clay crystals and humus in an exchangeable and available form for use by plants—not so with anions. Nitrates, sulfates, phosphates, borates, and molybdates are nutrient anions and therefore are not held by clay and humus by the same mechanism.

Available anions exist either in the soil solution or in some chemical compound in equilibrium with the soil solution. Phosphorus, sulfur, boron, and molybdenum are a part of the crystal structure of such minerals as apatite, tourmaline, and pyrite. But the nutrient elements in these minerals are so slowly soluble as to supply too few anions for normal plant nutrition, even though the minerals are finely ground.

It is soil organic matter that serves as the principal storehouse for the anions. Through the decomposition of organic matter by bacteria, fungi,

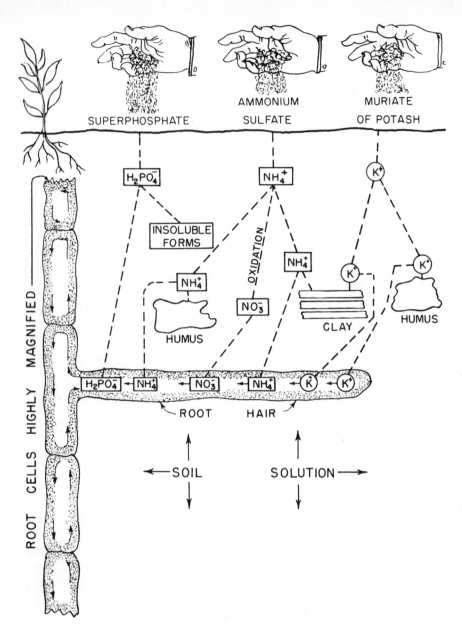

FIG. 11.3. The mechanism of uptake by a plant root of nitrogen, phosphorus, and potassium from common fertilizers.

and actinomycetes and subsequent oxidation, the anions are made available to growing plants.

More than 95 per cent of the nitrogen is held in the soil in the organic matter. Organic phosphorus may account for from 5 to 60 per cent of the total soil phosphorus. Up to 80 per cent of soil sulfur is reported to be held in the organic form. Boron and molybdenum reserves seem to be stored both in organic matter and in clay.

There is good evidence to support the theory that anions are taken into the plant only from the soil solution. But the amounts of anions in the soil solution at any one time are not enough to supply the plant for the entire growing season. The only logical answer is that, as the anions in the soil solution are taken up by the plant, other anions are released to the soil solution by organic decomposition or from slowly soluble compounds. It is also generally agreed that, as plant roots take in nutrient anions, the roots release OH^- or HCO_3^- ions in exchange.

For maximum production, crop plants usually require greater amounts of nutrients than the soil solution contains at any one time. As plants remove nutrient ions from the soil solution, the soil solution is being replenished with ions from clay minerals and humus by cationic exchange, by the slow decomposition of soil minerals, and by the more rapid decomposition of soil organic matter. It is seldom, however, that the rate of renewal for all essential elements is fast enough to achieve maximum crop production; for this reason, the use of fertilizers is required. The mechanism of plant nutrient uptake of N, P, and K from fertilizers is shown in Figure 11.3.

From the soil, plants accumulate nitrogen, phosphorus, and sulfur. In other words, plants nearly always contain a higher percentage of these elements than the soil in which the plants are growing. Conversely, soils almost always contain more iron, calcium, potassium, magnesium, and manganese than the plants growing in them.

NITROGEN

There are nearly 12 pounds of nitrogen above every square foot of the surface of the earth, yet nitrogen is one of our most critical elements because it is most frequently deficient for plant growth. The reason is that plants cannot utilize nitrogen as a gas; it must first be combined into some stable form.

Plants absorb nitrogen either as the ammonium or the nitrate ion. The ammonium ions can be held in an exchangeable and available form on the surfaces of clay crystals and humus, but bacteria soon transform the ammonium to nitrates, which are readily leachable. There is no good storehouse for available forms of nitrogen. The only storehouse of any kind for nitrogen is soil organic matter.

Soil organic matter is approximately 5 per cent nitrogen. It therefore follows that

$$\% \ N \times 20 = \% \ \text{organic matter}$$

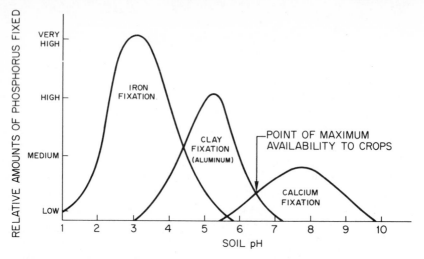

FIG. 11.4. The total phosphorus in many soils is medium to high, but phosphorus availability is often low. At a soil pH below 5.5, both iron and aluminum fix phosphorus and make it less available; whereas, at a pH above 7.0, calcium fixes phosphorus. The point of maximum phosphorus availability is pH 6.5 for mineral soils and 5.5 for organic soils. Lime on soils below pH 5.5 and gypsum on soils above pH 8.5 will make soil phosphorus more available to crops. *Source:* George D. Scarseth, *Man and His Earth,* Iowa State U. Press, Ames, Iowa, 1962, p. 143.

Nitrogen is a constituent of all living cells, and each molecule of chlorophyll contains four atoms of nitrogen. Nitrogen makes plants darker green and more succulent; it also makes cells larger with thinner cell walls. In addition, nitrogen increases the proportion of water and decreases the percentage of calcium in plant tissues.

Recent research in Arkansas has demonstrated that many crops when fertilized with nitrogen have an increased ability to absorb not only more nitrogen but also more phosphorus, potassium, and calcium. Nitrogen fertilization increases the cation-exchange capacity of plant roots and thus makes them more efficient in absorbing other nutrient ions.

PHOSPHORUS

Phosphorus nutrition is doubly critical; the total supply of phosphorus in most soils is usually low, and its relative availability is also low. Plants absorb phosphorus from the soil solution mostly in the form of $H_2PO_4^-$ ions.

The total phosphorus in an average arable soil is approximately 0.1 per cent, only an infinitesimal part of which at any one time is available to the plant. Under ideal conditions, as plants take in $H_2PO_4^-$ ions from the soil solution, other ions replace them from slowly soluble compounds in the soil. There is no efficient mechanism on clay crystals or on humus particles for holding exchangeable and available anions such as $H_2PO_4^-$.

Phosphorus availability is low in strongly acid soils because of the formation of iron and aluminum phosphates, from which phosphorus is very slowly available. In alkaline soils, tricalcium phosphate $[Ca_3(PO_4)_2]$ forms readily to reduce the availability of soil phosphorus (Figure 11.4).

The nucleus of each plant cell contains phosphorus; for that reason, cell division and growth are not possible without adequate phosphorus. Phosphorus is concentrated in cells near the most actively growing part of both roots and shoots, where cells are dividing rapidly.

In recent years it has been demonstrated that very large applications of a soluble phosphorus fertilizer on soils low in available zinc may cause zinc deficiency in such plants as corn, beans, and flax. The converse is also true; that is, on soils low in available phosphorus, applications of zinc tend to accentuate phosphorus deficiency.

A cold wet spring results in a retardation of plant growth, often accompanied by a purple color of the leaves caused by phosphorus deficiency. Plants growing in soils high in available phosphorus grow more rapidly in cold soils and their leaves do not exhibit a purple discoloration.[3]

Phosphorus deficiency was shown in 1968 to interfere with the normal opening of the stomata of certain plants, resulting in an increase in leaf temperature as much as 11°F higher during periods of sunshine than plant leaves that received adequate phosphorus.[4]

POTASSIUM

The amount of *total* potassium in most soils is sufficient to last forever; yet the money spent for potassium fertilizers is constantly on the increase. An explanation of this apparent contradiction lies in the fact that most of the potassium is a part of the molecule of very slowly soluble minerals such as orthoclase ($KAlSi_3O_8$). Soils may contain 2 per cent *total* potassium only one-fifth of which at any one time is in a readily available (exchangeable) form. However, during the growing season, approximately half of the potassium absorbed by the plant may come from the exchangeable form and the other half from relatively insoluble minerals that decompose and thereby release their potassium.

Until a few years ago it was thought that all potassium in the plant stayed in a mobile form. Radioactive potassium techniques have demonstrated that as much as one-third of plant potassium is fixed as a part of

---

[3] *Source:* C. D. Sutton, "Effect of Low Soil Temperature on Phosphate Nutrition of Plants—A Review," *Journal of the Science of Food and Agriculture,* Vol. 20, No. 1 (Jan., 1969), pp. 1–3.

[4] A. Wallace and A. Deutsch, "Phosphorus Deficiency Decreases Stomatal Activity and Water Use of Plants," *California Agriculture,* Vol. 22, No. 8 (Aug., 1968), p. 15.

plant proteins. Potassium also helps to maintain cell permeability, aids in the translocation of carbohydrates, keeps iron more mobile in the plant, and increases the resistance of plants to certain diseases.

Upon drying, all soils fix potassium in forms unavailable to plants. The amount of potassium rendered unavailable to plants because of fixation varies considerably, depending upon soil texture, kind of clay, intensity of drying, and the level of initial exchangeable potassium. The process of fixation is very rapid; in 10 minutes, 80 to 90 per cent of the potassium is fixed that eventually will become fixed.

The principal sources and forms of potassium in the soil that supply potassium to growing plants are given in the following diagram.[5]

Total Potassium

Relatively unavailable potassium	Fixed or slowly available potassium	Readily available potassium
Feldspars $KAlSi_3O_8$ Muscovite H_2KAl_3	Biotite $(H,K)_2 \cdot (Mg,Fe)_2Al_2(SiO_4)_3$ Hydrous mica Illite	Exchangeable potassium and water soluble potassium

CALCIUM

Calcium in the soil may average 1 per cent for the United States. Variations are great, however, for calcium minerals are fairly soluble, resulting in low-calcium soils in humid regions.

Most of the reserve calcium in the soil is in the form of calcium carbonate (limestone) or, in the West, a mixture of calcium carbonate and calcium sulfate. When compared with most potassium minerals, calcium-bearing minerals are usually much more soluble. Consequently there is nearly always more exchangeable calcium on the clay crystals and on the humus particles than there is potassium.

Calcium tends to make cells more selective in their absorption, since it is a constituent of the middle lamella of each cell wall. Rapidly growing root tips are especially high in calcium, indicating that calcium is needed in large quantities for cell division.

---

[5] J. C. Shickluna, "The Relationship of pH, Available Phosphorus, Potassium, and Magnesium to Soil Management Groups," *Michigan State University Quarterly Bulletin,* Vol. 45, No. 13 (1962), p. 140.

MAGNESIUM

Chlorophyll contains one atom of magnesium in each molecule; therefore, there could be no green plants without magnesium. There is almost as much magnesium as calcium in many soils, yet there are soils in the Northeast that are extremely deficient in magnesium.

Reserve magnesium occurs mostly in dolomitic limestone, a rock that consists of a mixture of calcium and magnesium carbonate. Dolomitic limestone is not so readily decomposed as is calcic limestone; for that reason, the amount of exchangeable magnesium is usually less than that of exchangeable calcium.

It has been demonstrated that magnesium aids in the uptake of phosphorus. This fact is of particular importance to the livestock industry, because the minimum phosphorus content desirable for the growth of forage plants is inadequate for satisfactory animal nutrition.

Buckwheat, spinach, cauliflower, muskmelon, Irish potato, pea, oat, and corn apparently require larger amounts of magnesium than most other crops.

When magnesium deficiency occurs, the plant foliage may be sprayed with a solution of magnesium sulfate in a concentration of 15 pounds of magnesium sulfate in 100 gallons of water. More long-lasting results may be obtained by applying to the soil 50 to 100 pounds of actual magnesium (Mg) per acre in the form of dolomitic limestone (12 per cent Mg), magnesium sulfate (chemical grade—20 per cent Mg; gypsum grade—10 per cent Mg), magnesium oxide (60 per cent Mg), sulfate of potash-magnesia (10 per cent Mg), or magnesium ammonium phosphate (15 per cent Mg). Of these materials dolomitic limestone is the slowest-acting but also the least expensive.

SULFUR

Sulfur is found in small amounts in the soil, averaging perhaps 0.15 per cent in a typical soil. A large part of the sulfur that plants use comes from sulfates, which are a by-product in superphosphate fertilizer, and through biological release from decomposing organic matter.

Many plant proteins contain sulfur, as does an oil produced by members of the cabbage family. Sulfur is also required for the synthesis of certain vitamins in plants.

MICRONUTRIENTS

Iron, manganese, zinc, copper, boron, chlorine, and molybdenum are listed as micronutrients because they are used by plants in such small amounts. Their exact function is not known. These elements may limit plant growth either because there may not be a sufficient amount of them in the soil, or, as is more often the case, because some condition in the soil reduces their availability.

All the micronutrients except molybdenum are more soluble in an acid soil; molybdenum solubility increases with liming.

Iron is not a constituent of chlorophyll, but it is essential for its formation (Figure 11.5). Manganese is related to oxidation-reduction balances in the plant, especially in connection with iron and nitrogen metabolism. Zinc is needed by plants in some of their enzyme systems. Copper activates a group of oxidizing enzymes and is a constituent of certain proteins. The function of boron in the plant is still obscure, although a lack of boron tends to increase the loss of phosphorus from plant roots. Molybdenum seems to take part in nitrogen metabolism in the plant and in nitrogen fixation by bacteria. The exact function of chlorine in plant nutrition is not known.

FOLIAR NUTRITION OF PLANTS [6]

It has been known for many years that plants are able to absorb essential elements through their leaves. The absorption takes place through the stomata of the leaves and also through the epidermis. Movement of elements is usually faster through the stomata, but the total absorption may be as great through the epidermis. Plants are also able to absorb nutrients through their bark.

The following elements have been successfully used to supply nutrients for plant growth by applying them as foliar sprays to the leaves:

Primary Nutrients	Secondary Nutrients	Micronutrients
Nitrogen	Magnesium	Iron
Phosphorus	Calcium	Zinc
Potassium	Sulfur	Boron
		Copper
		Molybdenum

[6] Damon Boynton, "Nutrition by Foliar Application," *Annual Review of Plant Physiology,* Vol. 5 (1954), pp. 31–54.

Iron	11	18	27	32	43
Chlorophyll	.3	.7	1.3	1.6	1.8

FIG. 11.5. A deficiency of iron causes chlorosis (whitening) of citrus leaves in California. The amount of iron (parts per million of dry weight) is associated with the amount of chlorophyll (in milligrams per gram of fresh weight)—the more iron, the more chlorophyll. *Source:* Ellis F. Wallihan, U. of Calif.

One difficulty in using foliar sprays to supply essential elements to crops is that translocation of the applied element may not be rapid enough for increasing crop yields. With some plants this problem is more difficult than with others. For example, the relative mobility of essential nutrients in bean plants, when applied as a foliar spray, in order of decreasing mobility, was as follows: [7]

Mobile	Partially Mobile	Immobile
Potassium	Zinc	Magnesium
Phosphorus	Copper	Calcium
Chlorine	Manganese	
Sulfur	Molybdenum	

Nitrogen fertilizer compounds have been used for several years as foliar sprays. Sodium nitrate, ammonium sulfate, potassium nitrate, and urea have all been used experimentally, but only urea gives satisfactory results. The other fertilizers cause the burning of leaves, due partly to the high osmotic concentration of the spray solution.

Urea has been successfully sprayed on apple trees, tomatoes, celery, lima beans, potatoes, cantaloupes, cucumbers, and sugar cane. Amounts up to 15 pounds of urea per acre at one spraying have been used with beneficial results on apple trees. Higher concentrations burn the leaves. The

---

[7] M. J. Bukovac and S. H. Wittwer, "Absorption and Distribution of Foliar Applied Mineral Nutrients as Determined with Radioisotopes," as reported in Walter Reuther (ed.), *Plant Analysis and Fertilizer Problems.* Pub. No. 8, American Institute of Biological Science, Washington, D. C. (1961).

usual concentration for apple trees is 5 pounds of urea per 100 gallons of water. This is commonly mixed and applied with the regular spray materials at weekly intervals early in the growing season.

The application of urea fertilizer to leaves of plants has given response approximately equal to that of fertilizer applied to the soil. The uptake of urea is faster when it is sprayed on the leaves, but it is cheaper to apply it to the soil.

Phosphorus is capable of being utilized by the plant when it is sprayed on the leaves. Although the practice is not common, there are many good reasons for predicting that there may be an increase in the foliar application of phosphorus.

One reason is that in most soils only a small percentage of phosphorus fertilizers is recovered by the plant (averaging about 20 per cent for the first year); whereas, when phosphorus is sprayed on the leaves, nearly all of it is absorbed. In one experiment, approximately 3 pounds of P_2O_5 sprayed on tomato leaves gave a greater early growth than did 135 pounds of P_2O_5 applied to the soil. The yield of tomatoes, however, was 12 per cent greater when the 135 pounds of P_2O_5 was applied to the soil than when 3 pounds of P_2O_5 was sprayed on the leaves.

Potassium applications as foliar sprays have been made, using potassium sulfate fertilizer. Some leaf injury resulted, and the conclusion was reached that soil application are far more satisfactory.

Magnesium is now commonly applied to plant foliage as solutions of magnesium sulfate (Epsom salts). One reason for the popularity of the practice is that soil applications of magnesium commonly take 3 years to correct magnesium-deficiency symptoms of such perennials as apple trees, whereas foliar sprays are effective within a few days after application.

A foliar application of a 2 per cent solution of $MgSO_4$ to tomatoes, oranges, and apples has relieved magnesium deficiency and has increased crop yields.

Calcium is seldom applied as a foliar spray because it can be efficiently applied to the soil. If $CaCO_3$ is too slow in reaction, then CaO or $Ca(OH)_2$ can be applied. But there is no good technical reason why calcium cannot be applied as a foliar spray.

Sulfur sprayed on leaves is readily absorbed by the plants. This fact was demonstrated, however, in connection with the study of the influence of certain sulfur sprays when used as a fungicide. Although there have been no reports of a sulfur deficiency being relieved by sulfur sprays, the practice may become established because it is physiologically sound.

Iron has been sprayed on foliage since about 1916 to relieve chlorosis. The first of such research work was carried out with chlorotic pineapples growing on highly alkaline soils in Hawaii. Periodic sprays of 5 per cent ferrous sulfate are now common practice on Hawaiian pineapple plantations. The biggest obstacle to this practice is the fact that, even though the iron moves readily into the leaves, it is translocated very slowly. As a result, after spraying with ferrous sulfate, chlorotic spots may still be in evidence in places which did not receive some of the iron spray. Iron chelates have also been successfully used as a spray.

The *leaves* of chlorotic grain sorghum on calcareous soil in Tulare County, California, were sprayed with 40 gallons per acre of 3-per cent ferrous sulfate solution about 1 week before heading, at a cost for materials of 50 cents per acre. The yield of grain sorghum was increased from 540 pounds of grain on the untreated plot to 1774 pounds on the treated plot, an increase of 222 per cent. Applications on the *soil* of more than 3000 pounds per acre of ferrous sulfate were required to accomplish similar increases in yields [8] (Figure 11.6).

Iron chlorosis in grain and forage sorghums is being cured in Texas with one or two foliar sprays of a 2.5-per cent iron sulfate (copperas) solution, made by dissolving 10 pounds of iron sulfate in 50 gallons of water.[9]

Iron deficiency on sorghums in the Great Plains can be successfully corrected by drilling at least 500 pounds per acre of ferrous sulfate or sulfuric acid under the seed zone before planting.

Manganese. Although soil manganese becomes less available in alkaline soils, many states in more humid regions of the country often report manganese deficiencies in fibric, hemic, and sapric soils and in local areas of alkaline soils. Manganese deficiencies are frequently corrected by spray applications of manganese sulfate, usually 5 to 10 pounds per acre. Manganese sulfate is also applied to the soil at rates of from 20 to 150 pounds per acre. Manganous oxide is also used to correct manganese deficiencies. In alkaline soils an acid-forming material, usually fertilizer, is applied to prevent fixation of the applied manganese.

Zinc is often sprayed on the leaves of apple and pear trees to relieve "leaf rosetting," a symptom of zinc deficiency. Approximately 25 pounds of zinc sulfate in 100 gallons of water (roughly a 3-per cent solution) applied to apple trees just before the buds open has corrected zinc deficiency. Zinc sulfide, zinc oxide, and zinc carbonate have all been successfully used as sprays. Driving galvanized (zinc-coated) nails in trees also relieves zinc deficiency (Figure 11.7).

---

[8] B. A. Krantz, A. L. Brown, B. B. Fischer, W. E. Pendery, and V. W. Brown, "Foliage Sprays Correct Iron Chlorosis in Grain Sorghum," *California Agriculture,* Vol. 16, No. 5 (1962), pp. 5–6.

[9] Charles Welch, "Foliar Application of Iron Sulfate Corrects Iron Deficiency in Sorghum," *Crops and Soils Magazine,* Vol. 21, No. 6 (March, 1969), pp. 20–21.

FIG. 11.6. A field of chlorotic grain sorghum on calcareous soil in California was sprayed with 40 gallons per acre of a 3 per cent solution of ferrous sulfate at a cost of 50 cents per acre and resulted in an increase in yield of 222 per cent. *Source:* B. A. Krantz, U. of Calif.

Boron, as boric acid or borax (sodium tetraborate), used as a foliar spray has proved to be a successful method of application. Internal cork of apples has been controlled by spraying the foliage with 8 pounds of borax in 100 gallons of water. As little as 2 pounds of borax per 100 gallons of water has checked "cracked stem" of celery. Boron has been satisfactorily applied to the soil, either alone or in mixed fertilizers.

Copper deficiency has been controlled by spraying the leaves with a mixture of 8 pounds of $CuSO_4$ plus 8 pounds of $Ca(OH)_2$, in 100 gallons

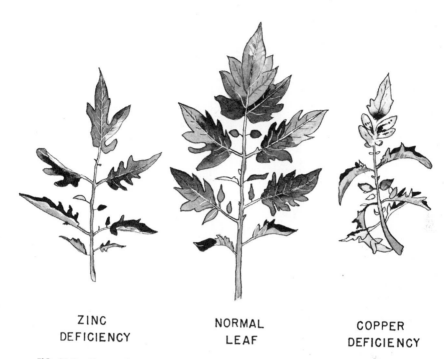

ZINC   NORMAL   COPPER
DEFICIENCY   LEAF   DEFICIENCY

FIG. 11.7. Tomato leaves. *Left:* zinc deficiency. *Center:* normal. *Right:* copper deficiency. Spraying to correct deficiencies should be done in the earliest stages of the deficiency symptoms in order to obtain maximum results. *Source: Calif. Agr. Exp. Sta.*

of water. Without the calcium hydroxide, the copper sulfate injures the foliage. Copper oxide has also been used successfully as a spray.

Molybdenum, as sodium molybdate, 1 ounce in 100 gallons of water, has eliminated deficiency symptoms in citrus trees. Somewhat like iron, however, molybdenum does not seem to be readily translocated within the plant. Spraying only the lower half of a citrus tree that showed molybdenum deficiency did not cure the deficiency symptoms on the upper half of the tree.

In highly acid soils, molybdenum is sometimes fixed in an unavailable form, thus causing deficiencies, particularly for legumes. The amount of molybdenum in soils and the amount required by plants is very small. In addition to sodium molybdate soil application of 0.5 to 2 pounds per acre, a commercial seed-coating preparation (Molygro) for some legumes, applied at about 2 ounces per acre, is used to correct deficiencies. Broadcast applications are best mixed with limestone on very acid soils to prevent fixation.

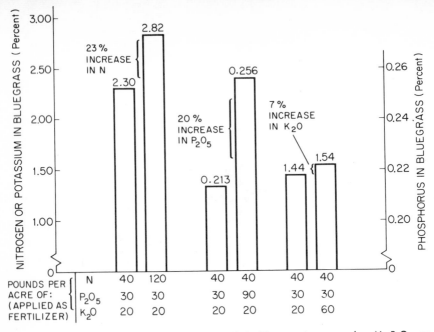

FIG. 11.8. Amount of N, P₂O₅, or K₂O in bluegrass increases when N, P₂O₅, or K₂O in fertilizer applied increases. *Source:* Conrad B. Kresge, "N–K Partners in Forage Grass Production," *Better Crops with Plant Food*, American Potash Institute, March–April, 1965, pp. 22–6.

FERTILIZER TO INCREASE NUTRIENT PERCENTAGE
IN PLANTS

It is generally true that the percentage composition of nutrients in plants is increased by the application of that nutrient in fertilizers. Evidence to support this statement is given for bluegrass in Figure 11.8.

Keeping two nutrients at a constant level and varying separately the N, P₂O₅, and K₂O, resulted in:

1. A 23 per cent increase in N in the bluegrass when the N in fertilizer was increased from 40 to 120 pounds per acre
2. A 20 per cent increase in P₂O₅ in bluegrass when the P₂O₅ in fertilizer was increased from 30 to 90 pounds per acre
3. A 7 per cent increase in K₂O in the plant when the K₂O in fertilizer was increased from 20 to 60 pounds per acre

Other research on bluegrass has indicated that when high levels of straight phosphorus fertilizers are applied (but no N or K₂O) and the bluegrass is analyzed 25 years later (compared with the no-phosphorus plot), nitrogen percentage increased by 117 per cent, phosphorus percentage by 200 per cent and K₂O by 93 per cent. The extra N came from

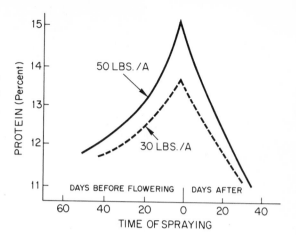

FIG. 11.9. By spraying a solution of urea on wheat plants at the time the wheat was starting to bloom (flower), protein content of the grain was increased a maximum of 3 per cent (from 12 to 15 per cent protein). The rates used were 30 and 50 pounds of N per acre and the wheat variety was Pawnee, a hard red winter type. (Kansas.) *Source:* Floyd W. Smith, "Fertilizing Wheat for Profit," *Plant Food Review,* Winter, 1964, National Plant Food Institute, Wash., D.C.

legumes that increased because of the phosphorus fertilizer, and the extra K in the bluegrass came from the more rapid release of potassium from a larger volume of soil minerals induced by the prolongation of roots that were stimulated by the phosphorus fertilizer.[10]

FERTILIZER TO INCREASE PROTEIN CONTENT OF WHEAT

During the winter of 1964, the experiment station at Kansas State University sprayed urea solutions on Pawnee hard red winter wheat plants in amounts of 30 and 50 pounds of N per acre at various times before and after flowering.

The purpose of the experiment was to determine if the protein content of the wheat kernels could be increased. The greatest increase in protein content of the wheat grain was achieved with both rates when the urea was applied at the time the wheat was starting to flower. The 30-pound rate per acre of N was responsible for an increase of from 12 to 13.5 per cent protein, and the 50-pound rate from 12 to 15 per cent protein (Figure 11.9).

Fertilizers have been applied experimentally to many crops by a large number of research scientists, with varying successes and failures. The Ontario Agricultural College in Canada concludes that the foliar application of nitrogen, phosphorus, and potassium does not result in any increase in yield of general farm crops.[11]

---

[10] *Crops and Soils Magazine,* American Society of Agronomy (April–May, 1965), p. 21.

Note: The soil was Carbo silt loam.

[11] *Foliar Fertilization for Field Crops in Ontario,* Publication No. 325, Ontario Department of Agriculture and Food, Toronto, Canada (1968), 3 pp.

FIG. 11.10. Surface crusts restrict plant growth by reducing the supply of oxygen for the roots. *Source:* Ben Osborn, Soil Conservation Service, U.S. Dept. of Agr.

To save money, fertilizers should be applied according to the research findings of the agricultural college in the state where the soil is located.

SOIL PHYSICAL CONDITIONS AND PLANT NUTRITION

Adequate amounts of nutrients for desirable plant growth may be available according to any chemical test but actually deficient because of soil physical conditions. Tillage pans, cloddy surface soils, surface crusts, or lack of desirable soil aggregation may reduce nutrient availability for one of these reasons:

1. By physically restricting root elongation, especially deep root penetration. This reduces the volume of soil in contact with plant roots, and the total amount of water and nutrients absorbed by the roots will therefore be less.
2. By restricting the exchange of oxygen and carbon dioxide in the soil, the ability of plant roots to translocate nutrients to the leaves is reduced [12] (Figure 11.10).
3. By retarding the growth of nitrifying bacteria, the soil nitrogen may

---

[12] Recent research has shown that low oxygen concentration in the soil results in an increase in nutrient absorption by the roots but a decrease in translocation of nutrients to the leaves.

remain as unavailable protein instead of breaking down to release available ammonium and nitrate ions.

4. By reducing water infiltration, water may become the first limiting factor in plant growth even though all nutrients are present in adequate amounts.

SUMMARY

Plants are now known to need only 16 of the 90 or more elements which they contain. Animals require 15 of the same 16 elements and, in addition, sodium, cobalt, selenium, and iodine. Nearly all plant nutrients are taken in by the plant in ionic forms. A plant root releases cations when it absorbs cations, and exchanges anions for other anions in the soil solution.

Nearly all nutrients can also be absorbed by the leaves of plants, through the stomata.

The physical condition of the soil must be satisfactory so as not to hinder plant nutrition. Tillage pans restrict root penetration, reduce oxygen exchange, retard nitrifying bacteria, and reduce available water.

Organic matter is the only good storehouse for soil nitrogen. Phosphorus is in short supply because the total amount in soils is low, and because it becomes quickly unavailable in acid and alkaline soils. There is plenty of total potassium in all soils, but it does not become available fast enough for normal crop growth. Calcium reserves are easily lost by leaching; as a consequence, soils in humid regions are acid. Magnesium is the only mineral on the chlorophyll molecule. Sulfur nutrition is related to organic-matter decomposition and to the addition of by-product sulfates in superphosphate fertilizer. All micronutrients except molybdenum are more readily available in acid soils.

The percentage of N, P, or K in a plant will increase with increasing applications of the respective element. Furthermore, protein content of wheat grain can be increased by timely sprays of urea on the wheat plants.

QUESTIONS

1. Name the elements essential for plants and for animals.
2. In what forms are nutrients absorbed by plant roots?
3. Describe the mechanism of a plant root absorbing potassium.
4. How do leaves absorb nutrients?
5. In what ways may a poor physical condition of the soil reduce plant growth?

REFERENCES

Bear, Firman E. (ed.), *Chemistry of the Soil* (2nd ed.). New York: Reinhold Publishing Corp., 1964.

Bear, Firman E., *Soils in Relation to Crop Growth.* New York: Reinhold Publishing Corp., 1965.

Berger, Kermit C., *Introductory Soils.* New York: The Macmillan Company, 1965.

Brown, A. L., B. A. Krantz, and J. L. Eddings, "Zinc–Phosphorus Interactions as Measured by Plant Response and Soil Analysis," *Soil Science,* Vol. 110, No. 6 (Dec., 1970), pp. 415–20.

Fox, H. R., "The Effect of Calcium and pH on Boron Uptake from High Concentrations of Boron by Cotton and Alfalfa," *Soil Science,* Vol. 106, No. 6, 1968, pp. 435–39.

Gauch, Hugh G., "Soilless Plant Culture," *Plant Food Review* (Fall, 1964), pp. 13–16.

Millar, C. E., L. M. Turk, and H. D. Foth, *Fundamentals of Soil Science* (4th ed.). New York: John Wiley & Sons, Inc., 1965.

Rajapakse, W. P., and G. Norton, *"The Role of Manganese in Plant Metabolism," University of Nottingham Reporter,* 1968, pp. 79–85.

Reuther, Walter (ed.), *Plant Analysis and Fertilizer Problems,* Publication No. 8. Washington: American Institute of Biological Science, 1961.

Russell, Sir E. John, and E. Walter Russell, *Soil Conditions and Plant Growth* (9th ed.). London: Longmans, Green & Company, Ltd., 1961.

Schütte, Karl H., *The Biology of the Trace Elements—Their Role in Nutrition.* Philadelphia: J. B. Lippincott Co., 1964, 228 pp.

Shaw, Earl J., *Western Fertilizer Handbook* (4th ed.). Sacramento, Calif.: California Fertilizer Association, 1965.

Shickluna, J. C., *A Survey on Micronutrient Deficiencies in the U.S.A. and Means of Correcting Them.* Madison, Wisc.: The Soil Testing Committee, Soil Science Society of America, 1965.

Sprague, Howard B. (ed.), *Hunger Signs in Crops.* New York: David McKay Co., Inc., 1964.

Tisdale, Samuel L., and Werner L. Nelson, *Soil Fertility and Fertilizers* (2nd ed.). New York: The Macmillan Company, 1965.

Truog, Emil (ed.), *Mineral Nutrition of Plants.* Madison, Wisc.: The University of Wisconsin Press, 1951.

CHAPTER **12**

SOILS OF THE TROPICS [1]

*There is no sharp geographic line of division between soils
of tropical regions and those of temperate climates.**

Below about 5000 feet in elevation and lying between the Tropic of
Cancer (23° 27′ north latitude) and the Tropic of Capricorn (23° 27′
south latitude) are the Tropics, sometimes called the Torrid Zone.

When people of the Temperate Zone become frustrated with ice and
snow, routine work, ringing telephones, crowded highways, and the competitive prestige of ownership of useless things, they dream of living in the
Tropics where the environment and human values allow time to enjoy the
luxury of leisure. A major equation in this dream is the assumption that
soils of the Tropics are so fertile and that rainfall is sufficient to permit
abundant food production with maximum human ease.

The primary differences in soil characteristics in the Tropics and those
in temperate regions are differences resulting from contrasts in temperature. In the Tropics, soil temperatures are high every day in the year,
whereas in most temperate regions, freezing of the soil interrupts the
weathering of minerals and soil profile development. Furthermore, there are

---

[1] This chapter name, Soils of the Tropics, will also include the areas between
the Tropic of Cancer and the Tropic of Capricorn that are above approximately
5000 feet whose climates are in fact *subtropical* and whose area may be approximately 10 per cent of the total of the Equatorial Region.

* James Thorp and Mark Baldwin in *"Laterite in Relation to Soils of the
Tropics,"* Annals of the Association of American Geographers, Vol. 30, No. 3
(1940), pp. 163–94.

a larger percentage of land surfaces in the Tropics that are older than those in temperate regions. The net results are that in the humid Tropics, weathering has been faster and more intense on *all* soils, and on *some* soils for longer periods of time.

On the oldest land surfaces, soils of the humid Tropics (Torrid Zone) and Subtropics, compared with soils of humid Temperate and Frigid Zones, have: [2]

1. More deeply and more intensely weathered profiles (Figure 12.1)
2. A lower cation exchange capacity
3. A higher anion exchange capacity
4. A lower buffer capacity, a slightly lower total water-holding capacity, but *a much lower available water-holding capacity.* Stated in another way, soils of the Tropics and Subtropics, with silt and clay percentages similar to those in the Temperate Zone, have chemical and physical properties more characteristic of coarser textured soils.
5. A plinthite layer in some soils that hardens by crystallization of the iron oxides, on continuous exposure to cycles of wetting and drying
6. A larger percentage of kaolinite and a smaller percentage of montmorillonite and illite clay
7. A lower percentage of silicon and weatherable minerals
8. A higher percentage of iron, aluminum, and often of titanium oxides, as Gibbsite [$Al(OH_3)$] and Goethite ($FeOOH$)
9. A higher degree of friability
10. Less accumulation of leaf litter and other organic debris
11. A lower degree of plasticity
12. A lower level of fertility.

As early as 1878, it was recognized by geologists and soil scientists that the process of decomposition of rocks to make soil is faster and extends to a greater depth in equatorial than in temperate and frigid latitudes. One of the first scientists to write on this subject was Stubbs.[3]

In soils of the Tropics, many particles the size of sand consist of altered minerals cemented by iron. This contrasts with the composition and struc-

---

[2] *Primary sources:* (1) Guy D. Smith, "Lectures on Soil Classification," *Pedologie,* No. 4 (1965), 134 pp.; (2) S. J. Toth, "The Physical Chemistry of Soils," in Firman E. Bear (ed.), *Chemistry of the Soil,* 2nd ed. (New York: Reinhold Publishing Corp., 1964), pp. 142–62.

[3] William C. Stubbs, *The Soils of Alabama, Berney's Hand-book of Alabama* (Mobile, Ala.: 1878), pp. 197–220.

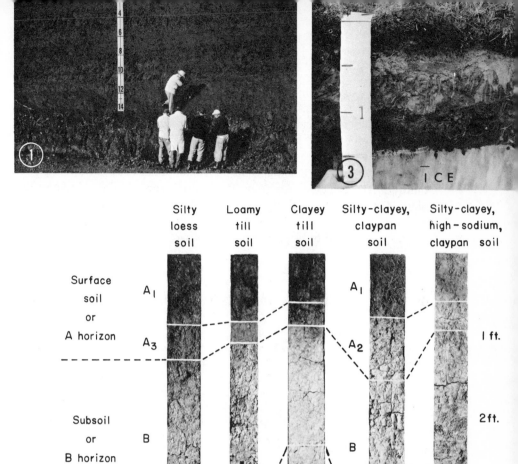

FIG. 12.1. Soils in the humid Tropics are more deeply and more intensely weathered than those in humid Temperate and Frigid Zones. This is true of upland, well-drained soils. Evidence of this may be seen in (1) from the State of Parana in southern Brazil, a deeply weathered profile of red soil more than 14 feet deep. Compare this soil of the Tropics with five representative soil profiles (2) from Illinois, in the Temperate Zone, where weathering has not extended as far as 4 feet, and with (3) from Alaska, where weathering has been restricted by perpetual ice which starts at 1½ feet. (Scales are in feet.) *Sources:* (1), Roy W. Simonson, Soil Survey Staff, Soil Conservation Service, U.S. Dept. of Agr.; (2), J. B. Fehrenbacher, B. W. Ray, and J. D. Alexander, U. of Ill.; and (3), J. C. F. Tedrow, Rutgers U.

ture of sand particles in temperate and arctic regions, which are mostly primary minerals such as quartz or feldspars.

Increasing yields per unit of area and increasing the acres in food crops have a good potential in the Tropics. But since the early part of the twentieth century, some of the more knowledgeable and concerned soil scientists have realized that soils and their manipulation to maximize yields require different techniques from those used successfully in temperate regions. One such soil scientist was E. W. Hilgard, who made the following comments: "It is quite obvious that a different standard of interpretation must be applied to tropical soils as compared with either the temperate humid, or the arid regions. . . ." [4]

SOIL FORMATION AND CLASSIFICATION

Not until the morphology and genesis of a soil are known can research to discover new and improved management systems be planned most effectively. Without such organized knowledge, purely empirical mass plot work alone must be resorted to with the hope that something will work. *This is the situation now with many tropical soils. The Ground-Water Laterite soils are an example. Until their genesis is worked out, finding practical systems of soil management by empirical plot trials alone seems nearly hopeless.*[5]

Soils at low elevations in the Tropics have all weathered under a continuously high temperature but under sharp contrasts in rainfall, from less than 5 to more than 400 inches a year. These great variations in rainfall have been responsible for the development of soils with continuous dryness (deserts) as well as soils that have developed under continuous wetness (tropical rain forests) and all intergrades.

With continuous dryness, weathering processes are primarily physical. This means in the *arid* Tropics mostly a mechanical breakdown of rocks and minerals by heating and cooling from day to night and abrasion by wind-blown particles. In a tropical climate with some *wet* months, rocks and minerals also weather chemically (oxidation, carbonation, hydrolysis, hydration, and solution). Simultaneously, some soil minerals decompose,

---

[4] E. W. Hilgard, *Soils: Their Formation, Properties, Composition, and Relations to Climate and Plant Growth in the Humid and Arid Regions* (New York: The Macmillan Company, 1911), p. 410.

[5] *Soil Survey Manual* (Washington: Agricultural Research Administration, U.S.D.A., 1962), p. 5.

Note: Italics are those of Roy L. Donahue.

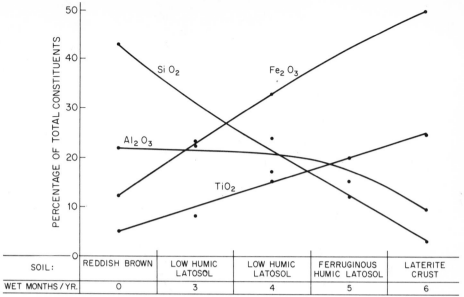

SOIL:	REDDISH BROWN	LOW HUMIC LATOSOL	LOW HUMIC LATOSOL	FERRUGINOUS HUMIC LATOSOL	LATERITE CRUST
WET MONTHS / YR.	0	3	4	5	6

FIG. 12.2

FIG. 12.2. A comparison of the percentages of iron (Fe_2O_3), aluminum (Al_2O_3), and titanium (TiO_2) in the A horizon of tropical soils with increasing numbers of wet months a year. (In the Tropics a "wet" month is usually defined as one which receives more than 4 inches of rain.) *Source:* Donald G. Sherman, "Factors Influencing the Development of Lateritic and Laterite Soils in the Hawaiian Islands," *Pacific Science,* Vol. 3, No. 4, Oct., 1949.

and new soil minerals crystallize. With more wet months, there is a destruction and transformation of many clay crystals into free oxides of iron, aluminum, and sometimes titanium; silicon percentage decreases.

When the annual rainfall in the Tropics is between 25 and 35 inches a year (3 to 4 "wet" months [6] a year), a maximum formation of kaolinite takes place. With increasing wetness, kaolinite decomposes and free oxides concentrate in the A and B horizons. Silicon in the kaolinite moves out of the soil profile, aluminum remains almost constant, and iron and titanium (if present) remain in the surface soil but increase in percentage because of the loss of other soil constituents.

With 12 "wet" months a year, and with water constantly moving downward and laterally through the soil profile, the ultimate soil in the Tropics is very high in iron oxides, aluminum oxides, and sometimes titanium oxides, and very low in silicon and all plant nutrients except iron. Such soils are often sufficiently rich in iron to be used as iron ore, or rich in aluminum to be mined as aluminum ore (Figures 12.2 to 12.4).

---

[6] A "wet" month in the Tropics is defined by Mohr as a month that receives more than 4 inches (100 millimeters) of rainfall. *Source:* E. C. J. Mohr, *The Soils of the Equatorial Regions* (Ann Arbor, Mich.: Edwards Brothers, 1944), 766 pp.

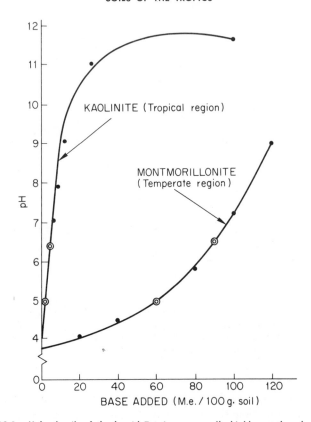

FIG. 12.3. Upland soils of the humid Tropics are usually highly weathered and contain *kaolinite* (2-layer) clay as the predominant clay mineral plus free oxides of iron and aluminum. These soils are characterized by a low exchange capacity and a low buffering capacity, as indicated when a base (such as lime) is added. Upland soils of humid Temperate Regions by contrast usually have a high percentage of 3-layer clays such as *montmorillonite*, a high exchange capacity, and a high buffer capacity. From these titration curves it is possible to approximate the lime requirement by using these values: 1 milliequivalent (m.e.) of base per 100 grams of dry soil $=$ 1,000 pounds of lime ($CaCO_3$) per acre. For example, to change the pH of the kaolinite (Tropical) soil from 5.0 to 6.5 (shown by circles) requires 4 m.e. of base or 4,000 pounds of calcium carbonate. By contrast, the same change in pH in the montmorillonite (Temperate) soil takes 30 m.e. of base or 30,000 pounds of calcium carbonate. *Sources:* Yoshito Matsusaka and G. Donald Sherman, "Lime Requirements of Hawaiian Soils," *Hawaii Farm Science*, Vol. 13, No. 3, Aug., 1964, and N. T. Coleman and A. Mehlich, *The Chemistry of Soil pH: The Yearbook of Agriculture* (1957), U.S. Dept. of Agr., p. 76.

Soils in equatorial regions of the world are more heterogeneous than those in temperate regions. At least five mountain ranges in the Equatorial Region have peaks above 15,000 feet, with perpetual ice and snow. By contrast, low elevations and hot deserts exist, as well as areas with more than 400 inches of rainfall a year. Histosols (bog soils) also are present

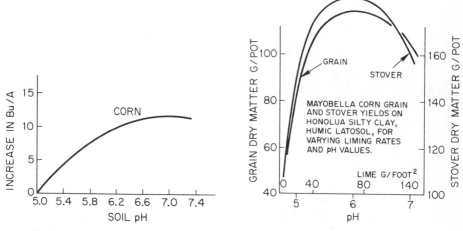

FIG. 12.4. *Left:* Corn response on a Temperate Region soil (Iowa) with varying amounts of lime, as measured by changes in soil pH. Corn yields increase gradually from pH 5.0 to about pH 6.5 and then increase negligibly. *Right:* Corn on a limed Tropical soil (as monitored by soil pH) in Hawaii responds more sharply, reaches a maximum yield at a lower pH (6.2), and shows a greater depression in yield above pH 6.2 than corn on a Temperate Region soil in Iowa. *Sources: Left,* Iowa State U. Pamphlet No. 315, 1965; *right,* Hawaii Farm Science, Special Issue on Liming, Vol. 13, No. 3, Aug. 1964.

in the Equatorial Region. Because of greater variation in the factors of soil formation, it is logical to assume that when more scientific information is available, there will likely be a larger number of soil series in the Equatorial Region than in the Temperate Region.

Venezuela, South America, lies at a latitude from 0° 45′ north to 12° 26′ north. Soils at the low elevations are therefore tropical. Eight of the ten soil orders are represented by large acreages. Histosols occur to a limited extent along tidal marshes, and if there were mountain peaks above about 7000 feet, Spodosols would probably be present. Six of the soil orders are listed in sequence, from the least intensely weathered to the most intensely weathered, as follows:

1. Mollisols
2. Inceptisols
3. Entisols (nonsandy, nonplinthitic)
4. Alfisols
5. Ultisols
6. Oxisols

Only Vertisols and Aridisols, although they occur in Venezuela, appear to be outside the weathering sequence—the former because of very fine tex-

(A), Dark red Oxisol from central Brazil, South America, described as follows: A horizon: 0–12 inches (disturbed), dark reddish brown (10R 3/2, moist); at first feels like a loam but after some manipulation feels like clay loam; weak fine granular structure; soft, very friable, non-sticky; roots present; smooth, gradual boundary. B horizon: 12–40+ inches, dark red (10R 3/6, moist; 10R 3/4, dry) clay loam; no discernible structure; soft, very friable, non-sticky to slightly sticky; no clay skins observed; few large roots; many pores visible under 10x lens. (Soil not examined below 40 inches.) Oxisols comprise about 50 per cent of the soils of Brazil.

(C), A deep, black Vertisol developed from river alluvium in eastern Ethiopia, on which irrigated cotton is being grown successfully.

(B), Light brownish gray Oxisol from Ivory Coast, western Africa. Oxisols comprise approximately 18 per cent of the soils of Africa.

(D), A shallow Vertisol in Central India.

(E), Inceptisol (Oxic), in Congo (Kinshasa), Central Africa.

(F), A deep sandy soil along the Atlantic Ocean in Ivory Coast, western Africa, supporting coconut trees.

FIG. 12.5. Soils of the Tropics and Subtropics are more variable than those in Temperate and Frigid Regions, as documented by these photographs from 5 countries on 4 continents. (Scales are in feet.) Sources: (A), Dirk van der Voet, Soil Conservation Service, U.S. Dept. of Agr.; (E), Charles E. Kellogg, Soil Conservation Service, U.S. Dept. of Agr.; (B), (C), (D), and (F), Roy L. Donahue; description of dark red Oxisol in Brazil by A. C. Orvedal, Asst. Director, Soil Survey Interpretation, Soil Conservation Service, U.S. Dept. of Agr.

ture and montmorillonite-type of clay and the latter because of low rainfall.[7]

Most people have visualized soils of the Tropics as red, well-drained, highly leached, and infertile—a fairly accurate characterization of the soil order of Oxisols. However, Oxisols comprise only 18 per cent of the soils of tropical Puerto Rico and the Virgin Islands. The complete listing of

[7] *Sources:* (1) F. C. Westin, Justo Avilan, A. Bustamante, and M. Mariño, "Characteristics of Some Venezuelan Soils," *Soil Science,* Vol. 105, No. 2 (Feb., 1968), pp. 92–102; (2) Fred C. Westin and Julia G. de Brito, "Phosphorus Fractions of Some Venezuelan Soils as Related to Their Stage of Weathering," *Soil Science,* Vol. 107, No. 3 (March, 1969), pp. 194–202.

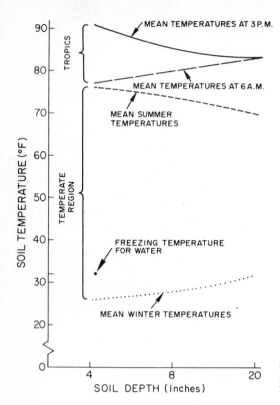

FIG. 12.6. A comparison of soil temperatures at soil depths of 4, 8, and 20 inches in the Tropics and in the Temperate Region.

Soil Order	Percentage of Total
Ultisols	26
Mollisols	19
Oxisols	18
Inceptisols	16
Alfisols	11
Vertisols	10
Total	100

soil orders and their percentage in Puerto Rico and the Virgin Islands are as follows [8] (see also Figure 12.5):

An example may be cited from Puerto Rico and the Virgin Islands in support of the statement that soils of the Tropics are more variable than those in temperate regions. The 93 pedons sampled to characterize the soils mapped by the United States Soil Conservation Service represent six of the ten soil orders in the new U.S. system of soil taxonomy. A closer in-

---

[8] *Soil Classification—A Comprehensive System: Seventh Approximation,* Soil Survey Staff, Soil Conservation Service, U.S.D.A. (Aug., 1960), and 1967 *Supplement.* Reference: *Soil Survey Laboratory Data and Descriptions of Some Soils of Puerto Rico and the Virgin Islands,* Soil Conservation Service, U.S.D.A. Soil Survey Investigation Report No. 12, in cooperation with Puerto Rico Agricultural Experiment Station (Aug., 1967), 191 pp.; in Foreword.

spection of soils of Puerto Rico reveals that these six soil orders may occur on the *same farm,* in contrast to about three soil orders on a farm in Kansas (Temperate Region).

SOIL TEMPERATURE

The thicker the clouds the thicker the soil (*Anonymous*).

Soil development in the Tropics is greatly influenced by the solubility, mobility, and chemical and biological precipitation of bases, silicon, iron, aluminum, and other minerals. Most directly determining the speed and direction of these soil-forming processes are rainfall (and resultant soil moisture) and soil temperature, which is always decreased by tropical rains.

Soil temperatures are compared in the Tropics near the Equator and in the Temperate Region in Figure 12.6.[9] Near the Equator there are no seasons resulting from differences in temperature; the primary temperature variations throughout the year occur throughout each day the sun is shining as compared with night. For this reason the soil temperature data for the Tropics are given in the figure for 6 A.M. and 3 P.M., the coolest and hottest soil temperatures, respectively, throughout the day and year. For comparison soil temperature data for the Temperate Region are given at comparable depths as mean summer temperatures and mean winter temperatures.

Figure 12.6 indicates that at a depth of 4 inches, the soil temperature in the Tropics at 6 A.M. is approximately equal to the mean summer temperature at the same depth in the Temperate Region. No other similarities exist. In the Tropics, the maximum difference in mean soil temperatures at any time is 14°F at a 4-inch depth at 6 A.M. and 3 P.M. In the Temperate Region, the maximum difference between any two mean soil temperatures in summer and winter is 50°F, at a soil depth of 4 inches.

In the Tropics, soil temperatures are equalized at a depth of 20 inches, compared with an extrapolated depth of 76 inches in the Temperate Re-

---

[9] Soil temperatures representing the Tropics are from Yangambi (Leopoldville), Congo (Kinshasa); latitude 0° 45′ north and longitude 24° 29′ east; elevation 1200 feet. Reference: J. L. D'Hoore, *Soil Map of Africa,* Pub. No. 93, Joint Project No. 11, Lagos, Nigeria (1964) pp. 34, 35. Soil Temperatures for mid-latitudes are at Ames, Iowa, U.S.A., latitude 42° north and longitude 93° 35′ west; elevation 1000 feet. *Reference:* Jen-Hu Chang, *Ground Temperature, II,* Blue Hill Meteorological Observatory, Harvard Univ., Milton, Mass. (1958), 196 pp. *Note:* An excellent discussion on the relevance of soil temperatures to soil genesis and classification is in Guy D. Smith, Franklin Newhall, Luther Robinson, and Dwight Swanson, *Soil-Temperature Regimes—Their Characteristics and Predictability,* U.S.D.A., Soil Conservation Service, T.P.–144 (April, 1964), 14 pp.

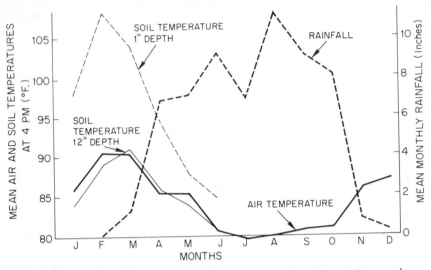

FIG. 12.7. A comparison of mean monthly air and soil temperatures at 4 p.m. and mean monthly rainfall in southeastern Nigeria. Data on rainfall and air temperature are for the period July, 1963 through June, 1964. Data on soil temperature are for the period Jan. through June, 1964. *Note:* No additional data are available for this station because the project was abandoned. *Source:* Kirk Lawton, L. Uwemedimo, A. Akubanjo, and L. Ekanem, *Basic Data,* U. of Nigeria Agro-Meteorological Sta., 0203 NSU, Nsukka, Nigeria.

gion. In other words, the soil temperature does not change in the Tropics at (and presumably below) a depth of 20 inches, or in the Temperate Region below 76 inches.[10]

Soil temperatures were recorded in southeastern Nigeria (latitude 7° north) for a period of 6 months. These soil temperatures are compared with air temperatures and rainfall at the same location (Figure 12.7). The most striking comparisons are the following:

1. February is dry (no rain) and is the month of highest mean monthly air temperature (91°F) and mean monthly soil temperature at 1-inch depth (108°F).
2. From January to June, the mean monthly air temperature and the mean monthly soil temperature at a depth of 12 inches are approximately equal.

---

[10] The soil temperatures reported here for Yangambi, Congo are the only soil temperature data available at low elevations within one degree of latitude from the Equator. No air temperature data are available at the same station, but mean monthly air temperature data from three other stations in the Congo lying within one degree of latitude have a maximum variation of 2.7°F (74.7°F for July and August to 77.4°F for March). (Stations are at Coquilhatville, Boende, and Kisangani.) By contrast, the mean monthly air temperature at Ames, Iowa for the coldest month (January) is 20.0°F and for the hottest month (July) is 74.8°F, a variation of 54.8°F. This variation is more than 20 times that in the Congo.

3. When rains start, soil temperatures become lower, probably because of more cloudiness and the cooling effect of the rain water.

SOILS WITH LATERITE

. . . A well developed laterite indicates a senile, infertile soil.[11]

Between 1800 and 1801, F. Buchanan, a British geologist and soil scientist, was studying and describing soils in southwestern India in the present states of Mysore and Kerala. He described a soil material that he called "laterite" (from the Latin, *later,* brick), as follows:

> It is full of cavities and pores, and contains a very large quantity of iron in the form of red and yellow ochres. In the mass, *while excluded from the air,* it is so soft that any iron instrument readily cuts it, and it is dug into square masses with a pick-axe and immediately cut into the shape wanted with a trowel or large knife. *It very soon after becomes as hard as brick,* and resists the air and water much better than any bricks that I have seen in India.[12] (Figure 12.8.)

Laterites have been reported to occur extensively in India, Ceylon, Australia, Thailand, Malaysia, Burma, Singapore, Philippines, Indonesia, Brazil, the Guianas, Peru, Equador, Puerto Rico, and Cuba, and in the following 30 countries in Africa where Laterites and associated soils are estimated to occupy approximately 2 million square miles, 18 per cent of the total land surface of Africa: Angola, Botswana, Burundi, Cameroon, Central African Republic, Congo (Brazzaville), Congo (Kinshasa), Gabon, Ghana, Guinea, Ivory Coast, Kenya, Liberia, Malawi, Malagasy Republic, Mali, Mozambique, Nigeria, Portuguese Guinea, Republic of South Africa, Rhodesia, Rwanda, Senegal, Sierra Leone, Swaziland, Tanzania, Togo, Uganda, Upper Volta, and Zambia.

CHARACTERIZATION OF LATERITE IN SOILS

As identified and mapped in the field, Laterites consist of at least two materials of contrasting chronological age, as first differentiated by Oldham

[11] Robert L. Pendleton, "Laterite and Its Structural Uses in Thailand and Cambodia," *Geographical Review,* Vol. 31 (1941), pp. 177–202.

[12] J. A. Prescott and Robert L. Pendleton, *Laterite and Lateritic Soils,* Technical Communication No. 47, Commonwealth Bureau of Soil Science, Commonwealth Agricultural Bureaux, Farnham Royal, Bucks, England (1966), p. 1.

FIG. 12.8. Ground-water Laterites have developed in the Tropics by the inflow of iron at the foot of seepage slopes. Under a forest cover the iron layer remains soft, but when the forest is cleared the increased intensity of wetting and drying and heating and cooling hastens crystallization and hardening of the iron. When the Laterite layer exists at the same level where it was formed, some scientists call it "low-level" Laterite; when uplifted, it is known as "high-level" Laterite. Ground-water Laterite is common but usually occurs in small patches in India (A), (B), (C), and (D), Brazil (E), and in Equatorial Africa (F) and (G). (J. L. D'Hoore in *Soil Map of Africa*, p. 38, states, "In northwest Africa sheets of old ground-water laterite now may be seen on high levels where dryness of climate resists their development as well as prevents their break up by a forest vegetation.") (A) and (D), High-level Ground-water Laterite in Central India. (Scale in A is in feet.) (B) and (C), Low-level Ground-water Laterite, while still reasonably soft, is being chiseled into building stone and used for building construction. (E), Low-level Ground-water Laterite in northeastern Brazil, with cambic horizon indicated by arrows. (Scale is in feet.) The official description of Ground-water Laterite from Brazil is as follows: 0–1 inches, Root mat of grasses and sedges; 1–8 inches, dark gray (N/2 moist, N/4 dry) silt loam; moderate, medium granular structure; moderately friable; clear, smooth boundary; 8–13 inches, very dark grayish brown (10YR 3/2 moist) silt loam (color lightens somewhat with depth); moderate, medium granular structure; friable to firm; clear, smooth boundary; 13–20 inches, brown (10YR 5/2, moist) matrix with red mottles, the reddest is 2.5 YR 4/8 moist; silty clay loam; weak, subangular blocky structure with peds of varying sizes; gradual, smooth boundary; 20–28+ inches, reddish gray (10R 6/1 moist) matrix with reddish plinthite spheres, the reddest is 10R 3/4 moist; some yellow mottles; massive; firm to very firm, the plinthite being more firm than the grey material; the grey material is sticky and plastic when wet. The taxonomy of this soil (E) is as follows: *Order*—Inceptisol; *Suborder*—Aquepts; *Great Group*—Tropaquepts; *Subgroup*—Plinthic Tropaquepts; *Family, Series, and Type*—not known. (Mouth of Amazon River in northeastern Brazil.) *Notes:* 1. This soil is representative of the soil on which the landing strip is made at Ipean, Ilha de Marajo. The soil does not flood but at the end of the rainy season in March and April the landing strip is too wet to use. 2. Soil profile description was made on Aug. 31, 1965 by Clifford Orvedal, Asst. Director, Soil Survey Interpretation, Soil Conservation Service, U.S. Dept. of Agr. 3. The soil classification was made by Dirk van der Voet, Soil Conservation Service, U.S. Dept. of Agr. (F), Low-level Ground-water Laterite in north central Ghana. (Scale is in inches.) (G), Low-level Ground-water Laterite in central Ghana, showing the ground water at a depth of 5½ feet. (Numbers are in feet and sections on range pole are 6 inches in width.) This soil has recently been tentatively classified by Henry B. Obeng, in charge of soil survey in Ghana, as follows: *Order*—Alfisol; *Suborder*—Ustalf; *Great Group*—Plinthustalf; *Subgroup* and *Family*—not yet determined; *Series*—Cobena; *Type*—Cobena loam. *Sources:* (E), Dirk van der Voet, Soil Conservation Service, U.S. Dept. of Agr.; all other photos by Roy L. Donahue.

in 1893 [13]: "high level" and "low level." The high-level form of Laterite occurs as hardened ironstone caps, usually of a meter or more in thickness, and appearing as an erosion-resistant surface of hilltops, with the edges consisting of broken chunks of Laterite that have rolled or slid down the slope. The low-level Laterite that has hardened on exposure is usually indistinguishable from the high-level form except that low-level Laterite occurs at a lower topographic level, either at the surface as hardened secondary ironstone or as softer material, at variable depths from the surface of the soil. In general, the deeper the Laterite layer the softer it is; however, even the softest and deepest material may harden almost irreversibly when exposed to the air for several months.

Most authorities agree, however, that the high-level and low-level kinds of Laterite were both formed in low-lying areas where seepage waters moved soluble iron down the slope to concentrate at ground-water level in what is now called the Laterite layer. The high-level Laterite was subsequently uplifted, the surface soil eroded, and the secondary ironstone layer hardened. The hardened Laterite has persisted because of its resistance to weathering and erosion.

Low-level Laterites occur on old land surfaces under conditions of forest or forest-grass vegetation with approximately 40 or more inches of annual rainfall and including a definite dry season. They form in place on the intermittent water table by the precipitation of iron from iron-laden acid solutions on many kinds of slowly permeable rocks and parent materials. Laterites form more slowly on alluvium, volcanic ash, calcareous soil materials, and on deep and slowly permeable clay soils.

The secondary ironstone in low-level Laterite may occur on the surface or at a depth of as much as 10 feet, and vary from a thickness of a few inches to as much as 6 feet. The ironstone layer of both the low-level and high-level forms may be pisolitic (composed of pea-sized particles) or vesicular (with irregular cavities).[14]

Chemically, soft and hardened Laterite consists of materials of varying composition but primarily hydrated oxides of iron with varying amounts of aluminum and silica. The primary difference between soft and hardened Laterite is that in hardened Laterite there is more iron in crystalline form.

---

[13] R. D. Oldham, "Laterite," Chap. 15 in *A Manual of Geology in India*, 2nd ed. (Calcutta, India: 1893), pp. 369–90, as reported in J. A. Prescott and R. L. Pendleton, *Laterite and Lateritic Soils*, Technical Communication No. 47, Commonwealth Bureau of Soil Science, Commonwealth Agricultural Bureaux, Farnham Royal, Bucks, England (1966), p. 8.

[14] Vesicular Laterite in physical appearance resembles a coal cinder and also volcanic lava; it is sometimes confused with the latter. Laterite, however, has a much greater density than a coal cinder or lava.

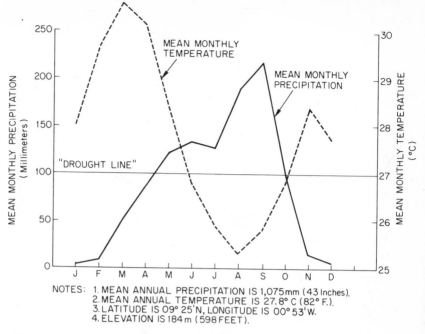

FIG. 12.9. Mean monthly precipitation and mean monthly temperature in Tamale, Ghana (western Africa). (10-year average.) *Notes:* 1. Mean annual precipitation is 1,075 mm (43 inches). 2. Mean annual temperature is 27.8°C (82°F). 3. Latitude is 09° 25'N, and longitude is 00° 53'W. 4. Elevation is 184 m (598 feet).

Three conditions are necessary for formation and hardening of Laterite:

1. A fairly level land surface situated at the foot of a seepage slope
2. An adequate supply of soluble iron, either by an inflow of iron from surrounding areas or by an outflow of other constituents and a resultant concentration of existing iron
3. An alternating wet season and dry season of approximately equal duration, and sufficient rain during the wet season to continuously saturate the zone of iron segregation. (Typical rainfall and temperature for the development and hardening of ground-water (low-level) Laterite is seen in Figure 12.9).

Soils with Laterite are usually highly leached, strongly acid, and have a low level of fertility. Nitrogen is probably the first nutrient element that limits plant growth. Phosphorus is usually rendered unavailable because of the high fixation as iron phosphate, and the availability of zinc, molybdenum, and copper are also low. Furthermore, the hardened ironstone layer restricts root elongation, decreases available water for plant growth during dry periods, and is conducive to an excess of water for plant growth during the rainy season. For these and perhaps other reasons, it is seldom that a satisfactory response is obtained with the use of fertilizers on most field crops growing on soils with Laterite.

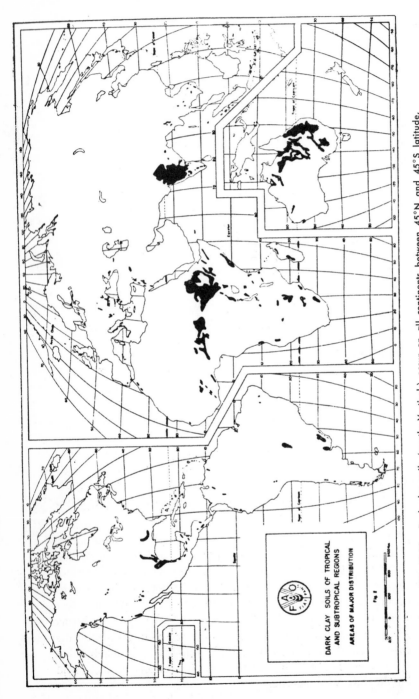

FIG. 12.10. Dark clay soils (mostly Vertisols) occur on all continents between 45°N and 45°S latitude, but mostly in Tropical Regions. The total world area of dark clay soils is 635 million acres, approximately one third of which exists in Africa, India, and Australia. Smaller areas exist in North America and South America. Source: R. Dudal (ed.), "Dark Clay Soils of Tropical and Subtropical Regions," FAO Agr. Dev. Paper, No. 83, 1965, p. 6.

DARK CLAY SOILS OF TROPICAL
AND SUBTROPICAL REGIONS

AREAS OF MAJOR DISTRIBUTION

Whereas the yields of most field crops on soils with Laterite are low, yields of certain crops, such as tobacco, sugarcane, turmeric, ginger, and sweet potatoes, are usually satisfactory. Of special significance is the fact that many tree crops, such as tamarind, mango, cashew, tapioca (cassava), coconut, banana, and teak, and many other forest trees, are well adapted to soils with Laterite.[15]

Continuous cultivation in the Tropics invariably results in losses of soil organic matter, except in instances where forest trees are left as shade or especially when banana is included in the crop rotation. The large number of banana leaves produced and the shading effect brings about an increase in soil organic matter [16] that helps to prevent hardening of the Laterite layer. The scientific explanation for this phenomenon is that soil organic matter and its decomposition products, such as citrates, retard the crystallization and therefore the hardening of amorphous ferric hydroxide.[17]

DARK CLAY SOILS [18]

Dark clay soils, classified mostly as Vertisols, occur in the Tropics and Subtropics on an estimated 635 million acres (Figure 12.10).

Dark clay soils have the following environments and characteristics:

1. An annual rainfall of 25 to 40 inches, with about 6 dry months
2. A mean annual temperature between 60 and 80° F
3. A clay soil texture throughout the profile
4. A slope less than 5 per cent
5. An elevation below 1000 feet
6. A parent material very high in calcium and sometimes magnesium
7. Very high power requirements if the soil is plowed with a turning plow but low power requirements if *only* a disc harrow is used

---

[15] J. A. Prescott and R. L. Pendleton, *Laterite and Lateritic Soils,* Technical Communication No. 47, Commonwealth Bureau of Soil Science, Commonwealth Agricultural Bureaux, Farnham Royal, Bucks, England (1966), p. 42.

[16] P. H. Nye and D. J. Greenland, *The Soil Under Shifting Cultivation,* Technical Communication No. 51, Commonwealth Bureau of Soils, Harpenden, Commonwealth Agricultural Bureaux, Farnham Royal, Bucks, England (1960), pp. 98–107.

[17] U. Schwertmann, W. R. Fischer, and H. Papendorf, "The Influence of Organic Compounds on the Formation of Iron Oxides," *Transactions, 9th International Congress of Soil Sciences,* Adelaide, Australia, Vol. I, Paper No. 66 (1968), pp. 645–55.

[18] *Principal reference:* R. Dudal (ed.), *Dark Clay Soils of Tropical and Subtropical Regions,* F. A. O. Agricultural Development Paper No. 83 (1965), 161 pp.

FIG. 12.11. Dark clay soils of the Tropics and Subtropics are classified mostly in the soil order of Vertisols. Vertisols are common in Texas, Alabama, Mississippi, California, and Puerto Rico. Very extensive areas of Vertisols also occur in India, Australia, Sudan, Ethiopia, Mexico, Uruguay, Argentina, and in Brazil. Representative Vertisols from 3 continents are portrayed by these photographs: (A) from northeastern Brazil, (B) and (C) from northwestern Ethiopia, (D) and (E) from central western India. (Scales are in feet.) Note the dark color, the wide and deep cracks which are self-mulching (at arrows), and the nodules of secondary limestone (L) in (E), where erosion has exposed the secondary limestone nodules. Vertisols are used primarily for growing cotton (B), grain sorghum (C), and Sesame. *Sources:* (A), Dirk van der Voet, Soil Conservation Service, U.S. Dept. of Agr.; all other photos by Roy L. Donahue.

8. Great difficulty in maintaining a ridge type of terrace but less difficulty when a channel type is used
9. Wide and deep crack upon drying (Figure 12.11)
10. Self-mulching, by surface peds falling into dry-weather cracks (Figure 12.11)
11. A rapid initial infiltration rate for rainfall, because of the cracks, but a very slow rate 24 hours later when the clay swells and the cracks are closed
12. When wet, extreme stickiness, and swelling of 25 to 50 per cent by volume. The soil surface is so sticky that roads and airfields cannot be used by motorized vehicles during the wet season
13. A low soil organic matter level, although the soils are deceptively dark in color
14. A high test for available potassium but usually a low test for available phosphorus and nitrogen
15. A native vegetation usually of tall grasses (prairie) or tall grasses and short trees (savanna)
16. A nonexistent water table and few permanent streams.

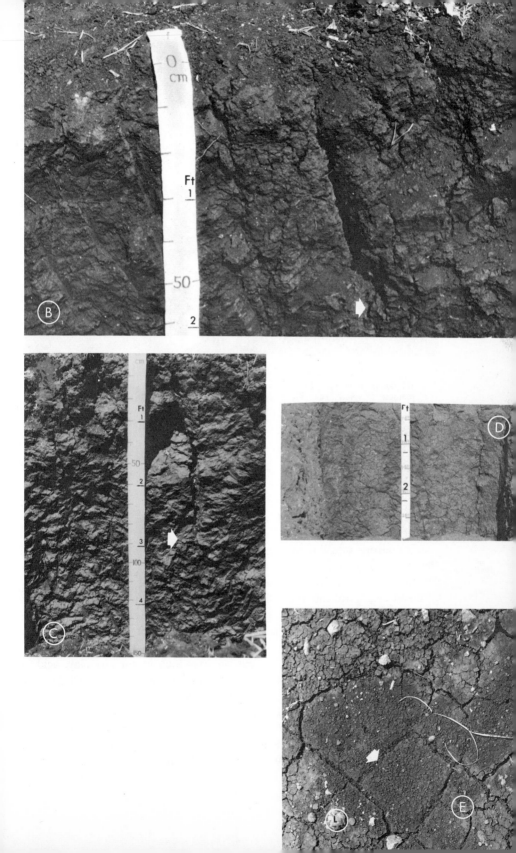

FIG. 12.12. Desirable strong granular soil structure in the Tropics, even at high elevations, is ephemeral. *Left:* Virgin sod soil being disk-plowed for the first time for planting to wheat. Note the strong granular soil peds (natural soil aggregates) and the numerous, permeating grass roots. *Right:* A similar soil nearby that has produced 2 crops of wheat. The desirable strong granular soil structure has deteriorated to a structureless powder that is subject to water and wind erosion. Because of declining yields of wheat, the field will be abandoned to weeds and grass. (25 miles southeast of Narok, Kenya. Annual rainfall is 40 inches; elevation is 9,000 feet; Masai Wheat Scheme.) *Photos by Roy L. Donahue.*

TILLAGE AND SOIL STRUCTURE

Raindrops splash bare, fine-textured soils into flowing mud, which seals the pores and cracks and results in less water entering the soil, more water flowing over the surface and causing soil erosion. When the mud dries, surface soil crusts are induced. This dry crust hinders the emergence of seedlings. Seedling emergence is further hindered in the Tropics because of the variable depth of planting resulting from the usual practice of planting by hand.

Crust formation is most serious on soils high in silt and clay, acid in reaction, low in organic matter, and on soils having a weak structure—the typical tropical soil (Figures 12.12 and 12.13).

Soils under continuous tillage tend to form surface crusts more readily because the natural peds have been destroyed and the percentage of organic matter has decreased. This statement is supported by recent research in tropical Australia by Arndt, who states, "Cultivation increases the tendency

FIG. 12.13. Soils under continuous tillage in the Tropics lose organic matter quickly because of rapid decomposition and because crop residues are often gathered for use as fuel (*left*). The result is a hard soil that is difficult to plow (*right*) and that may have to be beaten with sticks to prepare a desirable seed bed (*bottom*). (India.) *Photos by Roy L. Donahue.*

of soils to form seals and this can be the first serious consequence of continuous cropping." [19]

Almost no research in the Tropics has been conducted on the relationships among the related factors of continuous tillage, soil structure, soil organic matter, surface crust formation, and seedling emergence.

Maintaining desirable soil structure on red lateritic loam in equatorial Africa by *any method of clean tillage* appears to be impossible.[20]

SHIFTING CULTIVATION

The term "shifting cultivation" (also known as "bush fallow") means clearing forest or shrub (bush) land, planting crops for a few years, then abandoning the fields and allowing the trees to reseed the area and sprout again, while a fresh patch of bush is cleared. Any one field would be planted to cultivated crops for a period of 2 to 3 years and then would revert to wild trees, forbs, and grasses for a period of 10 to 20 years or more before it was cropped again.

[19] W. Arndt, *Resistance of Soil to Shoots and Roots,* Rural Research in CSIRO, East Melbourne, Victoria, Australia (Sept., 1966), p. 11.

[20] H. C. Pereira and P. A. Jones, "A Tillage Study in Kenya Coffee, Part 2, The Effects of Tillage Practices on the Structure of the Soil," *Empire Journal of Experimental Agriculture,* Vol. 22, No. 88 (1954), pp. 323–31.

FIG. 12.14. Success of the practice of shifting cultivation depends upon leaving sufficient trees and sprouting stumps to regenerate a forest quickly when the cropping cycle has terminated. *Top left:* Guineacorn in a proper system of shifting cultivation, where some trees are left. *Bottom left:* Not enough trees were left to regenerate and consequently grasses have dominated the vegetation in the bush fallow cycle. Under grasses, the soil becomes hotter and cooler and wetter and drier, and any laterite layer near the surface will harden irreversibly (*top right*). (Tape is in inches.) (Ghana, western Africa.) *Photos by Roy L. Donahue.*

Shifting cultivation is a common practice in the tropical rain forests and savannas of Asia, Africa, and in Central and South America. This cropping system is usually practiced in remote mountainous areas where rainfall is high and consequently the soils are highly leached and not plentifully supplied with essential plant nutrients. The practice is also common on level lands where hand-powered agriculture and abundant land prevail (Figures 12.14 and 12.15).

There are several explanations offered for the adoption of the practice of shifting cultivation:

1. Soils are low in available nutrients and are therefore able to support only a few years of satisfactory crop yields. The facts are that crop yields do decline usually after the second year. A low nutrient supply of one or more essential elements however, may or may not be the cause. Desirable structure of soils in the Tropics also deteriorates rapidly under tillage.
2. Many soils that occur where shifting cultivation is practiced are classified as "Laterites" or "Ground-Water Laterites." One characteristic of Laterites is that when cleared of trees and planted to crops, the soil becomes subjected to a greater fluctuation in heating and cooling and in wetting and drying. In addition, less organic matter is returned to the surface of the soil under field crop culture than under bush fallow. The net result is a *hardening of the Laterite layer in the soil.*

FIG. 12.15. The practice of shifting cultivation is the dominant system of agriculture in Tropical rain forests and savannas in Asia, Africa, and Central and South America. These 3 photographs from Costa Rica, Central America, illustrate: *top:* cutting the trees, but leaving the largest and most valuable ones to regenerate quickly when cropping is abandoned; *center:* burning the trees and brush as a cheap and rapid means of clearing the land for cultivation; *bottom:* cultivating small patches of land until crop yields decline, then abandoning the fields to trees again to rejuvenate soil structure and soil fertility. *Source:* Ernst Griffin, Mich. State U.

3. Weeds increase in numbers each year the land is cultivated, and weed competition may be one cause of decreasing crop yields.
4. Insects and diseases that damage crops increase in numbers and severity each year that a given crop occupies the same field. Allowing the field to "fallow in bush" breaks the cycle of specific insects and diseases.

5. Burning the forest trees and grasses also reduces the number of hiding and breeding places for the mosquito (*Anopheles*) that causes malaria in man, and for the tsetse fly (*Glossina* species) that causes sleeping sickness (*trypanosomiasis*) in animals and man. Burning also reduces the number of hiding places for lions, leopards, tigers, and jaguar that prey on animals and man.
6. Burning the forest encourages grasses to compete more favorably with trees and thus to provide more grasses for grazing by domestic animals.

The true reason or reasons for the traditional practice of shifting cultivation await the research scientists with costly, painstaking, and long-time studies. Even though the causes are not known, the results of the practice can be readily observed, as follows:

1. Shifting cultivation encourages migration among farmers, who must constantly be prepared to move when crop yields decline and a new field must be cleared.
2. The farmers who clear the forests do so with very simple axes and sometimes with inferior saws. The work is very hard and the yield from 2 to 3 years of crops is the only compensation. Rarely do the trees that must be cut have any commercial value because of their small size and the inaccessibility of markets.
3. With so much input of labor and so little output of crops, it is inevitable that the farmers have only enough productivity for a bare subsistence. A low income and a migratory population result in few opportunities, few health facilities, and usually great human misery.
4. Soil erosion is accelerated by the practice of clearing forest trees and planting crops. For it is a fact that forests offer better protection to the soil than any other form of soil cover, and that bare soil and most row crops afford the least protection.
5. Most trees are cut and burned in the system of shifting cultivation. If the forests were not so destroyed, many of the areas with commercial tree species could be managed in large tracts and could be logged scientifically by harvesting selectively over decades by constructing access roads and by establishing saw mills.
6. When *all* trees are cut and sprouts from the tree stumps are not permitted to live during the period of cultivation of field crops, wild grasses dominate the vegetation during the subsequent fallow periods. The result is a degeneration of soil productivity and much lower crop yields. The reasons for lower yields are that grasses re-

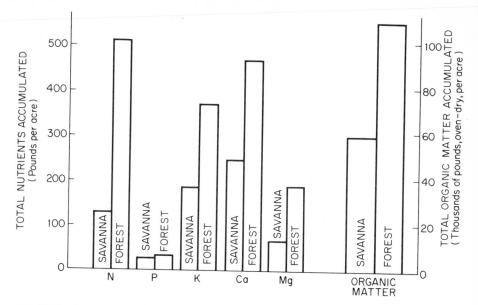

FIG. 12.16. Forest versus savanna: A comparison of total accumulations of organic matter and nutrients over a 20-year period in central Ghana. *Source:* P. H. Nye and D. J. Greenland, "The Soil Under Shifting Cultivation," *Tech. Comm.,* No. 51, Commonwealth Bur. of Soils, Harpenden, Eng., 1960, pp. 24, 25.

turn less organic matter, nitrogen, phosphorus, potassium, calcium, and magnesium to the soil than do trees (Figure 12.16).

SOIL FERTILITY AND FERTILIZER USE

Vine [21] challenges the pessimistic statements in the books of Stamp [22] and of Gourou [23] on the low fertility of soils of the Tropics and the near-hopelessness of making them very productive. Stamp cites excessive leaching, formation of hardened Laterite, abrasive quartz sand grains that grind away plows, and alkali formation in irrigated soils. Gourou suggests that the principal salvation of the Tropics is the production of lowland rice and perennial tree crops, such as cocoa, oil palm, and coffee.

Vine cites several examples in Nigeria to refute the pessimism of Stamp and Gourou, as follows:

1. Areas of rainfall above 80 inches (and, by inference, the areas of highly leached soils) are very small.

---

[21] H. Vine, *Is the Lack of Fertility of Tropical African Soils Exaggerated?* [Leopoldville, Congo (Kinshasa): Second Inter-African Soils Conference, 1954], Document No. 26, pp. 389–409.

[22] L. D. Stamp, *Our Undeveloped World* (London: Faber and Faber, 1953).

[23] P. Gourou, *The Tropical World* (London: Longmans, Ltd., 1953), trans. from French by E. D. Laborde.

2. Laterite has hardened in past geological ages but is not hardening now.[24]
3. Hardened steel can be used in making plows to withstand the abrasive action of sharp quartz sand.
4. Yields of corn and yams can be fairly high if leguminous, annual, green-manure crops, such as *Mucuna deeringiana,* animal manures, or small amounts of chemical fertilizers are used.

Because of soil heterogeneity in the Tropics, it is in reality almost impossible to generalize on statements of soil fertility. However, specific results of fertilizer trials on wheat, barley, and *teff* in Ethiopia are summarized as follows.

During the crop season of 1967–1968, the F.A.O. sponsored 273 field trials on *teff,*[25] wheat, and barley in six provinces in the highlands of Ethiopia (mostly dark clay soils, Vertisols). The results are summarized as follows:

> These overall results confirm the generally deficient nature of the plateau soils of Ethiopia with regard to phosphorus and nitrogen and the presence of adequate potassium.[26]

Phosphorus alone gave the greatest response, based upon the value of crop to cost of fertilizer (V:C) ratio. The next greatest response was nitrogen and phosphorus, and the third was nitrogen only. The use of potassium depressed the yields of all crops.

SOIL NITROGEN

When a dry soil is incubated in a moist condition, nitrates are produced in very large amounts during the initial 5- to 10-day period ("flush"). If the soil is again dried and rewetted, a second but lower grade of flush occurs during each wetting cycle following a drying cycle. The drier the soil and the greater the total amount of soil organic matter, the greater

---

[24] *Note:* The author of this chapter has not found statements from any other soil scientist to agree with this.

[25] Teff: *Eragrostis abyssinica.*

[26] F. A. O., *F. F. H. C. FAO Fertilizer Program,* Imperial Ethiopian Government, Ministry of Agriculture, Extension Service (April, 1968), 6 pp. plus 20 pp. of unpaged yield data, mimeographed.

Note: Where a fertilizer nutrient is indicated, 40 kilograms per hectare of N, P_2O_5, and/or K_2O were applied.

the flush of nitrates upon wetting. This phenomenon is characteristic of soil conditions in the wet-dry Tropics occurring in perhaps 90 per cent of the equatorial regions. It can also be deduced that in the wet-dry Tropics, organic-matter levels reach a very low equilibrium level under continuous cultivation.[27]

The early flush of nitrogen is relevant to plant growth in at least two respects:

1. Crops should be planted as soon as possible after the first rain or irrigation to absorb the flush of available nitrogen before nitrates are leached below root level.
2. Any nitrogenous fertilizer applied to the crop is most efficiently added as a side dressing or top dressing *after* the nitrogen flush and when the plants are several weeks old and the need for nitrogen is high, rather than as a preplant application.
3. Phosphorus is usually needed in tropical soils, and it should be applied preplant because it tends to increase the absorption of nitrogen by the plant and thus reduce leaching losses of early flush nitrogen.

Since most of the Tropics have a wet-dry climatic cycle that is conducive to nitrate production, and since nitrates are not adsorbed by any mechanism in the soil, nitrates move with soil water. Any leaching or runoff losses of water would therefore also carry nitrates away from a field. But nitrates (and soil nitrogen) in soils of the Tropics are not as deficient as many scientists predict. In fact, there is research evidence to show that nitrates move upward toward the surface during the dry season on an East African Kikuyu red loam soil planted to coffee.[28]

Soils of the cultivated fields of the Tropics *gain* nitrogen by additions of chemical fertilizers and animal manures; seeds; biological fixation; rainfall; and from fixed nitrogen in decomposing rocks. Forests and grasslands *gain* nitrogen by biological fixation, rainfall, and from rock decomposition. Soils of cultivated fields *lose* nitrogen by crop removal, leaching, and ero-

---

[27] References: (1) A. M. Kabaara, "Nitrate Studies in a Fertilizer Experiment in Sukumaland, Tanganyika" [now Tanzania], *East African Agricultural and Forestry Journal* (Oct., 1964), pp. 142–48; (2) H. F. Birch, "The Effect of Soil Drying on Humus Decomposition and Nitrogen Availability," *Plant and Soil,* Vol. 10 (1958), p. 9; Vol. 11 (1959), p. 262; Vol. 12 (1960), p. 81; also, "Soil Drying and Soil Fertility," *Tropical Agriculture,* Vol. 37 (Trinidad: 1960), p. 3.

[28] J. B. D. Robinson and P. Gacoka, "Evidence of the Upward Movement of Nitrate During the Dry Season in the Kikuyu Red Loam Coffee Soil," *Journal of Soil Science,* Vol. 13, No. 1 (1962), p. 133.

sion, whereas, forests and grasslands *lose* nitrogen by burning, leaching, and erosion.

Biological fixation of atmospheric nitrogen in the Tropics may be outlined as follows: [29]

    I. Symbiotic Fixation of Nitrogen
        A. Leguminous Bacteria
        B. Nonleguminous Bacteria
        C. Leaf-Gland Algae and Bacteria
            1. Blue-Green Algae
            2. Bacteria
        D. Lichens
    II. Nonsymbiotic Fixation of Nitrogen
        A. Bacteria
        B. Algae
        C. Fungi

In tropical India, a 52-year field study of the nitrogen cycle indicated that the only losses were from crop removal. All nitrogen fertilizer added was accounted for by being present in the soil or by crop removal. In fact, there was a slight *excess of nitrogen, which,* ". . . *May be attributed to bacterial fixation.*" [30]

Phosphorus added as fertilizer was also all accounted for by crop removal and an increase of phosphorus in the soil.

SUMMARY

Soils of the Tropics are more variable than soils in temperate regions. Soils on old uplands in the humid Tropics are more deeply and more intensely weathered and have a lower cation exchange but a higher anion exchange capacity, a lower buffer capacity, a lower available water capacity, a larger percentage of kaolinite but a smaller percentage of montmorillonite, a higher percentage of iron and aluminum oxides, a higher degree of friability, and a lower level of fertility.

---

[29] *Principal reference:* D. O. Norris, "The Biology of Nitrogen Fixation," in *A Review of Nitrogen in the Tropics with Particular Reference to Pastures—A Symposium,* Bulletin 46, Commonwealth Agricultural Bureaux (1962), pp. 113–29.

[30] A. Mariakulandai and S. R. Thyagarajan, "Longterm Manurial Experiments at Coimbatore," *Journal of the Indian Society of Soil Science,* Vol. 7, No. 4 (1959), p. 271. *Note:* Italics are those of Roy L. Donahue.

One of the most unique and troublesome characteristics of some soils of the Tropics is the presence of a Laterite layer that hardens upon continuous exposure to the sun and rain. Shifting cultivation (bush fallow) is one adaptation to reduce the hazard of hardening.

Desirable soil structure deteriorates very rapidly in the Tropics. Shifting cultivation has so far been the only practical solution to maintaining good soil structure. But unlike temperate regions, in the Tropics, forests are more effective than grasslands or grass-trees (savanna) in maintaining desirable soil structure as well as the level of soil productivity.

QUESTIONS

1. Soils on old uplands in the Tropics are highly weathered. Does this mean that they are useless for agriculture?
2. Soils on young land surfaces in the Tropics may be *more* productive than some of the best in the Temperate Region. Explain.
3. What is one practical solution to the problem of preventing the hardening of Laterite?
4. Briefly describe the practice of shifting cultivation.
5. Soils on old land surfaces in the Tropics are very highly leached. Why is nitrogen not always the first limiting factor in plant growth?

REFERENCES

Arakeri, H. R., G. V. Chalam, and P. Satyanarayana, in collaboration with Roy L. Donahue, *Soil Management in India* (2nd ed.). Bombay, India: Asia Publishing House, 1962, 609 pp.

Buringh, P., *Introduction to the Study of Soils in Tropical and Subtropical Regions.* Wageningen, Netherlands: Centre for Agricultural Publishing and Documentation, 1968, 118 pp.

Chinzei, T., K. Oya, and Z. Koja, in collaboration with Roy L. Donahue and John C. Shickluna, *Soils and Land Use in the Ryukyu Islands.* Naha, Okinawa: College of Agriculture, University of the Ryukyus, 1967, 187 pp.

Donahue, Roy L., *Estimates of Fertilizer Consumption in India in 1970–71.* New Delhi, India: The Fertilizer Asssociation of India, 1966, 140 pp.

Donahue, Roy L., *Soils of Equatorial Africa and Their Relevance to Rational Agricultural Development.* Research Report No. 7, Institute of International Agriculture, Michigan State University, 1970, 52 pp.

Donahue, Roy L., *Soils of Ethiopia.* Addis Ababa, Ethiopia: Institute of Agricultural Research, in press, 1971.

Fittkau, E. J., J. Illies, H. Klinge, G. H. Schwabe, and H. Sioli (eds.), *Biogeography and Ecology in South America*, Vol. I. The Hague: Dr. W. Junk N. V., Publishers, 1968, 445 pp.

Fitts, James Walter, *The Fertilizer Requirements of Countries in Latin America.* Raleigh, N.C.: International Soil Testing, North Carolina State University (Dec., 1968), 73 pp.

Kline, C. K., D. A. G. Green, Roy L. Donahue, and B. A. Stout, *Agricultural Mechanization in Equatorial Africa.* Institute of International Agriculture, Research Report No. 6, Michigan State University (Dec., 1969), 648 pp.

Kumar, L. S. S., A. C. Aggarwala, H. R. Arakeri, M. G. Kamath, Earl N. Moore, and Roy L. Donahue, *Agriculture in India, Vol. I, General.* Bombay, India: Asia Publishing House, 1963, 252 pp.

Mohr, E. C. J., and F. A. van Baren, *Tropical Soils.* New York: Interscience Publishers, 1954, 498 pp.

Moss, R. P. (ed.), *The Soil Resources of Tropical Africa.* London: Cambridge University Press, 1968, 226 pp.

Muhr, Gilbert R., N. P. Datta, H. Sankarasubramoney, V. K. Leley, and Roy L. Donahue, *Soil Testing in India* (2nd ed.). New Delhi, India: United States Agency for International Development, 1965, 120 pp.

Sherman, G. Donald, *Gibbsite-Rich Soils of the Hawaiian Islands.* University of Hawaii Agricultural Experiment Station Bulletin 116, 1958, 23 pp.

Tamhane, R. V., D. P. Motiramani, and Y. P. Bali, in collaboration with Roy L. Donahue, *Soils: Their Chemistry and Fertility in Tropical Asia* (3rd printing). New Delhi, India: Prentice-Hall of India (Private), Ltd., 1970, 475 pp.

Tropical Abstracts. Amsterdam, Netherlands: Department of Agricultural Research, Royal Tropical Institute, monthly publication.

PART **TWO**

**APPLICATIONS OF SOIL SCIENCE
TO PLANT GROWTH**

CHAPTER 13

LIME CHARACTERISTICS AND USE

In 45 A.D., *Columella, a Roman philosopher, used lime to increase the productivity of the soil for plants.*

Although the use of wood ashes, burned limestone, and marl (unconsolidated calcium carbonate) on the soil to increase plant growth is an ancient practice, Edmund Ruffin, a Virginia farmer-scientist, from 1825 to 1845, may have been the first person to apply lime on the soil to correct a condition which he said was soil acidity.

CONSUMPTION OF LIME

The use of ground limestone on agricultural crops in the United States has become a popular practice on acid soils in humid regions. In 1968, a total of 30,535,501 tons of limestone was reported as being used in 41 states. Tonnages used varied from 1500 in Idaho to more than 4 million in Illinois. Other states consuming more than 1 million tons in 1968 were Missouri, Iowa, Kentucky, Tennessee, and Ohio, in decreasing order. Only eight states reported that their farmers in 1968 had used 50 per cent or more of the optimum amount (Figure 13.1 and Table 13.1).

WHY SOILS ARE ACID

Soils are acid for one or more of the following reasons:

1. Some soils have developed from parent materials which are acid.

277

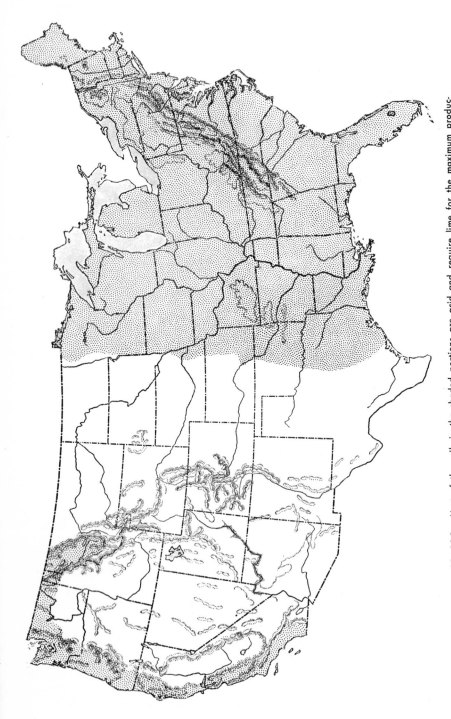

FIG. 13.1. Most of the soils in the shaded portions are acid and require lime for the maximum production of crops such as alfalfa, barley, and sweet clover. However, a soil test should be made before lime is applied. Source: About Agricultural Limestone, The National Limestone Institute.

TABLE 13.1 LIMESTONE USED BY FARMERS IN 1968 AND THE PERCENTAGE
USED OF THE OPTIMUM AMOUNT NEEDED *

State	Used in 1968 (Short Tons)	Percentage Used of Optimum Amount Needed (%)
Alabama	765,000	58
Arkansas	627,827	50
California	47,906	31
Connecticut	35,965	20
Delaware	81,296	57
Florida	834,610	63
Georgia	993,693	42
Idaho	1,500	†
Illinois	4,015,000	50
Indiana	1,299,143	29
Iowa	3,484,158	45
Kansas	590,620	25
Kentucky	1,971,505	42
Louisiana	285,377	31
Maine	127,015	30
Maryland	270,000	40
Massachusetts	49,091	18
Michigan	456,307	12
Minnesota	343,567	10
Mississippi	616,800	30
Missouri	4,070,216	41
Nebraska	143,460	11
New Hampshire	26,500	11
New Jersey	198,758	55
New York	678,525	28
North Carolina	964,711	41
Ohio	1,675,128	34
Oklahoma	281,880	15
Oregon	61,440	12
Pennsylvania	972,257	28
Rhode Island	14,867	55
South Carolina	410,000	24
South Dakota	4,200	†
Tennessee	1,711,071	65
Texas	248,303	13
Vermont	80,138	22
Virginia	805,361	30
Washington	66,997	25
West Virginia	242,381	44
Wisconsin	979,928	22
Wyoming	3,000	33
Total	30,535,501	35

* Source: National Limestone Institute, Inc., Washington, D.C.
† Insufficient data to calculate.

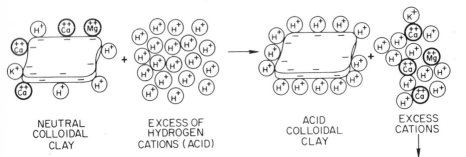

NEUTRAL
COLLOIDAL
CLAY

EXCESS OF
HYDROGEN
CATIONS (ACID)

ACID
COLLOIDAL
CLAY

EXCESS
CATIONS

FIG. 13.2. Soils that are permeable and that receive more than 40 inches of rainfall are liable to become acid because of leaching of calcium (lime) and other bases. In this illustration, hydrogen in the soil replaces calcium and magnesium (which leach out) leaving a hydrogen-saturated clay which is acid.

FIG. 13.3. Soils are acid partly because acid nitrogen fertilizers are used on them. The center plot received 200 pounds per acre of ammonium sulfate per year for 21 years. (Alabama.) (Compare with Fig. 13.4.) Source: Howard T. Rogers, Ala. Agr. Exp. Sta.

FIG. 13.4. The center plot received 200 pounds per acre per year of ammonium sulfate for 21 years, the same as in Fig. 13.3, but, in addition, it received 230 pounds of limestone per acre per year for the same period. (Alabama.) (Compare with Fig. 13.3.) Source: Howard T. Rogers, Ala. Agr. Exp. Sta.

2. Rain leaches lime downward, carrying some of it out of reach of plant roots (Figure 13.2).
3. Plant roots secrete hydrogen ions, which help to make soils more acid. ,
4. Most nitrogen carriers are acid and their use makes the soil more acid (Figures 13.3 and 13.4).
5. Sulfur is an ingredient of some fungicides, and its use creates acid conditions.

There are two forms of hydrogen in the soil that cause soil acidity, *active* and *exchangeable*. Active hydrogen is measured in pH units and exchangeable hydrogen is determined by measuring the amount of hydrogen that is held on the surfaces of clay and humus particles. Exchangeable aluminum that occurs in some soils is also a source of soil acidity.

When lime is added to a soil, both active and exchangeable acidity are partially neutralized. For this reason, the soil pH is not as good an indicator of lime requirement as a test for exchangeable hydrogen and aluminum.

All states have a soil testing service where the lime requirement of the soil for specific crops can be made; this service should be used.

ACID SOILS NOT PRODUCTIVE

Most crop plants do not yield their maximum potential on soils that are strongly acid. Principal exceptions are blueberries, cranberries, watermelons, and white potatoes. By contrast, alfalfa and sweet clover yield their maximum only when the soils are nearly neutral to slightly alkaline.

On strongly acid soils, the majority of crop plants produce yields less than their potential for one or more of the following reasons:

1. Aluminum toxicity
2. Manganese toxicity
3. Calcium deficiency
4. Magnesium deficiency
5. Molybdenum deficiency.

WHAT LIME DOES IN THE SOIL

Strongly acid soils are not productive for most crops. To increase the productivity of acid soils, the addition of lime is essential for these reasons:

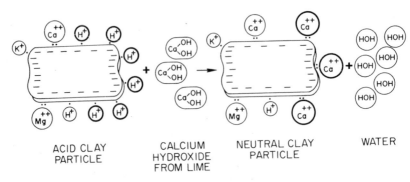

ACID CLAY	CALCIUM	NEUTRAL CLAY	WATER
PARTICLE	HYDROXIDE	PARTICLE	
	FROM LIME		

FIG. 13.5. Lime adds calcium to acid soils. Calcium is an essential nutrient for the growth of all higher plants.

1. Lime monitors the physiological balance of plant nutrients in the soil.
2. Lime adds essential calcium to acid soils for greater plant growth (Figure 13.5).
3. Lime makes phosphorus more available. In acid soils, iron and aluminum phosphates are relatively insoluble. Liming reduces the solubility of the iron and aluminum, and therefore less phosphorus is held in these slowly soluble and relatively unavailable forms.
4. Lime makes potassium more efficient in plant nutrition. When it is plentiful, all plants absorb more potassium than they need. Lime reduces the excessive uptake of potassium. Nutritionally and economically, this is a sound practice. When lime is abundant, plants take up more calcium and less potassium. Since calcium is often deficient in animal rations and potassium is in excess, it is desirable to increase the percentage of calcium in the plant. Economically the practice of liming is desirable because the plant absorbs more cheap calcium and less of the expensive potassium.
5. Lime increases the availability of nitrogen by hastening the decomposition of organic matter.
6. Lime furnishes calcium and magnesium (if the lime is dolomitic) for plant nutrition. These are 2 of the 16 elements essential for plant growth.
7. Beneficial soil bacteria are encouraged by adequate supplies of lime in the soil.
8. Harmful aluminum, manganese, and iron are rendered less soluble and harmless when a soil is well supplied with lime.
9. Over a period of years, a good liming program improves the physical condition of the soil by decreasing its bulk density, increasing its infiltration capacity, and increasing its rate of percola-

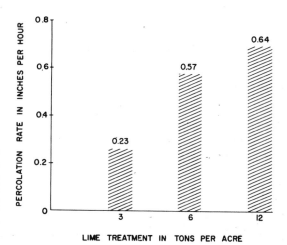

FIG. 13.6. The relationship between the amount of lime applied and the rate of percolation. *Source:* C. A. Van Doren and A. A. Klingebiel, "Effect of Management on Soil Permeability," *Soil Science Society of America Proceedings*, 1952.

tion of water. Figure 13.6 presents information showing that the more lime applied, the greater the rate of percolation.

10. There is less soil erosion following an adequate liming program. This result is due primarily to the increased vigor and density of plants following the application of lime.

11. Lime reduces the loss of nitrogen from soils. Research in New Jersey over a 40-year period demonstrated that soil adequately limed lost 22 per cent *less* nitrogen than comparable unlimed plots.

12. Liming acid soils in a California experiment reduced to about one-fifth the amount of Strontium-90 taken up by plants. Strontium-90 is one of the radioactive by-products of nuclear fission that is harmful to the human body.

CROP RESPONSE TO LIME

On mineral soils below pH 5.0, most crops respond to the judicious use of lime, but crop response is not uniform. Figure 13.7 indicates that

FIG. 13.7. Oat yields are increased as a result of the use of lime (as measured by soil pH), but corn responds even more. *Source:* Iowa State U. Pamphlet No. 315, 1965.

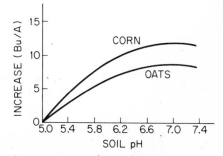

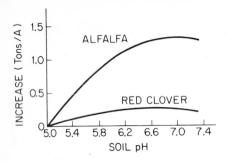

FIG. 13.8. Red clover yields are sluggish in response to a change in soil pH caused by increasing increments of lime, but alfalfa response is sharp. *Source:* Iowa State U. Pamphlet No. 315, 1965.

both corn and oats are increased in yield, but at a differential rate, when a soil with a pH of 5 is limed to about 6.5. Above 6.5, both crops barely respond. Red clover (Figure 13.8) shows only a slight response to lime at any pH between 5.0 and 7.4, whereas alfalfa yields are increased greatly from pH 5.0 to pH 7.0.[1]

Similar data are shown for Kansas in Figure 13.9, which compares the yields of limed and unlimed wheat, oats, corn, and alfalfa. Of these crops, alfalfa shows the greatest percentage increase in yield as a result of liming.

[1] *Primary source:* R. D. Voss, J. J. Hanway, J. T. Pesek, and L. C. Dumenil, *A New Approach to Liming Acid Soils,* Iowa State University Pamphlet 315 (March, 1965), 11 pp.

FIG. 13.9. The effect of lime on the yield of four crops in Kansas. *Source:* Roscoe Ellis, Jr., *Liming Kansas Soils,* Kans. Agr. Exp. Sta. Cir. No. 313, rev. 1958.

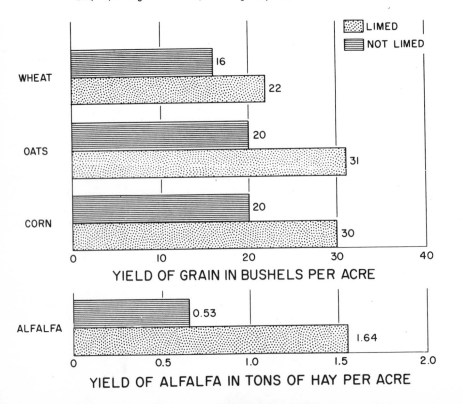

FIG. 13.10. Lime on the peanuts on the left increased the grade by two points and the yield by 100 pounds per acre. Gypsum gave a similar response. (Twenty-inch rainfall belt in Central Texas.) *Source:* N. H. Hunt, County Agricultural Agent, Pearsall, Texas.

On adequately fertilized soils in Maryland, the proper amount of lime applied on a soil with a pH of 5.6 gave increased corn yields valued at $3.95 for each dollar spent on lime. Soybeans on a soil with a pH of 6.0 gave a return of $6.16, and alfalfa on a soil with a pH of 6.4 gave a return of $8.85 for each dollar spent on lime.

The total dry matter in crops produced over a 40-year period in New Jersey was increased by 35 per cent as a result of the use of lime.[2]

Both yield and quality of peanuts were increased with the use of lime in the 20-inch rainfall belt of central Texas (Figure 13.10).

LIMING MATERIALS

More than 90 per cent of agricultural lime is calcium carbonate; some is calcium and magnesium carbonate, and a much smaller quantity is calcium oxide or calcium hydroxide. To a chemist, lime is calcium oxide, but to a farmer and an agronomist or soil scientist, lime means calcium carbonate equivalent.

The common liming materials are:

1. Calcic limestone ($CaCO_3$), which is ground limestone.
2. Dolomitic limestone [$CaMg(CO_3)_2$], from ground limestone high in magnesium.
3. Quicklime (CaO), which is burned limestone.
4. Hydrated (slaked) lime [$Ca(OH)_2$], from quicklime that has changed to the hydroxide form as a result of reactions with water.
5. Marl ($CaCO_3$), from the bottom of small ponds in areas where the soils are high in lime. The lime has accumulated by precipitation from drainage waters high in lime. Some marls contain many shell remains from marine animals.
6. Chalk ($CaCO_3$), resulting from soft limestone.

---

[2] A. L. Prince, S. J. Toth, A. W. Blair, and F. E. Bear, "Forty-Year Studies of Nitrogen Fertilizers," *Soil Science,* Vol. 52, No. 4 (Oct., 1941).

7. Blast-furnace slag ($CaSiO_3$ and Ca_2SiO_4), a by-product of the iron industry. Some slags contain phosphorus and a mixture of CaO and Ca $(OH)_2$. This product is called *basic slag* and is used primarily for its phosphorus content.

8. Miscellaneous sources, such as ground oystershell, wood ashes, and by-product lime resulting from papermills, sugar beet plants, tanneries, and water-softening plants.

Gypsum ($CaSO_4$) is sometimes added to the soil to supply calcium, but it has no influence on soil pH and therefore is not considered to be a liming material.

All the liming materials mentioned have value for supplying either calcium or both calcium and magnesium, raising the pH and making aluminum, manganese, and iron less toxic. The choice of a particular liming material is determined by the cost in relation to its purity, the ease of handling, and the speed with which the lime reacts in the soil.

CHEMICAL GUARANTEES OF LIME

There are several methods of expressing the chemical guarantee of lime. The most common ones are the following four.

1. Calcium carbonate equivalent, sometimes known as the *neutralizing power*. If chemically pure calcium carbonate were present in a lime, the calcium carbonate equivalent would be 100. If all of the lime were in the calcium carbonate form but it was 95 per cent pure, the calcium carbonate equivalent would be 95. Other forms of lime can be converted to the calcium carbonate equivalent by the use of atomic and molecular weights. One example will be shown. Assume that it is desired to calculate the calcium carbonate equivalent of chemically pure calcium oxide, CaO.

$$\% \ CaCO_3 \ \text{equivalent} = \frac{\text{Molecular wt. of } CaCO_3}{\text{Molecular wt. of CaO}} \times 100 =$$

$$\frac{100}{56} \times 100 = 178.6$$

The calcium carbonate equivalent of any quantity of CaO can be obtained by multiplying the pounds of pure CaO by 178.6 per cent.

2. Calcium oxide equivalent. This form of chemical guarantee is obtained also by the use of molecular weights. For example, if pure calcium carbonate were converted to its calcium oxide equivalent, the calculations would be:

$$\% \text{ CaO equivalent} = \frac{\text{Molecular wt. of CaO}}{\text{Molecular wt. of CaCO}_3} \times 100 =$$

$$\frac{56}{100} \times 100 = 56.0$$

To obtain the CaO equivalent of magnesium carbonate, the calculations are:

$$\% \text{ CaO equivalent} = \frac{\text{Molecular wt. of CaO}}{\text{Molecular wt. of MgCO}_3} \times 100 =$$

$$\frac{56}{84} \times 100 = 66.7$$

3. Conventional oxides. This form of lime guarantee consists of converting the calcium to calcium oxide, the magnesium to magnesium oxide, and combining the two.

4. Elemental percentage of calcium and/or magnesium. This method of expressing lime guarantees is determined in a similar way. If pure $CaCO_3$ were to be reported as elemental calcium, the calculations would be:

$$\% \text{ Ca} = \frac{\text{Atomic wt. of Ca}}{\text{Molecular wt. of CaCO}_3} \times 100 =$$

$$\frac{40}{100} \times 100 = 40$$

PHYSICAL GUARANTEES OF LIME

The chemical activity of liming material is determined by the solubility of the chemical compounds in the lime. For example, calcium oxide is more soluble than calcium carbonate, whereas calcic limestone is more soluble than dolomitic limestone. Calcium silicate is the least soluble of the liming materials.

It is obvious that the finer the lime particles, the faster they react in the soil to become available to plants (Figure 13.11).

There are no national laws governing the physical guarantee of lime, this regulation being left to the states. Thirty-nine states have lime laws, which are difficult to summarize. In general, the physical guarantees of lime in the respective states may be roughly averaged in this way: 85 per cent must pass through a 15-mesh sieve, and 30 per cent must pass through a 100-mesh sieve.

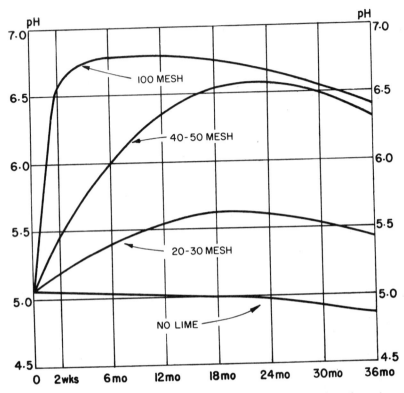

FIG. 13.11. The finer the limestone, the more quickly it reacts with the soil to raise the pH and the quicker the calcium becomes available to the plant. However, since plant roots grow toward and into coarse particles of limestone, a satisfactory fineness is: "All through an 8-mesh sieve, ¼ through a 100-mesh sieve (with all fines left in)." *Source:* National Plant Food Institute and Purdue U.

LIME REQUIREMENT OF CROPS

To arrive at a satisfactory solution to the problem of how much lime to apply, the requirement of the crop is a good starting place. Since soil acidity and lime level have a fairly good correlation in humid regions, the pH of the soil is used as an index of the lime needs of various crops.

The relative lime requirement of selected crops can be studied from Table 13.2, which indicates that alfalfa, barley, cotton, sugar beets, and sweet clover have the highest; corn, tobacco, and wheat have a medium requirement; buckwheat, potatoes, rice, and rye have a low requirement; and blueberries and cranberries have the lowest requirement.

TABLE 13.2 RELATIVE LIME REQUIREMENT OF SELECTED CROPS *

High Lime Requirement	Low Lime Requirement
Alfalfa	Buckwheat
Asparagus	Oats
Barley	Potatoes
Beans	Raspberries
Cotton	Rice
Peas	Rye
Red clover	Strawberries
Soy beans	Vetch
Sugar beets	
Sunflower	Very Low Lime Requirement
Sweet clover	
	Blueberries
Medium Lime Requirement	Cranberries
Corn	
Grain sorghum	
Grasses (most)	
Peanut	
Sweet potato	
Tobacco	
Trefoil	
Wheat	

* Primary source: E. C. Doll, Lime for Michigan Soils, Michigan State University Extension Bulletin 471 (1966).

EFFECT OF LIME ON SOIL pH

After the pH requirement of the crop is known, it is then necessary to test the pH of the soil. The soil texture and humus content must also be estimated to arrive at the relative buffer capacity.

The relationships between texture and the buffer capacity (resistance to a change in pH) are shown in Figure 13.12. The more clay and organic matter there is in a soil, the more limestone is needed to change the pH. If soils are fairly high in organic matter, as they are in the northern states, it will take more lime to accomplish the same change in pH. Conversely, in southern soils, which are lower in organic matter, less lime will be needed to accomplish the same pH change.

There is also a variation in lime needs to change the pH, depending upon the type of clay present as well as the range of the pH change desired. As the pH change is closer to 7.0, the amount of lime required to effect

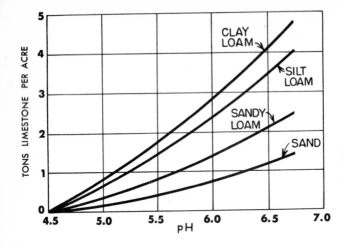

FIG. 13.12. Approximate tons of limestone required to raise the pH of a 7-inch layer of soil. *Source: Liming Soils, An Aid to Better Farming,* U.S. Dept. of Agr. Farmers' Bul. No. 2124, 1959.

the same pH change is greater. For example, in a silt loam in Figure 13.12 it takes 30 per cent more lime to change the soil pH from 5.5 to 6.5 than it does to change it from 4.5 to 5.5.

SOIL pH AND NUTRIENT AVAILABILITY

The general relationship between soil pH and plant nutrient availability is shown in Figure 13.13. From this chart, it is obvious that the primary nutrients—nitrogen, phosphorus, and potassium—as well as the secondary nutrients—sulfur, calcium, and magnesium—are as available or more available at a pH of 6.5 than at any other pH. Molybdenum, manganese, and boron availability is also similar to that of the primary and secondary nutrients.

The minor elements—iron, manganese, boron, copper, chlorine, and zinc—are less available at a pH of 6.5 than at more acid reactions. In general, however, minor elements are sufficiently available so that plant growth is not limited at a pH of 6.5. A pH from 6.5 to 7.0 is therefore considered to be the pH range in which most nutrients are desirably available to plants. It can also be concluded that at this soil pH range, commercial fertilizers are most readily available.

AMOUNTS OF LIME TO APPLY

When the lime requirement has been determined for a particular crop on a particular soil, the amount of lime to apply can be calculated only after knowing the fineness and neutralizing value of the limestone or the marl or refuse lime to be used. Table 13.3 was prepared to assist in this determination.

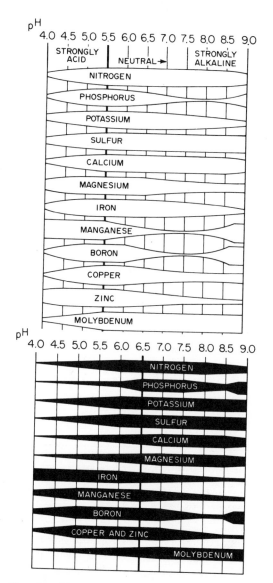

FIG. 13.13. The relationship between soil pH and relative plant nutrient availability (the wider the bar the more the availability). *Above:* fibric, hemic, and sapric soils (peat and muck). Note that the pH for greatest availability for most nutrients is about 5.5. *Below:* mineral soils. Note that the pH for greatest availability for almost all nutrients is about 6.5, which is one pH unit higher than for organic soils. Maintaining pH of 5.5 for organic soils and pH of 6.5 for mineral soils does not, however, guarantee adequate plant nutrients, because the total supply of one or more nutrients may be low. *Sources: Above,* R. E. Lucas and J. F. Davis, Dept. of Soil Science, Mich. State U.; *below,* National Plant Food Institute, Wash., D.C.

TABLE 13.3   AGRICULTURAL LIME CONVERSION *

Limestone

Grade Name	Per Cent Passing Through 100-Mesh Sieve	Neutralizing Value	Factor
Superfine	80+	90+	0.80
Pulverized	60–79	90+	0.85
Agricultural ground	40–59	90+	1.00
Agricultural ground	40–59	80–89	1.1
Agricultural ground	40–59	70–79	1.3
Fine meal	30–39	90+	1.2
Fine meal	30–39	80–89	1.35
Fine meal	30–39	70–79	1.55
Coarse meal	20–29	90+	1.3
Coarse meal	20–29	80–89	1.45
Coarse meal	20–29	70–79	1.65

Marl and Refuse Lime

Pounds of $CaCO_3$ Equivalent per Cubic Yard	Factor
800–999	2.5
1000–1199	2.0
1200–1399	1.7
1400–1599	1.5
1600+	1.25

* Multiply the "lime requirement" in tons per acre by the "factor" for the liming material being used. Note: "Lime requirement" is an expression of the tons of limestone having a neutralizing value of 90 per cent or higher with 40 to 59 per cent passing through a 100-mesh sieve, that is required to raise the pH of soil to 6.6.

If the soil testing laboratory reported a lime requirement of 4 tons per acre and the reader has ready access to "fine meal" limestone (column 1, line 6) with 30 to 39 per cent passing through a 100-mesh sieve and a neutralizing value of 90+, multiply 4 tons by 1.2, getting 4.8 tons. Since it is difficult to spread exactly 4.8 tons per acre, it is better practice to plan to use 5 tons per acre.

With the same recommended lime requirement of 4 tons per acre, and if marl is readily available with a purity of 1000 to 1199 pounds of $CaCO_3$ equivalent per cubic yard (bottom part of Table 13.3, column 1, line 2), multiply 4 by 2.0, getting 8.0 cubic yards of marl.

METHODS OF APPLYING LIME

The most efficient way to use lime is to apply small amounts every year or every 2 years. But this liming program increases the cost of application. The usual liming practice consists of a compromise between what is most effective and what is the cheapest per ton of lime applied. Lime can be applied to advantage at any stage in the cropping system, but normally it is best applied 6 to 12 months in advance of seeding a legume.

The rate of lime application should always be determined by means of soil testing. Applying 2 tons per acre on a field that needs 4 tons is short-sighted economy, since there may be little or no return on a considerable cash outlay for legume seed, fertilizer, and lime. Similarly, liming a field that needs no lime will give no benefit at all and may be injurious.

It is desirable that newly spread lime be well mixed with the whole plow layer. On strongly acid soils, where 3 to 6 tons or more per acre of lime are required, it is recommended that one-half the dose be applied before plowing, and the other half be applied and disked in after plowing.

When not more than 2 tons per acre are needed, the entire amount can be applied and disked in after plowing and before seeding the legume or legume-grass mixture.

The mechanics of getting lime on the field has changed in recent years. In early American days, farmers spread lime on the land with a shovel from the back of a wagon. Then crude spreaders were devised. Later, commercial hopper-type spreaders pulled by a tractor were available. Now the most common method of spreading lime is to have it spread by a truck with a specially built, V-shaped bed and a spreading mechanism in the rear (Figures 13.14 and 13.15).

Where equipment for spreading is not readily available, it is satisfactory to put lime in the gutters behind the dairy cows or to spread lime on the loaded manure spreader. Spreading the manure automatically spreads the lime with it.

FIG. 13.14. The most accurate method of spreading lime is with the use of a tractor-drawn hopper spreader. Source: N.H. Agr. Exp. Sta.

FIG. 13.15. Spreading lime or phosphate can be done very rapidly with this V-shaped bed mounted on a truck. *Source: Texas Agricultural Limestone Association.*

OVERLIMING

When excessively large amounts of lime are applied to sandy soils low in humus, injury to plant growth sometimes occurs. Injury to plant growth may be due to any one or a combination of these causes:

1. Boron deficiency
2. Iron, manganese, copper, or zinc deficiency
3. Phosphorus availability may be reduced to a critically low level.

Overliming injury may be reduced by the application of large amounts of manure, green-manure crops, compost, phosphorus fertilizers, boron, or a mixture of minor elements. Overliming injury, however, is not very common.

THE LIME BALANCE SHEET

Lime is lost from the soil by:

1. Leaching
2. Removal by harvested crops and other products sold, such as milk
3. Erosion
4. Neutralization by acid-forming fertilizers.

Some of the rainfall percolates downward in the soil, carrying lime with it. On the average, approximately 100 pounds per acre per year of calcium carbonate is lost in this way. Harvested crops sold and products sold, such as milk, remove another 100 pounds of lime. Erosion may take another 100 pounds per acre per year. Finally, acid fertilizers neutralize approximately 100 pounds of lime an acre each year.

This loss totals 400 pounds on many humid crop-acres each year.

Simply to add enough lime to maintain the lime level in the soil would require the application of 1 ton of lime every 5 years. An adequate liming program on such soils involves the use of sufficient lime to raise the pH of the soil to the desired level for the crop to be grown and the application of 1 ton each 5 years for maintenance.

SUMMARY

On mineral soils, nearly all farm crops respond to lime applications when the soil pH is below 6.5. On organic soils, the corresponding threshold is pH 5.5.

The common liming materials are ground limestone, ground dolomitic (high-magnesium) limestone, burned lime, and hydrated lime. In some areas, marl, wood ashes, and oystershell are important sources of liming materials.

Chemical guarantees include the calcium carbonate equivalent, the calcium oxide equivalent, conventional oxides, and elemental guarantees.

A common physical guarantee for lime is 85 per cent through a 15-mesh sieve, and 30 per cent through a 100-mesh sieve. Alfalfa, sweetclover, barley, and sugar beets all have a high lime requirement.

Large amounts of lime should be applied in two applications: half before plowing and the other half disked in after plowing. Small applications can be disked into the surface of the soil after plowing. Overliming injury has been highly overemphasized. Maintenance rations of lime average approximately 1 ton per acre every 5 years.

QUESTIONS

1. Name four causes of acid soils.
2. Give five benefits to be dirived from lime when it is properly applied.
3. What kinds of crops respond best to liming?
4. Calculate the calcium carbonate equivalent of 50 pounds of pure calcium oxide.
5. Approximately how much limestone will it take to raise the pH from 5.5 to 6.5 in a sandy loam?

REFERENCES

Berger, Kermit C., *Introductory Soils.* The Macmillan Company, 1965.

Defense Against Radioactive Fallout on the Farm, Farmers' Bulletin No. 2107. United States Department of Agriculture, 1964.

Doll, E. C., *Lime for Michigan Soils.* Extension Bulletin 471, Michigan State University, 1966.

Green, R. E., *Evaluation of Liming Materials Derived From A Calcareous Beach Deposit in Hawaii,* Technical Bulletin No. 65. Hawaii Agriculture Experiment Station, University of Hawaii (Dec., 1968).

Hawaii Farm Science, Agricultural Progress Quarterly (special issue on liming). Hawaii Agriculture Experiment Station, College of Tropical Agriculture, University of Hawaii, Vol. 13, No. 3 (Aug., 1964).

Kamrath, E. J., *Soil Acidity and Response to Liming.* Bulletin 4 International Soil Testing Service, North Carolina State University Agricultural Experiment Station Technical Bulletin 4 (Oct., 1967), 20 pp.

Lucas, R. E., and J. F. Davis, "Relationships Between pH Values of Organic Soils and Availability of 12 Plant Nutrients," *Soil Science.* Vol. 92, No. 3 (Sept., 1961).

Pearson, R. W., and Fred Adams (eds.), *Soil Acidity and Liming,* Agronomy Monograph Series No. 12. Madison, Wisc.: American Society of Agronomy, 1967.

Shickluna, J. C., B. Watson, and E. P. Whiteside, "Plants in Calcium Soils Less Susceptible to Fallout," *Crops and Soils.* American Society of Agronomy (Mar., 1965), pp. 28, 29.

Shoemaker, Harold, "Fit Lime to Plowing Depth," *Better Crops with Plant Food Magazine.* American Potash Institute, 1964.

Voss, R. D., J. J. Hanway, J. T. Pesek, and L. C. Dumenil, *A New Approach to Liming Acid Soils.* Iowa State University Pamphlet 315 (Mar., 1965), 11 pp.

Way, Winston A., *The Whys and Hows of Liming.* University of Vermont Extension Service Brieflet 997, 1968.

Whitaker, Colin W., M. S. Anderson, and R. F. Reitemeier, *Liming Soils— An Aid to Better Farming,* Farmers' Bulletin No. 2124. United States Department of Agriculture, 1964.

Wolcott, A. R., "The Acidifying Effects of Nitrogen Carriers," *Agricultural Ammonia News.* Agricultural Ammonia Institute (July–Aug., 1964).

Wolcott, A. R., H. D. Foth, J. F. Davis, and J. C. Shickluna, "Nitrogen Carriers: I. Soil Effects," *Soil Science Society of America Proceedings.* Vol. 29, No. 4 (July–Aug., 1964), pp. 405–10.

Woodruff, C. M., "Crop Response to Lime in the Midwestern United States," in, Robert W. Pearson and Fred Adams (eds.), *Soil Acidity and Liming.* Madison, Wisc.: American Society of Agronomy, 1967, pp. 207–31.

CHAPTER 14

FERTILIZER CHARACTERISTICS

In 1842 John Bennet Lawes of England treated bones with sulfuric acid and patented the product under the name of "superphosphate."

Not many years ago the fertilizer business was a scavenger industry, dependent upon such packinghouse by-products as dried blood, tankage, and bone meal, and upon animal manures. Certain plant materials like cottonseed meal also contributed to the supply of commercial fertilizers. Now the fertilizer industry is a complex chemical business, the largest of the heavy chemical industries of the world. But not without good reason.

In 1850, the first chemical fertilizer in the United States was made with a mixture of guano (bird manure) and potash salts at Baltimore, Maryland. One hundred years later, the commercial fertilizer-manufacturing plants in our nation were valued at 1 billion dollars and were selling each year 900 million dollars' worth of fertilizer.

The scientific use of commercial fertilizers by the farmers has made it possible to feed a hungry world, reduce the cost of production, and reduce the amount of labor required to produce a bushel of corn or a peck of potatoes. More than that, the proper use of fertilizers makes it possible, on the average, for the farmer to get 3 dollars back in increased production for each dollar invested in fertilizer. Commercial fertilizers have thus increased the efficiency of the farmer and at the same time have reduced the cost of food to the consumer.

THE FERTILIZER INDUSTRY

The tonnage of fertilizer produced by industry in 1880 was approximately 1 million. In 1887, Congress passed the Hatch Act, which established agricultural experiment stations in each of the states. This act encouraged research on the use of fertilizers. Then the Smith–Lever Extension Act of 1914 gave impetus to an expanded use of commercial fertilizers by providing extension agronomists and county agricultural agents who worked directly with farmers in educational programs to improve agriculture. As a result of research, farm demonstrations, news releases, and bulletins and books on fertilizer use, fertilizer consumption rose to more than 14.8 million tons of N + P_2O_5 + K_2O in 1968.

MATERIALS SUPPLYING NITROGEN

The production of nitrogenous fertilizers has increased faster than that of any other chemical fertilizer. The principal nitrogenous fertilizers and their percentages of nitrogen are given in Table 14.1.

TABLE 14.1 PRINCIPAL NITROGENOUS MATERIALS

Material	Nitrogen Content (%)
Anhydrous ammonia	82
Urea	46
Ammonium nitrate	33.5
Aqua ammonia	20 to 24
Nitrogen solutions	20 to 41
Ammonium sulfate	21
Diammonium phosphate	21
(plus 53% available P_2O_5)	
Ammonium phosphate sulfate	16
(plus 20% available P_2O_5)	
Sodium nitrate	16
Organic products	1 to 12
(animal manures, meat meal, cottonseed meal, and fish meal)	

Ammonia is the principal nitrogenous fertilizer, with more than 90 per cent of all nitrogenous fertilizers consisting of ammonia or a fertilizer made from ammonia. Anhydrous ammonia, liquid ammonia, ammonium nitrate, urea, ammonium sulfate, synthetic sodium nitrate, and ammonium phos-

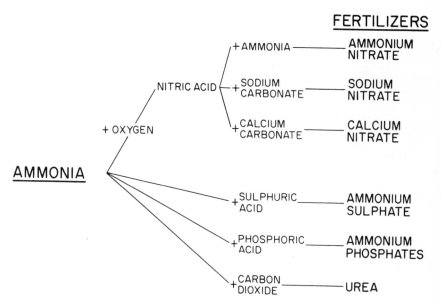

FIG. 14.1. More than 90 per cent of all nitrogenous fertilizers consist of ammonia or fertilizers made from ammonia. Source: *Fertilizer Salesman's Handbook*, National Plant Food Institute.

phate are made with ammonia as the source of nitrogen. Only cyanamid and Chilean nitrate of soda do not use ammonia in their manufacture (Figure 14.1).

Ammonia is a colorless gas containing one atom of nitrogen to three atoms of hydrogen (NH_3). Since a nitrogen atom weighs 14 times more than an atom of hydrogen, the per cent N in pure NH_3 is found in this way:

$$\% \ N = \frac{14}{14 + 3} \times 100 = 82.35$$

Commercial-grade ammonia is 99.5 per cent pure; therefore, it contains 82 per cent N.

In recent years, the use of ammonia for direct application to crops by injecting it into the soil has increased tremendously. It is supplied for this purpose in pressure tanks and reacts somewhat like butane gas; that is, it is a liquid when under tank pressure but a gas in atmospheric pressure. In this form it is called anhydrous ammonia. ("Anhydrous" means without water.)

Many safety precautions must be observed when handling anhydrous ammonia. Some of these precautions include the following:

FIG. 14.2. Storing and handling anhydrous ammonia requires specialized equipment capable of withstanding pressure of at least 250 pounds per square inch. *Source: Agricultural Ammonia Institute.*

1. Do not get a flame near a mixture of 16 to 25 per cent ammonia because these mixtures will burn.
2. Keep away from ammonia when it escapes into the atmosphere. It will cause severe irritations of the eyes, nose, throat, and lungs. The skin can also be easily burned with ammonia. Rubber gloves and goggles will give some protection.
3. Use only pressure tanks that are designed to withstand pressures of at least 250 pounds per square inch (Figure 14.2).
4. Paint all ammonia tanks white and store in a cool, shady place.
5. Arrange for an inspection of all tanks at least once a year.

In addition to the use of anhydrous ammonia as a low-cost source of nitrogenous fertilizer, it can also be used as a defoliant to hasten the shedding of leaves of cotton to facilitate mechanical harvest.[1] Experimentally, anhydrous ammonia has also been used as a source of a part of the protein feed for sheep and cattle. In 1968, anhydrous ammonia for the first time was transported by tanker ships across the Atlantic Ocean from the United States to Europe.

Anhydrous ammonia is now being transported several thousand miles from large, efficient plants that manufacture the ammonia from low-cost natural gas, through underground pipelines, from Texas to the Midwest.

---

[1] Max K. Buchmiller, *Anhydrous Doubles as Defoliant,* Doane's Agricultural Report Business Issue 31, Vol. 26, No. 14 (Sept. 8, 1968).

FIG. 14.3. Urea is the cheapest per pound of N of all solid nitrogenous fertilizers, but its manufacture requires more capital investment. This is a urea synthesis plant with a tall tower in the center background (at arrow) that is used to "pebble" the urea fertilizer by spraying a solution into the top. *Source:* Nitrogen Division, Allied Chemical Co.

Urea [$CO(NH_2)_2$] is now cheaper per pound of N than any other solid nitrogenous fertilizer, and it analyzes 46 per cent N. The nitrogen must be converted by bacteria to ammonium before plants can use it. Urea is a synthetic organic fertilizer (Figure 14.3).

Urea fertilizer is readily soluble and leachable when it is first applied to the soil, but when it changes to ammonium it is held by clay and humus in a form that is readily available to plants. Under favorable temperature and moisture conditions, urea hydrolyzes to ammonium carbonate and then to nitrate within less than a week.

When urea contains more than 1 per cent biuret, the biuret is toxic to such sensitive plants as tobacco.

Ammonium Nitrate (NH_4NO_3) is a good, cheap source of solid nitrogen, analyzing 33.5 per cent N. Half of the nitrogen is in the ammonium form and half is in the nitrate form.

Nitrogen solutions are water solutions of ammonium nitrate or urea or both. When the liquid is to be stored under pressure, some anhydrous ammonia is usually introduced.

An example of a nitrogen solution that contains 32 per cent N but that can be stored and used under atmospheric conditions is listed:

Material	Per Cent of Material	Per Cent N in N Solution
Ammonium nitrate (33% N)	48.5	16.0
Urea (46% N)	34.8	16.0
Water	16.7	—
Total	100.0	32.0

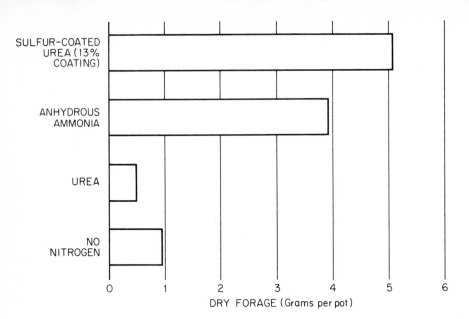

FIG. 14.4. Sulfur-coated urea is a new product that releases nitrogen slowly, is cheaper to produce than other slow-release nitrogenous fertilizers, and does not burn when top-dressed on grass as readily as does untreated urea. Bermuda grass forage yields were higher following applications of sulfur-coated urea as compared with applications of anhydrous ammonia and untreated urea. The urea not coated with sulfur burned the grass; and thus decreased yields below the "no nitrogen" pot. *Source:* S. E. Allen, D. A. Mays, and G. L. Terman, "Low-cost, slow-release fertilizer developed," *Crops and Soils Magazine,* Dec., 1968.

When higher concentrations of N are desired, varying amounts of anhydrous ammonia may be used along with ammonium nitrate. For example, a 37-per cent nitrogen solution can be made which has a pressure of 1 pound per square inch at 104°F. This product is made with:

Material	Per Cent of Material	Per Cent N in N Solution
Ammonium nitrate (33% N)	69.8	23.0
Anhydrous ammonia (82% N)	17.1	14.0
Water	13.1	—
Total	100.0	37.0

Some companies make even higher concentrations of N solutions; to do so, they add more anhydrous ammonia. This also increases the pressure of the solution.

Ammonium Sulfate $[(NH_4)_2SO_4]$ comes mostly from recovered coke–oven gases and contains 21 per cent N. Because of its relatively high cost, it is fast losing in sales to ammonium nitrate and urea. For use on rice, however, ammonium sulfate is the best form of nitrogenous fertilizer be-

302

cause the nitrogen in the ammonium form does not quickly change to nitrogen gas and is therefore not lost by denitrification.

Nitrate of soda (NaNO₃) is found as a natural impure product in the interior of Chile and is also made synthetically. Because of its low percentage of N, nitrate of soda is fast disappearing from use. There is also objection to its use on fine-textured soils, where the Na residue disperses the clay particles and causes puddling and crust formation.

SLOW RELEASE OF NITROGEN FROM FERTILIZER [2]

The nitrogen in most chemical fertilizers is readily soluble and available for use by plants. However, plants need nitrogen every day in their life rather than a surplus one day and a deficiency the next. To attempt to solve this problem, research scientists working for the Tennessee Valley Authority sprayed molten sulfur on granular urea. Then a wax was added and later a clay, to increase flowability. Solubility of *untreated* urea was *100 per cent within a few minutes.* By contrast *only 1 per cent* of the nitrogen *in the coated urea dissolved every 5 days.* Greenhouse and field tests have confirmed the greater efficiency of uptake of nitrogen by upland plants from the coated urea with a slow release of N (Figure 14.4).

However, sulfur-coated urea pellets are not effective on lowland rice because of a coating of iron sulfide that forms around each pellet, thereby rendering the nitrogen too slowly available.

MATERIALS SUPPLYING PHOSPHORUS

The phosphorus fertilizer ores used in the United States come largely from Florida, where ancient oceans have left millions of years of accumulation of marine shell organisms. This rock phosphate ore is refined and ground to produce the rock phosphate of commerce. Rock phosphate is acidulated with sulfuric acid to make 20 per cent superphosphate and with phosphoric acid to produce 45 per cent triple superphosphate. Other materials used in supplying phosphorus are diammonium phosphate (53 per cent P₂O₅ and 21 per cent N), monoammonium phosphate (48 per cent P₂O₅ and 11 per cent N), ammonium phosphate sulfate (20 per cent P₂O₅ and 16 per cent N), and basic slag (approximately 10 per cent P₂O₅).

---

[2] D. W. Rindt, G. M. Blouin, and J. G. Getsinger, "Sulfur Coating on Nitrogen Fertilizer to Reduce Dissolution Rate," *Journal of Agricultural and Food Chemistry,* Vol. 16, No. 5 (Sept.–Oct., 1968), pp. 773–78.

Since the price of sulfur, and therefore of sulfuric acid, in recent years has increased because of scarcity, there has been renewed interest in using nitric acid to make several kinds of nitrophosphates, the most common of which contains 20 per cent N and 20 per cent P_2O_5.[3]

Ordinary superphosphate is made by adding 10 parts by weight of rock phosphate to 9 parts of 70 per cent sulfuric acid in a mixer and thoroughly agitating it for 1 to 2 minutes. The acid causes intense heat, which aids in the drying process. The mixture may be dumped in a pile and allowed to cure for several months, or it may be conveyed in a continuous curing operation. If cured in a pile, at the end of the curing period the superphosphate must be blasted out, ground, sometimes pelleted, and used either directly as a fertilizer or put into mixed fertilizers.

Upon treatment with sulfuric acid, some of the tricalcium phosphate is changed to dicalcium phosphate and some to monocalcium phosphate. Monocalcium phosphate is water soluble and therefore readily available to plants. Dicalcium phosphate is not water soluble but is citrate soluble and available to plants. Tricalcium phosphate is very slowly soluble and slowly available to plants.

In the United States, phosphorus fertilizers move into agricultural trade channels on the basis of guaranteed *available* P_2O_5. The term "available" implies that the phosphorus so guaranteed can be absorbed and utilized by growing plants. Guarantees of available P_2O_5 are checked by state agricultural chemists in their laboratories by determining the amount of phosphorus that is soluble in neutral ammonium citrate.

The average composition of granulated 20 per cent superphosphate is as follows:

Item	Per Cent
Gypsum $CaSO_4$	48
Monocalcium phosphate $Ca(H_2PO_4)_2$	30
Dicalcium phosphate $Ca_2H_2(PO_4)_2$	9
Iron oxides, aluminum oxides, and silica	9
Tricalcium phosphate $Ca_3(PO_4)_2$	2
Moisture	2
Total	100

Triple superphosphorus is made by adding two parts by weight of rock phosphate to one part of 70 per cent phosphoric acid. The batch is mixed

---

[3] "Nitrophosphates Highlight 15th Fertilizer Round Table," *Agricultural Chemicals* (Dec., 1965), pp. 40–43, 92.

for 3 minutes, dried, ground, and used directly for field application or in mixed fertilizers.

A newer product, known as *nitric phosphates,* results in concentrations of nitrogen and phosphorus in chemical combination. A typical process consists of mixing rock phosphate, nitric acid, and phosphoric acid, then adding ammonia. The products can be varied, depending upon the proportion of the ingredients. An average analysis is 20–20–0. To these types of formulations can be added muriate of potash to produce a product such as 10–10–10.

For direct application to the soil, triple superphosphate supplies 52 per cent of the phosphorus; ordinary superphosphate, 29 per cent; calcium metaphosphate, 8 per cent; basic slag and phosphoric acid, 3.5 per cent each; natural organics, 3 per cent; and bone meal, 1 per cent.

MATERIALS SUPPLYING POTASSIUM

Muriate of potassium is the principal fertilizer supplying potassium; second in importance is sulfate of potash. Of increasing significance is sulfate of potash-magnesia and potassium nitrate.

Muriate of potash is usually 95 per cent pure KCl, equivalent to 60 per cent K_2O. It is mined in New Mexico, Utah, and California. In Saskatchewan, Canada, the largest potash reserves in the world are now being mined. When mined, it contains approximately 40 per cent KCl; purification increases the concentration to approximately 95 per cent KCl.

Sulfate of potash is found also as a salt in areas near the KCl. Average purified sulfate of potash is 95 per cent K_2SO_4, equivalent to approximately 51 per cent K_2O.

Sulfate of potash-magnesia is a naturally occurring double salt, found in the same general area with the other potash salts. It is a combination of potassium sulfate and magnesium sulfate. It analyzes approximately 18 per cent MgO and 22 per cent K_2O (Figure 14.5).

FIG. 14.5. The world's largest potash mine and refinery are at Esterhazy in western Canada and supply more than 10 per cent of the world's potash fertilizer. *Source:* International Minerals and Chemical Corp.

SOME NEW FERTILIZERS [4]

For many years, scientists at the Tennessee Valley Authority have been developing new fertilizers with higher analyses. One of the most recent fertilizers that has been made and used in large-scale field demonstrations by a few of the most progressive farmers is ammonium polyphosphate with an analysis (grade) of 15–60–0. Both the nitrogen and phosphorus are sufficiently soluble to be used for making liquid fertilizers. Also, the material may be used directly as a fertilizer or blended to make other compounded or mixed fertilizers.

Another new fertilizer is urea ammonium phosphate, with grades ranging from 34–17–0 to 29–29–0 to 25–35–0.

SECONDARY AND MICRONUTRIENT ELEMENTS

The secondary and micronutrient elements that are sometimes added to fertilizer formulations or used as a fertilizer either singly or in mixtures include the following:

Boron
Calcium (usually applied as a lime)
Copper
Iron
Manganese
Magnesium (often applied as a lime)
Molybdenum
Sulfur
Zinc

To certain specialty fertilizers are added secondary and micronutrient elements in varying amounts. For field-crop fertilization, the micronutrient elements are often added separately after a known deficiency occurs. As soils are cropped more heavily, the need for adding more micronutrient elements will no doubt increase.

Agricultural limestone contains varying amounts of secondary and micronutrient elements that contribute to the plant requirements when lime is applied to a field. As a result of analyzing 220 samples of agricultural

---

[4] A. B. Philips, "More New Fertilizers are Coming," *Hoard's Dairyman*, Vol. 113 (Oct. 10, 1968), pp. 1128–29.

limestone from 35 states, the average analyses for the following essential elements were reported: [5]

Calcium carbonate	75.4%
Magnesium carbonate	17.1%
Iron	0.4%
Potassium	0.2%
Sulfur	0.1%
Manganese	330 p.p.m.
Phosphorus	210 p.p.m.
Zinc	31 p.p.m.
Boron	4 p.p.m.
Copper	2.7 p.p.m.
Molybdenum	1.1 p.p.m.

FORMULATION OF MIXED FERTILIZERS

A complete fertilizer contains nitrogen, phosphorus, and potassium. The grade numbers in a complete fertilizer are the guaranteed analysis written as whole numbers. A 10–10–10 fertilizer, for example, is guaranteed to contain 10 per cent total nitrogen, 10 per cent available P_2O_5, and 10 per cent water-soluble K_2O. There is no actual P_2O_5 or K_2O in any fertilizer; these expressions are only the chemical equivalents of the actual chemical compounds in a fertilizer.

Calculations in the formulations of mixed fertilizers may be worthwhile as an example. Suppose that you were to mix a ton of a 10–10–10 fertilizer, using

21% ammonium sulfate
45% triple superphosphate
60% muriate of potash

A 10–10–10 would require per ton:

$$10\% \times 2000 = 200 \text{ lb of N}$$
$$10\% \times 2000 = 200 \text{ lb of } P_2O_5$$
$$10\% \times 2000 = 200 \text{ lb of } K_2O$$

Pure N, P_2O_5, and K_2O would require 200 pounds of each plus 1400 pounds of filler; but pure materials are not stable under atmospheric con-

---

[5] P. Chichilo and C. W. Whittaker, "Trace Elements in Agricultural Limestones of the United States," *Agronomy Journal*, Vol. 53, No. 3 (1961), pp. 143–44.

TABLE 14.2   OPEN FORMULA TO MAKE A TON OF 14–14–14 FERTILIZER *

Material	Pounds of Materials per Ton	Analysis (%)	Percentage in Mixture		
			Nitrogen (N)	Phosphorus (P$_2$O$_5$)	Potash (K$_2$O)
Ammonium sulfate	149	20.8 N	1.55	—	—
Ammonium nitrate	524	33.5 N	8.78	—	—
Nitrate solution (ammonium nitrate plus ammonia)	194	40.6 N	3.94	—	—
Superphosphate, triple	606	47.0 P$_2$O$_5$	—	14.24	—
Muriate of potash	467	61.0 K$_2$O	—	—	14.24
Clay to coat granules of fertilizer	60	—	—	—	—
Total	2000	—	14.27	14.24	14.24

* *Source:* Lime and Fertilizer Branch, U.S.D.A.

ditions and therefore cannot be used to make a fertilizer. To compute the pounds of an impure product needed to supply a given number of pounds of pure product, divide the latter by the former, in this way:

Ammonium sulfate needed $=$

$$\frac{200 \text{ (lb of N needed)}}{\% \text{ purity of N (in ammonium sulfate)}} = \frac{200}{0.21} = 952 \text{ lb}$$

Triple superphosphate needed $=$

$$\frac{200 \text{ (lb of P}_2\text{O}_5 \text{ needed)}}{\% \text{ purity of P}_2\text{O}_5 \text{ (in triple superphosphate)}} = \frac{200}{0.45} = 444 \text{ lb}$$

Muriate of potash needed $=$

$$\frac{200 \text{ (lb of K}_2\text{O needed)}}{\% \text{ purity of K}_2\text{O (in muriate of potash)}} = \frac{200}{0.6} = 333 \text{ lb}$$

Total	1729 lb
Filler	271 lb
Grand total	2000 lb

Although it is probably true that, because of the high cost, very few 10–10–10 fertilizers would be mixed in this way, the principle of calculation is the same regardless of the method or materials used. Modern fertilizers are usually made by ammoniating the superphosphate. As much ammonia as possible is used because it is cheaper than the solid nitrogenous fertilizers.

The example of a 10–10–10 fertilizer is called the *grade* of the fertilizer, whereas the relative amounts of N, P_2O_5, and K_2O are in the proportion of 1–1–1 and are known as the *ratio*.

The ingredients necessary to make a ton of 14–14–14 are shown in Table 14.2. Three sources of nitrogen are used: ammonium sulfate, ammonium nitrate, and nitrogen solution. Triple superphosphate is used to supply all of the phosphorus, and muriate of potash supplies all of the potash. Clay is used at the rate of 60 pounds per ton to coat the granules of fertilizer so that they will not absorb so much water from the atmosphere.

Note that the mixture is guaranteed to be a 14–14–14, but that it is formulated to contain 14.27 per cent nitrogen, 14.24 per cent available P_2O_5, and 14.24 per cent water-soluble K_2O. These excess amounts of nutrients are added to the mixture to allow for some loss of nitrogen as ammonia, and to allow for possible errors in mixing and in segregation before the fertilizer reaches the consumer.

PROBLEMS OF MANUFACTURING MIXED FERTILIZERS

Perhaps a generation ago, many farmers mixed their own fertilizers with the use of carriers such as ammonium sulfate, superphosphate, and muriate of potash. No longer can farmers afford to do this. One reason is that labor costs are too high; the other reason is that modern fertilizers are not simple mixtures of dry materials. Fertilizers of today are the result of complex and costly methods which increase the concentration of the fertilizers and reduce their unit cost per pound of plant nutrients.

There are five basic ingredients in modern fertilizers:

1. Carriers of N, P_2O_5, K_2O, and, at times, other materials
2. Conditioners, such as ground vermiculite, tobacco stems, rice hulls, or other similar material
3. Neutralizers of acidity, primarily dolomitic limestone
4. Fillers, such as sand, to "make weight" when a fertilizer formulation does not equal the required weight. Most modern, high-grade fertilizers, however, do not contain fillers.

5. Materials for specialty fertilizers, such as insecticides, fungicides, or herbicides [6]

The biggest problem in making mixed fertilizers is to maintain a desirable physical condition. Most fertilizer plants are overcoming this difficulty by pelleting the fertilizer and by using certain clays and waxes to coat the pellets. Improper mixing and inadequate curing also contribute to an undesirable physical condition of the fertilizer.

One special problem of long standing is the manufacture of nitrogenous fertilizers that release nitrogen slowly over the growing season of plants. A new and promising approach is the development of a sulfur-coated urea with wax and a microbiocide added.

ACIDITY OR BASICITY OF FERTILIZERS

Since most fertilizers are used on acid soils, their acidifying properties are of great concern. A glance at Tables 14.3 and 14.4 establishes the fact that most of the nitrogenous fertilizers are acid; the potassium and the phosphorus fertilizers are physiologically neutral.

There is a tremendous difference in the acidifying properties of the nitrogen materials. Sodium nitrate is alkaline, whereas the other nitrogenous carriers are acid. Listed in order of least acid forming to most acid forming *per pound of N* are:

Sodium nitrate (alkaline)
Urea
Ammonium nitrate
Anhydrous ammonia
Ammonium sulfate

SUMMARY

The chemical fertilizer industry has grown during the past 100 years to become the largest chemical industry, selling each year more than 900 million dollars' worth of fertilizer. The proper use of fertilizers makes it

---

[6] *Note:* Fertilizer-pesticide mixtures have now become so sophisticated that several fertilizer companies are offering for sale a material that fertilizes with long-lasting effects: kills crabgrass; kills grub worms but does not kill earthworms; and kills chickweed, clover, and all other broad-leaved weeds, such as dandelion.

TABLE 14.3 THE COMMON NITROGENOUS FERTILIZERS AND THEIR
ACIDITY OR BASICITY *

Material	N (%)	Equivalent Acidity		Equivalent Basicity	
		Per 1 lb of N	Per 100 lb of Material	Per 1 lb of N	Per 100 lb of Material
Sodium nitrate	16			1.8	29
Urea	46	1.6	74		
Ammonium nitrate	33	1.8	59		
Anhydrous ammonia	82	1.8	148		
Ammonium sulfate	21	5.2	109		

* In terms of pure calcium carbonate equivalent.

TABLE 14.4 THE COMMON FERTILIZERS SUPPLYING PHOSPHORUS AND
POTASSIUM AND THEIR PHYSIOLOGICAL REACTION IN THE SOIL

Material	Nutrient (%)	Physiological Reaction in Soil
	P_2O_5	
Superphosphate	20	Neutral
Triple superphosphate	45	Neutral
	K_2O	
Muriate of potash	60	Neutral
Sulfate of potash	50	Neutral

possible for farmers to average 3 dollars from increased yields for each dollar invested in fertilizer.

The Hatch Act of 1887, setting up agricultural experiment stations, and the Smith–Lever Act of 1914, establishing Extension Services, were significant in promoting the widespread use of fertilizers.

Most nitrogenous fertilizers are made with ammonia as the source of N. Nitrogen solutions for use in fertilizer formulations and for direct application to the soil have increased tremendously in recent years. There has also been a big increase in the direct use of anhydrous ammonia injected into the soil.

A sulfur-coated urea in pellets appears very favorable for furnishing a low-cost, slow-release nitrogenous fertilizer.

The principal material supplying phosphorus for commercial fertilizers is rock phosphate. Superphosphate is made by treating rock phosphate with sulfuric acid. Triple superphosphate comes from the acidulation of rock phosphate with phosphoric acid.

Two newer phosphatic fertilizers include ammonium polyphosphate (15–60–0) and urea ammonium phosphate, which ranges in grades from 34–17–0 to 29–29–0 to 25–35–0.

Potash salts are supplied by muriate of potash, sulfate of potash, and sulfate of potash-magnesia.

Problems encountered in the formulation of mixed fertilizers involve careful calculations of costs and efforts to overcome the hazard of a poor physical condition in the finished product.

QUESTIONS

1. Give three reasons for the rapid growth of the fertilizer industry.
2. Give some advantages of the direct use of anhydrous ammonia as a fertilizer.
3. What is the principal material supplying potassium?
4. Calculate the ingredients to make a ton of 8–16–16 fertilizer, using urea (46 per cent N), triple superphosphate (45 per cent P_2O_5), and muriate of potash (60 per cent K_2O).
5. What are the main problems encountered in the manufacture of fertilizers?

REFERENCES

Bear, Firman E. (ed.), *Chemistry of the Soil* (2nd ed.). New York: Reinhold Publishing Corp., 1964.

Beaton, J. D., and D. W. Bixby, *Sulphur-Containing Fertilizers: Properties and Applications.* Washington, D.C.: The Sulfur Institute, 1971.

Berger, Kermit C., *Introductory Soils.* New York: The Macmillan Company, 1965.

Changing Patterns in Fertilizer Use. Madison, Wisc.: Soil Science Society of America, 1968.

Engelstad, O. P., and G. L. Terman, "Importance of Water Solubility of Phosphorus Fertilizers," *Commercial Fertilizer.* Vol. 113, No. 6 (Dec., 1966), pp. 32, 33, 35.

———, "Fertilizer Nitrogen: Its Role in Determining Crop Yield Levels," *Agronomy Journal* (Sept.–Oct., 1966), Vol. 58, pp. 536–39.

Farber, Eduard, *History of Phosphorus,* Paper 40, The Museum of History and Technology. Washington: Smithsonian Institution, 1965, pp. 178–200.

Hignett, Travis P., "Bulk Blending of Fertilizers: Practices and Problems," *The Fertilizer Society of London Proceedings* (1965) No. 87.

Kennedy, F. M., E. A. Harre, T. P. Hignett, and D. L. McCune, *Estimated World Fertilizer Production Capacity as Related to Future Needs* 1967 *to* 1972–80. Muscle Shoals, Ala.: Tennessee Valley Authority, 1968.

Locke, Lowell F., and Harold V. Eck, *Iron Deficiency in Plants: How to Control It in Yards and Gardens,* Home and Garden Bulletin No. 102. United States Department of Agriculture, 1965.

McKelvey, V. E., *Phosphate Deposits,* Geological Survey Bulletin 1252–D. Washington: United States Department of the Interior, 1967, 21 pp.

McVickar, Malcolm H., *Using Commercial Fertilizers* (3rd ed.). Danville, Ill.: The Interstate Printers & Publishers, Inc., 1970.

McVickar, Malcolm H., et al. (eds.), *Agricultural Anhydrous Ammonia: Technology and Use.* Memphis: Agricultural Ammonia Institute, 1966.

McVickar, Malcolm H., G. L. Bridger, and Lewis B. Nelson, *Fertilizer Technology and Usage.* Madison, Wisc.: Soil Science Society of America, 1963.

Superphosphate: Its History, Chemistry, and Manufacture. Washington: United States Department of Agriculture and Tennessee Valley Authority, 1964.

Tisdale, Samuel L., and Werner L. Nelson, *Soil Fertility and Fertilizers* (2nd ed.). New York: The Macmillan Company, 1965.

See also: Fertilizer Publications. Agronomic, Chemical, and Engineering. Muscle Shoals, Ala.: Tennessee Valley Authority, National Fertilizer Center.

CHAPTER 15

FERTILIZER USE

And in the course of a long time, the bones became de-cayed, so that there was no strength in them, and they were reduced to dust; then they carried the remains and put it on the surface of the plowed land, and from that time they had astonishing crops of wheat and barley, and of every other grain for many years.*

Without commercial fertilizers, world populations would probably soon exceed the food supply. With fertilizers and other modern necessities, such as improved seed, better insecticides, and more effective fungicides, critical population pressures can be delayed perhaps indefinitely.

Based upon the records of production of fertilizers in the past, it is expected that the world production will increase approximately 8 per cent each year. This rate of fertilizer production should be sufficient to help feed a growing population and to prevent starvation. This is not enough, however, in a modern world where the application of science can help to achieve an abundance of food for everyone.

On many soils throughout the world the proper use of fertilizers, along with the adoption of other crop production practices, such as insect and disease control measures, can result in a doubling of existing crop yields.

---

* Circa first century, as quoted in: D. J. Brown, *The Field Book of Manures,* and the *American Muck Book: Treating of the Nature, Properties, Sources, History, and Operations of all the Principal Fertilizers and Manures in Common Use* (New York: C. M. Saxton and Co., 1856), pp. 234, 235.

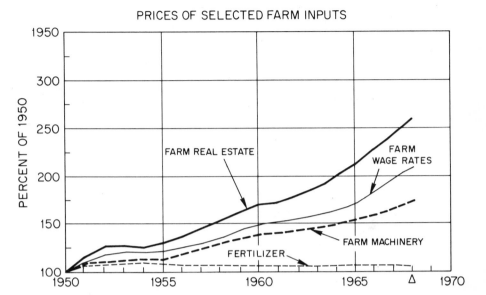

FIG. 15.1. Fertilizer consumption has continued to increase over the years partly because the relative price has remained fairly low; whereas, the price of farm real estate, farm wage rates, and the price of farm machinery have increased. *Source: Handbook of Agricultural Charts,* Economic Research Service, U.S. Dept. of Agr., 1968, p. 14.

FERTILIZER CONSUMPTION

Of all farm inputs, fertilizer has been one of the most profitable. Fertilizer-responsive plants, the rapid development of weedicides and pesticides, narrow rows, sophisticated fertilizer placement implements, the low cost of fertilizer, and an efficient soil testing service have all contributed to the economic response and the continuous popularity of fertilizers in the United States and in the world. The relatively low cost of fertilizers as compared with the cost of other farm inputs, such as land, wages, and farm machinery, have also contributed toward increasing fertilizer consumption (Figure 15.1).

The trends in the use of fertilizers have included:

1. Higher analysis of mixed fertilizers, year by year, as indicated in the following table:

Year	Percentage Plant Nutrient in Mixed Fertilizers		
	N	P_2O_5	K_2O
1900	2.0	9.4	2.5
1930	3.1	9.8	5.0
1950	4.0	10.9	8.3
1960	6.5	13.0	12.1
1968	8.6	16.7	12.9

Since 1900, the percentage of N in mixed fertilizers has increased 330 per cent; P_2O_5, by 78 per cent; and K_2O, by 416 per cent.

2. A larger percentage of the use of straight materials as compared with mixtures. In 1968, 55 per cent of all fertilizers consumed were mixtures and 45 per cent were straight materials, such as ammonium nitrate, superphosphate, or muriate of potash. In 1948, these percentages were 74 and 26, respectively.

3. The greatest increase in consumption of nitrogen fertilizer, followed by potassium and phosphorus at approximately the same rate.

4. A greater rate of increase of liquid fertilizers than of solid fertilizers.

5. A continuous increase in total fertilizer consumed ($N + P_2O_5 + K_2O$), indicated as follows:

Year	$N + P_2O_5 + K_2O$ * (In Millions of Short Tons)
1900	0.3
1930	1.5
1966	12.4
1970 (est.)	18.5
1980 (est.)	26.6

* Russell Coleman, "The Outlook for Fertilizers," *Chemical Engineering Progress,* Vol. 64, No. 7 (July, 1968), pp. 68–71.

The consumption of fertilizers by each of the 50 states in the United States and by each of the 10 provinces in Canada for the year 1968 and the percentage that this amount is of the potential use, are portrayed in Table 15.1.

For the 50 states, the 1968 consumption for N was 50 per cent of potential; for P_2O_5, 44 per cent of potential; and for K_2O, 47 per cent of potential. For the 10 provinces of Canada, the corresponding percentages were 24, 31, and 26, respectively.

TABLE 15.1 *UNITED STATES AND CANADA:* CONSUMPTION OF N, P_2O_5, AND K_2O FERTILIZER IN 1968 AS COMPARED WITH POTENTIAL CONSUMPTION, BY STATES AND PROVINCES * (THOUSAND SHORT TONS)

States in U.S.	N		P_2O_5		K_2O	
	Actual Consumption	Actual Is of Potential (%)	Actual Consumption	Actual Is of Potential (%)	Actual Consumption	Actual Is of Potential (%)
Alabama	125.5	76	105.4	76	103.8	73
Alaska	0.7	44	0.7	44	0.4	31
Arizona	84.5	40	28.9	28	1.1	275
Arkansas	94.4	48	61.3	33	68.5	34
California	407.7	55	151.6	27	55.0	45
Colorado	69.2	22	40.4	21	7.2	30
Connecticut	7.8	45	6.9	42	6.5	34
Delaware	11.6	56	12.1	11	14.7	15
Florida	147.3	41	102.0	50	189.2	48
Georgia	243.3	65	146.9	71	212.1	74
Hawaii	24.0	39	15.5	24	24.0	39
Idaho	72.9	41	46.4	41	4.2	102
Illinois	587.9	60	425.6	55	449.5	62
Indiana	324.4	70	269.8	87	317.0	71
Iowa	636.2	80	345.7	53	282.3	58
Kansas	357.1	77	158.3	52	34.1	31
Kentucky	85.1	52	87.1	38	99.2	39
Louisiana	105.2	55	56.9	38	50.1	34
Maine	19.3	64	23.5	84	22.7	70
Maryland	40.8	60	39.3	55	49.3	60
Massachusetts	7.7	39	7.5	39	7.0	36
Michigan	114.4	53	127.3	59	136.0	58
Minnesota	253.5	56	210.6	54	196.6	17
Mississippi	135.0	46	66.2	32	64.0	30
Missouri	278.2	59	165.0	45	159.6	36
Montana	30.8	15	43.5	27	1.5	21
Nebraska	404.8	87	106.4	35	28.6	33
Nevada	2.7	4	1.5	16	0.1	100
New Hampshire	1.9	26	1.8	24	1.9	24
New Jersey	23.8	70	23.4	62	23.4	50
New Mexico	29.1	11	13.7	8	1.1	367
New York	67.7	44	82.9	51	68.1	32
North Carolina	164.0	75	133.3	77	168.4	72
North Dakota	62.7	16	111.0	34	9.2	16
Ohio	192.1	51	208.6	63	221.7	63
Oklahoma	127.7	31	80.5	26	26.9	14
Oregon	90.4	74	51.9	42	11.3	16
Pennsylvania	68.3	28	83.6	38	71.0	25
Rhode Island	1.6	44	1.7	51	1.6	47
South Carolina	88.0	80	75.8	71	94.1	73
South Dakota	59.7	18	37.9	18	7.8	9
Tennessee	93.1	68	85.6	66	93.1	65
Texas	459.5	34	214.4	25	74.3	19
Utah	13.2	17	21.4	45	0.3	75
Vermont	4.0	18	5.8	22	6.1	11

TABLE 15.1   (Continued)

States in U.S.	N Actual Consumption	N Actual Is of Potential (%)	P_2O_5 Actual Consumption	P_2O_5 Actual Is of Potential (%)	K_2O Actual Consumption	K_2O Actual Is of Potential (%)
Virginia	70.8	69	79.6	62	87.2	62
Washington	143.2	66	48.6	39	24.8	39
West Virginia	6.1	9	9.0	18	7.0	13
Wisconsin	110.3	42	113.1	46	205.2	44
Wyoming	12.9	18	8.7	13	0.6	300
Total	6562.3	50	4344.5	44	3786.2	47

Provinces in Canada

	N Actual Consumption	N Actual Is of Potential (%)	P_2O_5 Actual Consumption	P_2O_5 Actual Is of Potential (%)	K_2O Actual Consumption	K_2O Actual Is of Potential (%)
Alberta	100.8	27	96.1	34	1.5	65
British Columbia	7.9	16	8.2	18	4.1	15
Manitoba	57.4	33	53.6	39	1.6	15
New Brunswick	6.2	23	10.5	45	8.9	28
Newfoundland	0.4	33	0.6	50	0.6	50
Nova Scotia	3.1	56	4.6	46	4.6	46
Ontario	108.5	38	131.0	48	111.8	48
Prince Edward Island	5.0	25	9.6	41	9.4	24
Quebec	19.7	11	38.9	19	40.1	12
Saskatchewan	44.6	13	96.9	23	0.7	24
Total	353.7	24	450.1	31	183.1	26
Grand Total For U.S. and Canada	6915.9	48	4794.6	42	3969.3	45

* J. D. Beaton and S. L. Tisdale, *Potential Plant Nutrient Consumption in North America,* Technical Bulletin 16, Sulphur Institute, Washington (July, 1969), 64 pp.

Note: Potential consumption is defined as the amounts of N, P_2O_5, and K_2O that would be consumed if all crops in 1967 had been fertilized with the rates recommended. Values used in these calculations were obtained from the preceding reference.

Only four states (but no provinces) have achieved 70 per cent or more of potential fertilizer use of all three nutrients. They are Alabama, Indiana, North Carolina, and South Carolina.[1]

---

[1] J. D. Beaton and S. L. Tisdale, *Potential Plant Nutrient Consumption in North America,* Technical Bulletin 16, Sulphur Institute, Washington (July, 1969), 64 pp.

THE FERTILIZER INFORMATION GAP [2]

The Harvest Publishing Company surveyed 1000 farmers in five states —Kansas, Missouri, Michigan, Ohio, and Pennsylvania (with a 35 per cent return) and discovered that farmers are not well informed on the efficient use of fertilizers. For example, fewer than 25 per cent bought fertilizer on the basis of a soil test, and approximately one-third did not know the meaning of the grade numbers on a bag of fertilizer but thought that all fertilizers were the same except for different brand names.

EFFICIENCY OF NITROGENOUS FERTILIZERS ON ALKALINE SOILS [3]

When various nitrogenous fertilizers are applied on the *surface* of *alkaline* (high-lime) soils and are *not* worked into the soil, the nitrate (NO_3) forms (ammonium nitrate or sodium nitrate) give a greater crop response than either the ammonium (NH_4) or amine (NH_2) (as in urea) forms of nitrogen. The explanation is that the lime in the soil reacts with the ammonium to release ammonia (NH_3) gas, which escapes into the atmosphere.

The relative efficiency of various forms of nitrogen on alkaline soils, using Johnsongrass as the test crop, are as follows (Table 15.2):

TABLE 15.2 THE RELATIVE YIELD OF JOHNSONGRASS ON ALKALINE SOILS WHEN FERTILIZED WITH VARIOUS SOURCES OF NITROGENOUS FERTILIZERS

Source of N	Relative Yield
Ammonium sulfate	37
Urea	53
Ammonium nitrate	89
Sodium nitrate	100

---

[2] Nancy Lee Harding, "The Fertilizer Information Gap," *Farm Store Merchant,* Vol. 11, No. 8 (August, 1968), pp. 94–96.

[3] *Source:* C. E. Scarsbrook, H. W. Grimes, and L. A. Smith, "Efficiency of Nitrogen Sources Varies on Alkaline Soils," *Highlights on Agricultural Research,* Vol. 11, No. 4 (Auburn, Ala.: Auburn University, 1964).

NUTRIENTS CONTAINED IN CROPS

Crops sold from the farm remove nutrients that must be replaced. Sometimes soil minerals break down fast enough to release a large part of the nutrients found in crops, but this process is too slow to provide all

FIG. 15.2. Major and secondary essential nutrients contained in the entire plant with the yield indicated. *Source: Magnesium–Sulfur, Essential Plant Nutrients,* International Minerals and Chemical Corp., Skokie, Ill., undated publication.

Crop	Crop Yield per Acre	Nitrogen N	Phosphate P_2O_5	Potash K_2O	Calcium Ca	Magnesium Mg	Sulfur S
				Nutrients Contained in Crop — Pounds per Acre per Crop			
Alfalfa	5 tons	250	60	225	160	25	23
Corn	150 bu	220	80	195	58	50	33
Cotton	1½ bales	95	50	60	28	8	4
Coastal Bermuda Grass	6 tons	150	60	180	33	22	40
Soybeans	40 bu	145	40	75	7	9	7
Rice	6500 lbs	185	51	18	20	15	18
Tobacco	2800 lbs	95	25	190	105	24	21
Wheat	60 bu	125	50	110	16	18	16
Oats	100 bu	100	40	120	14	20	20
Potatoes	400 bu	200	55	310	50	15	18
Peanuts	3000 lbs	220	45	120	105	28	25
Grain Sorghum	8000 lbs	260	110	220	45	36	38
Banana	1200 plants	400	400	1500	300	156	*
Coffee	1784 lbs	27	4	43	56	61	16
Oil Palm	13382 lbs	80	18	120	64	18	*
Pineapple	15000 plants	134	107	535	102	53	*

* No information

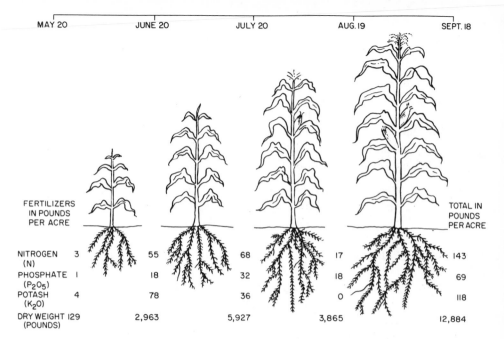

	MAY 20	JUNE 20	JULY 20	AUG.19	SEPT.18

FERTILIZERS IN POUNDS PER ACRE

TOTAL IN POUNDS PER ACRE

NITROGEN (N)	3	55	68	17	143
PHOSPHATE (P_2O_5)	1	18	32	18	69
POTASH (K_2O)	4	78	36	0	118
DRY WEIGHT (POUNDS)	129	2,963	5,927	3,865	12,884

FIG. 15.3. Corn needs the largest amount of K from June 20 to July 20, but it re-quires more of its N and P a month later. The seasonal demand of the plant must be kept in mind when deciding when to apply fertilizers. *Source: Plant Food Your Corn Absorbs,* Folder D-9, American Potash Institute, undated publication.

nutrients for modern crop yields. Legumes are capable of fixing atmospheric nitrogen in amounts satisfactory for crop yields of ten years ago, but are not sufficient for today's crop yields. The only logical conclusion for profit-able farming is that most of the nutrients removed by high-yielding crops must be replaced by commercial fertilizers (Figure 15.2).

Pounds per acre of nutrients removed by selected crops are shown in Figure 15.3. Banana removes more nitrogen, phosphorus, and potassium, whereas coffee removes the least, of the crops shown. Corn and cotton are intermediate in their chemical composition. Most of the nitrogen removed by alfalfa and other legumes is fixed by legume bacteria from atmospheric nitrogen.

The soil and fertilizers must not only supply the *total* nutrients needed by the growing crops, they must also supply these nutrients *in time* for adequate periodic growth. For example, Figure 15.3 indicates that the demands of corn for N and P are greatest from July 20 to August 19, but that the greatest demand for K is one month earlier.

TIME OF APPLYING FERTILIZERS

It has been traditional to apply fertilizer to row crops just prior to or at the time of planting. This practice is still satisfactory if the efficient use of fertilizer is the only objective. But farmers are the busiest at this time of year, and fertilizer costs more when it must be supplied to everyone at the same time. Fall applications of fertilizers are a possibility in some areas. In Ontario, Canada, no fall application of nitrogen is recommended.[4]

Studies over the United States of fall-applied fertilizers in relation to leaching losses indicate that:

1. Phosphorus leaches out of organic and mineral soils only in trace amounts. It is often fixed in some less soluble form, however, unless the soil pH is approximately 5.5 for organic soils or 6.5 for mineral soils and there is an abundance of fresh organic matter present.
2. Potassium leaches from *mineral soils* in small amounts, but this is not serious enough to limit the practicability of applying potash fertilizers in the fall. Leaching losses of potassium can be reduced by maintaining a crop on the land at all times.
3. Leaching losses of potassium on *organic soils* are very great; for this reason, it is not recommended to apply potassium fertilizers in the fall on fibric, hemic, or sapric (peat or muck) soils.[5]
4. Nitrogen is readily lost in the water that percolates through the soil. The nitrogen lost in percolating waters is almost entirely in the nitrate form; loss in the ammonium form is negligible because clay and humus particles adsorb the ammonium ion and hold it in a nonleachable form.

There is some possibility of losses of nitrogen by leaching immediately after urea fertilizer is applied to the soil; however, in approximately 1 day, the urea is usually changed by bacteria and chemical transformation into the nonleachable ammonium form of nitrogen. Therefore, the practicability of applying nitrogenous fertilizers in the *fall* lies primarily in minimizing the leaching losses of nitrogen. This problem can be solved by one or more of the following procedures:

1. Keep the nitrogen in the ammonium form by applying the NH_3 or NH_4 form to a cold soil (50°F or colder).
2. Do not apply nitrogen fertilizers in the fall in the areas where the percolation losses of water are great.

---

[4] C. S. Baldwin and C. K. Stevenson, "Fall-Applied Nitrogen? NO!" *Crops and Soils Magazine* (Aug.–Sept., 1968), pp. 15, 16.

[5] J. C. Shickluna, J. F. Davis, and R. E. Lucas, "Why Potatoes and Onions Need Phosphorus and Potassium on a Virgin Organic Soil," *Better Crops with Plant Food,* American Potash Institute (March–April, 1965).

3. Apply nitrogen fertilizer in the fall only where a cool-season crop is growing vigorously enough to absorb the nitrogen.
4. Apply nitrogen fertilizer in the fall on crop residues that will utilize the nitrogen for their decomposition and thereby retain the nitrogen from loss by leaching.
5. Apply ammonia nitrogen in the fall on acid soils where conversion to the nitrate form is slow.

Nitrogen in the ammonium form is oxidized to the nitrate form quite rapidly when the soil temperature is above 50°F. Applying the ammonia to the soil in the fall after the soil gets no warmer than 50°F will help in reducing the transformation of ammonia to the nitrate form. In this way, the losses of nitrogen by leaching will be reduced. In Iowa, soil temperatures usually drop below 50°F about November 1, and rise above 50°F in mid-April. Cool-weather applications of anhydrous ammonia can therefore be safely made in Iowa and at similar latitudes between November 1 and April 15.

When there is no percolation of water through the soil, there can be no leaching losses of nitrates. Research on clay soils at Temple, Texas, and Clarinda, Iowa, has shown that during the average year there are no percolation losses of water. This means that there can be no losses of nitrogen even though it is in the nitrate form. At La Crosse, Wisconsin, there is an average annual percolation loss of 2.3 inches of water. Here there will be a considerable loss of nitrates, and fall applications of ammonia should be made only when the soil temperature is below 50°F. By contrast, the water leached through the soil at Statesville, North Carolina, averages 16.7 inches a year. Add to this the factor of high soil temperatures, and the conclusion is reached that the application of ammonia to the soil in the fall under these conditions is not recommended.

There is one more factor to be considered in the question of applying ammonia in the fall. It has been shown by repeated research that when there is a vigorously growing crop on the land, there will be no leaching losses of nitrates because plant roots will absorb the nitrates as fast as they are produced.

If a large amount of crop residues is to be incorporated into the soil in the fall, an application of nitrogen fertilizer will hasten decomposition and at the same time the straw will reduce the loss of nitrogen by leaching.

On soils that are acid, conversion by bacteria of the nonleachable ammonia and ammonium forms of fertilizer to the leachable nitrate form is slow; thus, fall application of NH_3 and NH_4 is safer on acid soils than on soils with a pH above 7.0.

FIG. 15.4. Anhydrous ammonia can easily be applied by bubbling it into irrigation water, but distribution is usually not uniform. *Source:* Shell Chemical Corp.

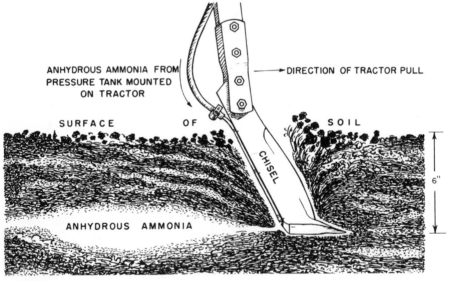

FIG. 15.5. Under pressure and behind a chisel at a depth of approximately 6 inches, anhydrous ammonia is released as a gas and is adsorbed on the surface of clay and humus particles for use by plants. *Source: Crops and Soils Magazine.*

FERTILIZER PLACEMENT

Phosphorus and potash fertilizers move downward in the soil very slowly. For this reason it was thought that surface applications on sod crops were not effective. Although these fertilizers do not move readily, plant roots move toward surface-applied phosphorus and potassium and absorb the fertilizers as long as the surface of the soil stays moist.

Ammonia must be applied deep enough to be absorbed by the soil, usually at a 6-inch depth. Ammonia can also be applied by bubbling it into irrigation water, but by this method distribution over the field is generally not uniform and loss by volatilization is usually great (Figures 15.4 and 15.5). Nitrates and other solid nitrogenous fertilizers are effective when applied on the surface because the next rain or application of irrigation water will leach them into the soil where plant roots are growing. However, there is one exception: solid urea fertilizer should not be applied on the surface of a pasture or lawn sod because of the hazard of loss of nitrogen as a gas. This loss can be prevented by first burning the sod, by placing the urea *in* the soil, or by sprinkling water on the sod immediately after application.[6]

For row crops, the mixed fertilizer is usually applied slightly below and to one side of the seed. Care must be used, however, that potash and nitrogen fertilizers are not placed too close to the seed, thereby injuring the sprouting seed. In sand soils and/or with large amounts of nitrogen or potash, there is a real hazard of seedling injury.

Normal and triple superphosphate may not cause injury to the sprouting seedling, but when applied with the seed, they can delay germination by desiccation by readily absorbing soil moisture. All ammonium phosphates, and particularly diammonium phosphate, have high salt effects and must therefore be placed away from the seed or plant (Figure 15.6).

A newer method of applying fertilizers is to put either a part or all of the fertilizer on the cover crop before it is turned under prior to planting a

---

[6] J. P. Vavra, "N Lost from Surface-Applied Fertilizer," *Crops and Soils Magazine,* Vol. 18, No. 7 (April–May, 1966), p. 22.

FIG. 15.6. Side placement of fertilizer in two bands placed 2 inches to each side and 2 inches below the level of lima bean seed. Source: U.S. Dept. of Agr.

row crop. In this way, the nutrients are transformed into the organic form, from which they are released by decomposition of the cover crop at about the same rate as they are needed by the next crop.

PROFIT FROM USING FERTILIZER

Profit from the use of fertilizer varies widely, but a model response curve will help to visualize the subject (Figure 15.7). From the application of the first few increments of fertilizer, the increase in yield is small ("A"); however, the next few increments of fertilizer ("B") are much greater; finally, as more fertilizer is added, the crop yields are depressed ("C"). The most profitable level is reached when the last increment of fertilizer added is just paid for by the value of the increased yield.

The use of fertilizers pays in many areas and with a wide variety of crops. The net profit in relation to the cost of the fertilizer is demonstrated in Figure 15.8 for the states of Washington and Ohio.

In Washington, when no fertilizer was used, a net profit of 5 dollars an acre was realized. When 8 dollars was spent for fertilizer, the net profit was 25 dollars. But when 16 dollars was spent on fertilizer, the net profit was 38 dollars.

FIG. 15.7. A small amount of fertilizer may give very little initial yield, then a rapid increase, shown in A. Additional units of fertilizer may result in a rapid increase in yield, then a leveling, as in B. Further increases in fertilizer may give declining yields, as shown in C. The most profitable level of fertilizer is reached when the last increment is just paid for by the value of the increased yield, which may be near the maximum yield. *Source: Plant Food Review*, Vol. 3, No. 1, 1957.

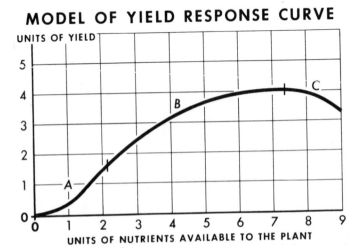

MODEL OF YIELD RESPONSE CURVE

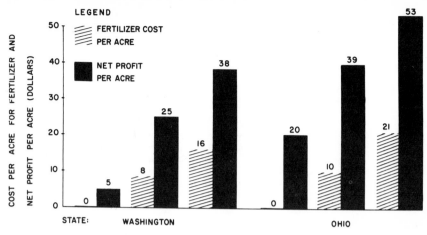

FIG. 15.8. The relationship between fertilizer costs and net profit per acre. *Source:* National Plant Food Institute.

Similar results were obtained in Ohio, where 20 dollars net per acre was realized when no fertilizer was used, 39 dollars when 10 dollars was spent on fertilizer, and 53 dollars per acre net profit when 21 dollars was spent on fertilizer.

It seems that even more money could be spent for fertilizer as long as the increased income continues to more than pay for the cost of the fertilizer.

FERTILIZERS INCREASE SOIL MOISTURE EFFICIENCY

The proper use of fertilizers on soils of low natural fertility makes it possible to grow a wider variety of crops. Widening the selection of crops can result in the use of more vigorous, efficient, and valuable cropping systems. The net result of the liberal use of fertilizers is greater efficiency in the utilization of land, labor, capital, and water.

Research in Arizona has demonstrated that the number of inches of water required to produce a ton of alfalfa hay varied from 6.8 to 12.8, depending upon the relative wetness of the soil and the rate of application of phosphate fertilizer. Water was most efficiently used when the soil was kept relatively wet and when the highest rate of phosphate fertilizer was used.[7]

---

[7] The "wet" plots were kept at a moisture tension of less than 175 centimeters (2.3 atmospheres) of tension at a depth of 12 inches, and the "highest rate" of phosphate fertilization was 600 pounds of P_2O_5 per acre, corresponding to 3000 pounds of 20 per cent superphosphate per acre. O. J. Kelley, "Requirement and Availability of Soil Water," *Advances in Agronomy,* Vol. 6 (1954), pp. 67–94.

TABLE 15.3   WATER USE EFFICIENCY OF WHEAT IS HIGHEST UNDER
MEDIUM IRRIGATION AND ADEQUATE FERTILIZATION *

Treatment	Water Level	Available Water to 5 Feet at Seeding (in.)	Rainfall (in.)	Available Water to 5 Feet at Harvest (in.)	Water Used (in.)	Yield (Bushels per Acre)	Water Use Efficiency (Bushels per Inch)
	Low	4.00	4.69	1.97	6.72	6.7	1.00
Without	Medium	6.87	4.69	3.56	8.00	12.8	1.60
fertilizer	High	7.27	4.69	3.82	8.14	14.6	1.79
	Mean						1.46
	Low	4.00	4.69	2.00	6.69	6.8	1.02
With	Medium	6.87	4.69	3.16	8.40	18.4	2.19
fertilizer	High	7.27	4.69	3.00	8.96	18.7	2.09
	Mean						1.77

* E. B. Norum, "Fertilized Grain Stretches Soil Moisture," *Moisture and Fertility*, American Potash Institute, Special Issue, 1963.

A corn-wheat-red clover experiment was conducted in Kentucky over a period of 12 years. During any one year, *regardless of the rainfall,* the highest yields were obtained on the plot that received lime, nitrogen, phosphorus, and potassium according to the results of a soil test.[8]

Studies of wheat yields in relation to available water in the surface 5 feet of soil, with adequate fertilization versus no fertilizer, have been made in North Dakota (Table 15.3). The greatest differences in yield due to fertilizer were 37 per cent at the "medium" water level.

FERTILIZERS IMPROVE SOIL STRUCTURE

Continuous tillage is usually conducive to a deterioration of desirable soil structure; however, when tillage is practiced with well-fertilized crops,

---

[8] O. P. Engelstad and E. C. Doll, "Crop Yield Response to Applied Phosphorus as Affected by Rainfall and Temperature Variables," *Agronomy Journal*, Vol. 53 (1961), pp. 389–92.

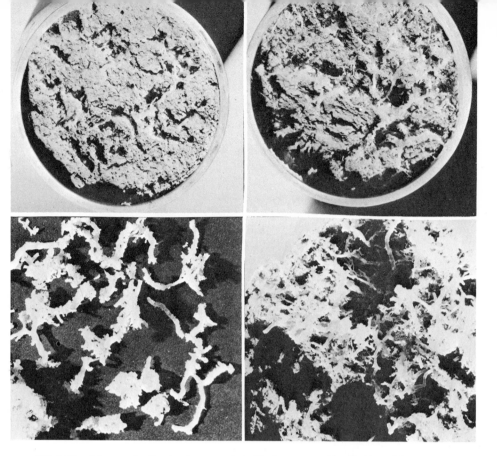

FIG. 15.9. Adequate fertilization increases desirable large pores in soil. *Above left and right:* Both soil cores came from field plots in Kentucky that for 16 years had the same rotation: corn-wheat-two years of grass + clover. The only difference was that the soil on the left had received no fertilizer, while the soil on the right had been adequately fertilized. *Below left:* Paraffin casts from the soil that had no fertilizer. *Below right:* Paraffin casts from the soil that had been adequately fertilized. *Note:* The paraffin was heated to a liquid and poured on each soil. The hot paraffin flowed into the larger pore spaces and solidified. *Source:* William Survant, Ky. Agr. Exp. Sta.

vigorously growing roots and abundant crop residues can improve soil structure.

Research in Kentucky compared the soil structure of two adjoining field plots that had had the same cropping system of corn-wheat-two years of grass plus clover for a period of 16 years; one plot was fertilized and the other was not. The results were as follows:

At the end of the 16-year period, soil cores were collected from each plot and hot paraffin was poured over them. The hot paraffin went into all large soil pores and hardened upon cooling. The soil was gently washed away with water, leaving a paraffin cast that represented the original large pores. The superior soil structure of the fertilized plot, as indicated by the more porous paraffin cast, may be seen in Figure 15.9.

SUMMARY

Commercial fertilizer is helping to keep a world population from outrunning its food supply. On a long-time average, fertilizer production is doubling every 15 years. The use of nitrogenous fertilizer is increasing faster than the use of either potassium or phosphorus, with potassium and phosphorus about equal. The increased use of ammonia for direct application to the soil and of nitrogen solutions for use in mixed fertilizers account for most of the rapid increase in the use of nitrogenous fertilizers.

For the 50 United States, consumption of N was 50 per cent of potential; P, 44 per cent; and K, 47 per cent.

Ammonia fertilizers may be safely applied in the fall when there is no leaching, when the soil stays colder than 50°F after application, or when there is a vigorous cool-season crop growing.

Anhydrous ammonia must be applied approximately 6 inches in depth to insure adequate adsorption. Solid nitrogenous fertilizers are usually applied on the surface of the soil. Both nitrogen and potassium fertilizers are toxic in large concentrations and must therefore be placed away from the seed. Surface-applied phosphorus appears to be a satisfactory practice on perennial grasses and legumes. Potassium must not be applied to organic soils in the fall because of serious losses by leaching.

Adequate liming and fertilization permit a wider selection of valuable crops, which increases infiltration and efficient use of water and results in an improvement in soil structure. Amounts of lime and N–P–K fertilizers to apply should be determined by a soil test. Fertilizers return approximately 3 dollars in increased yields for each dollar invested in fertilizer.

QUESTIONS

1. What is the cost of fertilizer relative to the cost of other production inputs?
2. Explain the fertilizer information gap.
3. Which form of N fertilizer is most efficient on alkaline soils?
4. Discuss the leaching losses of N.
5. What is the best way to determine the amount and grade of fertilizer to apply?

REFERENCES

Buckman, Harry O., and Nyle C. Brady, *The Nature and Properties of Soils* (7th ed.). New York: The Macmillan Company, 1969, 653 pp.

Fertilizer Recommendations for Vegetables and Field Crops in Michigan. Michigan State University Extension Bulletin E–550 (Feb., 1970), 24 pages. (*Note:* All states publish similar bulletins. To obtain a copy, contact the County Extension Director.)

Garman, William H. (ed.), *The Fertilizer Handbook.* Washington: National Plant Food Institute, 1963.

Hill, W. S., "The Need for Fertilizers," in *The Farmers' World: The Yearbook of Agriculture* (1964). United States Department of Agriculture, pp. 101–105.

Ignatieff, Vladimir, and Harold J. Page (eds.), *Efficient Use of Fertilizers.* Rome, Italy: Food and Agriculture Organization of the United Nations, 1962.

Knuti, Leo L., Milton Korpi, and J. C. Hide, *Profitable Soil Management.* Englewood Cliffs, N. J.: Prentice-Hall, Inc., 1962.

McVickar, Malcolm, *Using Commercial Fertilizers* (3rd ed.). Danville, Ill.: The Interstate Printers and Publishers, Inc., 1970.

McVickar, Malcolm H., *et al.* (eds.), *Agricultural Anhydrous Ammonia— Technology and Use.* Memphis: Agricultural Ammonia Institute, 1966.

Mehring, Arnon L., "Dictionary of Plant Foods," *Farm Chemicals Magazine.* Willoughby, Ohio: Meister Publishing Co., 1961.

Mullins, Troy, *Production Requirements and Estimated Costs and Returns for Rice and Beef Cattle Under Alternative Rotation Programs in the Coast Prairie, Texas,* Mp–801. Texas Agricultural Experiment Station, 1966.

Tisdale, Samuel L., and Werner L. Nelson, *Soil Fertility and Fertilizers* (2nd ed.). New York: The Macmillan Company, 1965.

Welch, L. F., P. E. Johnson, J. W. Pendleton, and L. B. Miller, "Efficiency of Fall- Versus Spring-Applied Nitrogen for Winter Wheat," *Agronomy Journal.* Vol. 58, No. 3, 1966.

CHAPTER 16

TILLAGE

Some folks begin their spring plowing when the county agent tells them to, some folks watch their neighbors, but we always wait till there's life in the ground.*

Soils must have the right proportion of air and water for plant roots to absorb adequate amounts of nutrients for luxuriant plant growth. Coarse-textured soils usually have good air and water relations but too small a capacity for supplying enough water and plant nutrients. Fine-textured soils normally have a satisfactory nutrient and water reserve but often contain too little air.

Plowing and cultivating coarse-textured soils do not seem to make them a better medium for plant growth. Tillage of fine-textured soils immediately improves the air-water relations for better plant growth (Figure 16.1).

---

* Mildred Walker, in *Winter Wheat* (New York: Harcourt, Brace, and World, Inc., 1944), p. 300. Used with permission.

FIG. 16.1. Houston Black Clay that has been bedded on the contour in preparation for planting cotton or grain sorghum. (Hill County, Texas.) *Classification:* Udic Pellustert; fine, montmorillonitic, thermic. *Source:* Curtis L. Godfrey, Tex. A&M U.

On all soils, the primary purpose of tillage is to control weeds. But with the use of new herbicides for the control of weeds, this age-old reason for tillage must be reexamined. Tillage temporarily aerates the soil and controls weeds, but year by year it destroys desirable soil structure and eventually reduces aeration.

In southeastern United States on a Norfolk sandy loam, a traffic pan that developed at a depth of 7 to 9 inches was too dense for cotton roots to penetrate.[1]

A virgin grassland soil is rich in organic matter and has a structure that is desirable for rapid plant growth. Under continuous tillage, the organic matter level decreases and the soil structure deteriorates. Continuous tillage, especially with the use of heavy machinery, tends to compact the soil and thus make it a less desirable medium for plant growth (Figures 16.2 through 16.4).

---

[1] J. L. Tackett and R. W. Pearson, "Oxygen Requirements for Cotton Seedling Root Penetration of Compacted Soil Cores," *Soil Science Society of America Proceedings*, Vol. 28 (1964), pp. 600–605.

FIG. 16.2. In our modern society, bigness has been equated with success. On farms, bigness means large acreages, more horsepower per tractor, and deeper and faster plowing. But heavy tractors pulling heavy-draft implements rapidly on clean-tilled soils has resulted in a destruction of desirable soil structure, more surface runoff, less infiltration of water for plant growth, less air in the soil, and a less favorable medium for seed germination, seedling growth, and reduced yields. *Above:* A 5-bottom turning plow as an example of twentieth century bigness, whose use often results in soil structure deterioration and a tillage pan. *Below:* Once-over corn planting in last year's grain stubble leaves the soil between the rows rough enough to absorb more rainfall (note at arrow), reduces surface soil crusting, reduces erosion by water and wind, and enhances plant growth. *Note:* All major machinery companies now promote "once-over" planting and have the equipment available for this purpose. *Sources: Above,* John Deere Co.; *below,* Allis-Chalmers Co.

FIG. 16.3. *Left:* A virgin soil that has developed under a tall-grass prairie. (Scale is in inches.) *Right:* The same kind of soil nearby that has been cultivated for 50 years or more. (Texas.) *Source:* Tex. Agr. Exp. Sta.

FIG. 16.4. Both soils are Austin Clay. The compact soil on the left came from a field where a tractor had been used for several years, whereas the porous soil on the right had been tilled only with a hand spading fork for 20 years. Note the root coming out from the earthworm hole near the center. *Source:* D. O. Thompson, Tex. Agr. Exp. Sta.

TILLAGE AND PLANT GROWTH

In the arid regions, the method of tillage is usually related to available moisture and to hastening organic decomposition for the release of nitrogen for plant growth.

Experiments and demonstrations to find the minimum amount of tillage for maximum yields have been carried on in many places. Corn in many states has been grown successfully with only one cultivation. The system is usually to spray, in one or two operations, an insecticide such as Di-Syston, mixed with a liquid starter fertilizer such as 6–18–6, to plant 28,000 kernels per acre in 20-inch rows, and to spray a mixture of a weedicide and an insecticide, such as atrazine and aldrin. (In 1968, this system in the Corn Belt resulted in a yield of 200 bushels of corn per acre.) [2]

At three locations in Virginia, no-tillage corn grain yields averaged 18 per cent higher (with a range of from 12.5 to 23.7 per cent) when the seed was planted in annual ryegrass (*Lolium multiflorum*) that had been killed with a suitable herbicide and left on the soil as a standing dead mulch. This was in comparison with plots where the annual ryegrass had been removed for use in making grass silage.

Other researchers demonstrated a 20 to 63 per cent *decrease* in desirable surface soil aggregation following conventional methods of tillage to control weeds, as compared with heribicide treatments.[3]

DEEP TILLAGE

Land that is plowed at the same depth for many years usually develops a compacted layer where the bottom of the plow slides over and compresses the soil. Compacted layers are formed at greater depths by the use of heavy machinery on moist, fine-textured soils. Some type of chiseling on such soils has often increased crop yields (Figure 16.5).

Chiseling in Indiana in connection with deep placement of fertilizers has increased corn yields approximately 30 per cent in some tests. The use of the chisel to a depth of 20 inches resulted in:

1. Temporary elimination of the plow sole and the heavy-machinery pan that was developed below the plow sole

---

[2] *Farm Journal* (July, 1969), p. 24D.

[3] G. M. Shear, "The Development of the No-Tillage Concept in the United States," *Outlook on Agriculture,* Vol. 5, No. 6 (1968), pp. 247–51.

FIG. 16.5. Chiseling at a depth of 12 inches increased cotton yields in Mississippi almost 90 per cent over the conventional method of land preparation. *Source:* Caterpillar Tractor Co.

2. An increase in plant roots in this layer as a result of chiseling and fertilization
3. Greater movement of water into the soil and more water being stored for longer periods within reach of plant roots
4. An increase in crop yields.

Breaking the soil with various implements was studied in Mississippi in relation to the yield of cotton. One series of plots was listed 8 inches deep (conventional method). Another was chiseled 12 inches deep with the chisels spaced 20 inches apart. A third treatment consisted in plowing with a disk plow 12 inches deep, and a fourth treatment involved plowing 8 inches deep with a conventional plow to which was attached a chisel that shattered the soil 4 inches deeper.

The yields of seed cotton demonstrated that the disk plow and the moldboard plow plus chisel were equally effective in producing the highest yield, namely, 1800 pounds per acre. The next highest yield resulted from chiseling 12 inches deep, with 1700 pounds; and least effective was the conventional method of listing (middlebusting) to a depth of 8 inches, with a yield of 900 pounds of seed cotton per acre. This test demonstrated the desirability of tillage to a depth of 12 inches to shatter the tillage pan.

TILLAGE TO CONTROL INSECTS

The timely plowing under of crop residues is an effective means of controlling certain insects. The Hessian fly, which is serious in many wheat fields, can be controlled by plowing under infested wheat stubble and volunteer wheat. The wheat jointworm is held in check by destroying all volunteer grain. Plowing under corn stalks reduces next year's crop of European corn borers. Timely cultivation aids in reducing grasshopper infestations by drying out their eggs. The cotton boll weevil and the pink boll worm in cotton are held in check by the early destruction and plowing under of cotton stalks.

TILLAGE AND ORGANIC MATTER

All forms of tillage in the spring help to warm and aerate the soil and to hasten the decay of organic matter. In any one season this is desirable because current crops benefit from nutrients released from the decomposing organic matter. If organic matter is not replenished, however, there will be a net loss over the years, sometimes with a serious reduction in crop yields.

Total organic matter lost during 65 years of continuous tillage in South Dakota was 40 per cent of the original organic matter.

SOIL MOISTURE AND TILLAGE

In arid and semiarid regions, either fall plowing or early spring plowing permits more water to be stored in the soil for use by crops than does late spring plowing. There is also more storage of water in the soil when no crop is grown for a year. This practice is known as *fallow*. But always the hazard of wind and water erosion in fallowed fields must be equated against increases in crop yields resulting from more available soil water.

The results of tillage research over a period of 18 years at three locations in western Kansas is portrayed in Figure 16.6. The average annual precipitation at these stations was 20 inches. Late spring plowing resulted in an average of 1 inch of available water at seeding time and an average yield of 8.4 bushels of wheat per acre. By contrast, early spring plowing allowed 1.8 inches of available soil water to be stored, which produced 11.5 bushels of wheat. A yield of 20.3 bushels of wheat per acre was obtained following a year of fallow when there were 5.9 inches of available water at seeding time.

KINDS OF PAN FORMATIONS

Pan formation may be *natural,* caused by geological or pedological forces, or they may be *anthropic,* having been induced by man.

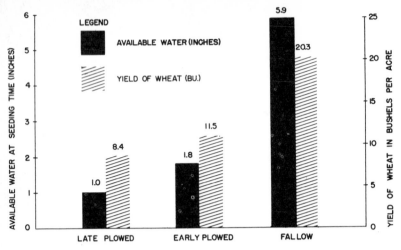

FIG. 16.6. The relationship between the time of plowing and fallow on available moisture at seeding time and the yield of wheat. *Source:* R. I. Throckmorton and H. E. Myers, *Summer Fallow in Kansas,* Kan. Agr. Exp. Sta. Bul. No. 293, 1941.

FIG. 16.7. Soil profile of a natural clay-pan soil (*Planosol*—soil on a plain) in Illinois. The high percentage of clay and its compaction in the B₂ horizon and the high percentage of compacted and acid, infertile silt in the A₂ horizon are the causes of the restricted root development of unfertilized corn in these two horizons. With no fertilization (1) corn yields averaged 42 bushels per acre; with proper fertilization (based upon a soil test) (2) corn yields were doubled, to 84 bushels. (Scale is in feet.) The soil classification of the profile is as follows: *Order*—Alfisol; *Suborder*—Aqualf; *Great Group*—Albaqualf; *Subgroup*—Mollic Albaqualf; *Family*—fine, montmorillonitic, mesic Mollic Albaqualf; *Series*—Cisne; *Type*—Cisne silt loam. (Fayette County, Illinois.) *Source:* R. T. Odell, U. of Ill.

FIG. 16.8. Tillage pans are created by plowing five-textured soils when too wet, and at the same depth each year. *Source:* Henry Corrow, Ext. Serv., U. of N.H.

Natural pans are of five types, depending upon the kind of cementation:

1. *Clay pans,* due to an accumulation of clay, usually in the B horizon (Figure 16.7)
2. *Silt pans,* caused by an excess of silt in or near the B horizon. These horizons are often called *fragipans.*
3. *Indurated horizons,* as a result of cementation by iron, aluminum, silica, calcium carbonate, calcium sulfate, or humus
4. *Dispersed horizons,* formed by the dispersing action of sodium on fine-textured soils
5. *Compacted horizons,* due to the application of forces during geologic deposition. One theory of their formation is that they were compressed by the weight of the glacial ice.

Anthropic pans, induced by man, are of three general types:

1. *Tillage pans,* caused by the compaction of heavy implements or by compression at the bottom of a plow. The latter often are called *plow pans* (Figure 16.8).
2. *Surface crusts,* created by the impact of falling raindrops on bare soil
3. *Hoof pans,* made by the packing action of grazing animals, especially on moist, fine-textured soils

339

TILLAGE PANS AND INFILTRATION RATES

The bottom of a plow sliding along a fine textured soil at the same depth year after year soon compacts the soil into a tillage pan. This compacted layer restricts root penetration, decreases permeability, and reduces the infiltration of water.

The effect of a tillage pan on the rate of infiltration in the southern Great Plains was studied. Thirty-five years of cropping to grain sorghums reduced the infiltration rate to 0.2 inches of water per hour. Below the plow pan, the infiltration was 9 inches per hour, or 45 times faster.

Probably the best cure for the plow pan is chiseling to at least 12 inches in depth, to be followed by perennial grasses and legumes in a regular cropping system (Figure 16.9).

CROPPING SYSTEMS AND SOIL STRUCTURE

Perennial grasses are probably the best vegetation for improving soil structure; continuous tillage seems to be the worst land use for maintaining desirable structure. Since the farmer usually makes more money from cultivated crops, there must be a compromise between continuous grass and continuous tillage (Figure 16.10).

FIG. 16.9. Tillage pans can be broken by some type of chisel set at a depth of about 12 inches. Note the greatest soil compaction at and below plow depth. *Source:* Caterpillar Tractor Co.

FIG. 16.10. Soil from various long-time cropping systems in Illinois: (1) virgin sod; (2) rotation of corn-oats-clover-wheat (clover); (3) rotation of corn-corn-corn-soybeans. *Source:* Ill. Agr. Exp. Sta.

The influence of various cropping systems on soil structure in Illinois is reported in Table 16.1. The cropping system of corn–oats–2 years of alfalfa–bromegrass resulted in 54 per cent soil aggregation and 68 bushels of corn per acre. Corn–oats–alfalfa was next most desirable in maintaining soil structure and corn yields. Other rotations, in order of effectiveness, were corn–oats (sweetclover); corn–oats; and 10 years of continuous corn.

Soil density, as measured by the weight of dry soil per cubic foot, is another way to report indirectly the results of cropping systems or other land use on the physical conditions of the soil. The surface foot of virgin soil averaged 66 pounds, whereas soil cropped for 40 years averaged 82 pounds per cubic foot on an oven-dry basis. The second-foot comparisons

TABLE 16.1 THE RELATIONSHIP BETWEEN CROPPING SYSTEM, PERCENTAGE AGGREGATION OF THE SOIL, AND YIELDS OF CORN *

Cropping System	Percentage of Plow Layer Aggregated	Corn Yield per Acre (Bu)
Corn–oats–2 years of alfalfa plus bromegrass	54	68
Corn–oats–alfalfa	53	59
Corn–oats (sweetclover)	45	47
Corn–oats	40	39
Continuous corn for 10 years	23	22

* R. S. Stauffer, *Tilth of Corn-Belt Soils,* Illinois Agricultural Experiment Station Circular 655 (1950). No fertilizer was used. Soil was dark-colored, level, clay.

FIG. 16.11. Tillage pans restrict deep root penetration, probably because of a reduction in soil oxygen and because there are too few crevices through which the roots can grow. *Source:* Caterpillar Tractor Co.

FIG. 16.12. Bulk density of virgin soils compared with that of corresponding tillage pans in four predominant soils in the Gulf Coastal Plains. These soils are classified as coarse-textured and any bulk density greater than 1.6 is reported to restrict plant root growth. *Source:* A. Kashirad, J. G. A. Fiskell, V. W. Carlisle, and C. E. Hutton, "Tillage Pan Characterization of Selected Coastal Plain Soils," *Soil Science Society of America Proceedings,* Vol. 31, No. 4, July–Aug., 1967, pp. 534–41.

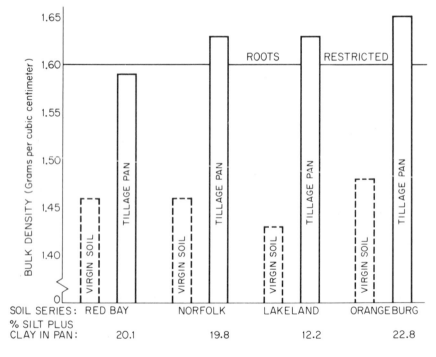

were 70 and 87, and the third-foot comparisons were 77 and 91 pounds per cubic foot, for virgin and cultivated soil, respectively. The 40 years of continuous cropping resulted in a soil compaction of more than 24 per cent in the surface 2 feet and 18 per cent in the third foot.

FACTORS RESTRICTING ROOT GROWTH IN COMPACT HORIZONS

Several soil scientists have discovered that normal root growth is severly restricted in fine-textured soil horizons when the bulk density is greater than 1.4, and in coarse-textured soil of bulk density above 1.6. Compact horizons may restrict root growth for several reasons:

1. There may not be enough channels through which roots can grow.
2. Oxygen may not be in sufficient concentration at all times for normal root respiration or for translocation of nutrients (Figures 16.11 and 16.12).
3. Water for plants may be in excess during rains and deficient between rains.
4. Available nutrients may be in too low a concentration.
5. Toxic substances may be present, such as a high concentration of manganese, carbon dioxide, or aluminum.

SUMMARY

Soils are cultivated to aerate and warm the soil and to control weeds. With the wise use of herbicides, tillage to control weeds is not now so essential as it was a few years ago. No-tillage crop production is popular.

In semiarid regions, early spring tillage is conducive to the storage of more available water and to higher crop yields.

Deep tillage, especially when combined with the establishment in the cropping system of perennial grasses and legumes, results in more desirable soil structure and greater crop yield.

Timely tillage is important in the control of certain insects, such as the Hessian fly, wheat jointworm, European corn borer, grasshopper, boll weevil, and pink boll worm.

There are five types of natural pan formations and three types of anthropic pans. The natural pans are clay pans, silt pans (fragipans), indurated horizons, dispersed horizons, and compacted horizons. Anthropic pans are tillage pans, surface crusts, and hoof pans. Plow pans decrease the infiltration rate of water as much as 45 times.

Continuous corn may have only 23 per cent of the surface soil aggregated, whereas a good cropping system consisting of corn–oats–2 years of alfalfa plus bromegrass may have 54 per cent of the surface soil aggregated. Continuous tillage for 40 years was responsible for compacting the surface 2 feet of soil more than 24 per cent, and the third foot was 18 per cent heavier as a result of tillage compaction.

QUESTIONS

1. Why is it necessary to cultivate less now than formerly?
2. Why does deep tillage often result in increased crop yields?
3. State the advantages and disadvantages of tillage.
4. Explain the formation of anthropic pans.
5. Explain how compact soil horizons may restrict root growth.

REFERENCES

Cook, R. L., *Soil Management for Conservation and Production.* New York: John Wiley & Sons, Inc., 1962.

Gerard, C. J., C. A. Burleson, W. R. Cowley, M. E. Bloodworth, and G. W. Kunze, *Hardpan Formation in Coarse and Medium Textured Soils in the Lower Rio Grande Valley of Texas.* Texas A&M University Bulletin No. B–1007, 1964.

Jones, J. N., J. E. Moody, G. M. Shear, W. W. Moschler, and J. H. Lillard, "The No-Tillage System for Corn (*Zea mays*)," *Agronomy Journal.* Vol. 60, No. 1 (Jan.–Feb., 1968), pp. 17–20.

Kashirad, A., J. G. A. Fiskell, V. W. Carlisle, and C. E. Hutton, "Tillage Pan Characterization of Selected Coastal Plain Soils," *Soil Science Society of America Proceedings.* Vol. 31, No. 4 (July–Aug., 1967), pp. 534–41.

Knuti, Leo L., Milton Korpi, and J. C. Hide, *Profitable Soil Management.* Englewood Cliffs, N. J.: Prentice-Hall, Inc., 1962.

McAlister, J. T., *Mulch Tillage in the Southeast.* United States Department of Agriculture Leaflet No. 512, 1962.

McKibben, George E., "No-Tillage Planting is Here," *Crops and Soils Magazine.* Madison, Wisc.: American Society of Agronomy, Apr.–May, 1968.

Millar, C. E., L. M. Turk, and H. D. Foth, *Fundamentals of Soil Science.* New York: John Wiley & Sons, Inc., 1965.

Power to Produce: The Yearbook of Agriculture (1960). United States Department of Agriculture.

Soil: The Yearbook of Agriculture (1957). United States Department of Agriculture.

Soils and Men: The Yearbook of Agriculture (1938). United States Department of Agriculture.

CHAPTER 17

WATER CONSERVATION

The ideal soil for crop plants, from the point of view of water supply, is one that takes in the water that falls, allows the excess to drain away, and holds enough for plants between periods of moistening. *

Certain principles have been established for efficient land use for water conservation.

1. Deep-rooted plants, such as trees and shrubs, use (transpire) more soil water than shallow-rooted plants, such as grasses. When the objective of land management is to increase the amount of water from a particular watershed, either by greater surface runoff or by greater deep percolation, it is therefore usually recommended to cut the deep-rooted trees and shrubs and to replace them with shallow-rooted grass.

2. Thinning dense coniferous stands of forest trees on north slopes of areas of heavy snowfall allows greater snow accumulation and therefore more water for runoff or deep percolation to increase streamflow.

In 1966, a detailed summary was made of runoff and erosion research in 24 states and Puerto Rico.[1] The conclusions are summarized as follows:

---

* Charles E. Kellogg in *Climate and Man: The Yearbook of Agriculture* (1941), U.S.D.A.

[1] The states that cooperated in this summary of erosion research were Arkansas, Georgia, Illinois, Indiana, Iowa, Kansas, Louisiana, Maine, Michigan, Mississippi, Missouri, Nebraska, New Jersey, New York, North Carolina, Ohio, Oklahoma, Pennsylvania, South Carolina, Tennessee, Texas, Virginia, Washington, and Wisconsin.

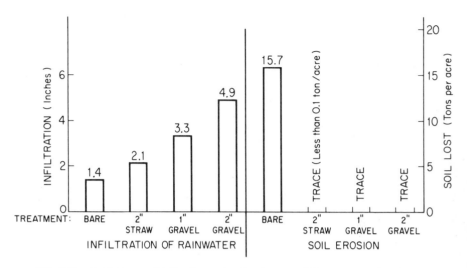

FIG. 17.1. Mulch increased infiltration and reduced erosion following an 8.49 inch rain on Austin Clay on Oct. 4, 1959, Temple, Texas. *Source:* John E. Adams, "Influence of Mulches on Runoff, Erosion, and Soil Moisture Depletion," *Soil Science Society of America Proceedings,* Vol. 30, 1966.

1. On the research plots that were cropped, the percentage of the total precipitation that did *not* soak into the soil (runoff) varied from 3 to 36. In absolute amounts, this represented from 1 to 14 inches of rainfall per year.
2. Runoff losses were reduced an average of 40 per cent by using recommended soil and crop management practices.
3. The greater the runoff the less the yield of all crops, especially corn.
4. Runoff losses were reduced slightly more than 15 per cent by contour tillage.
5. Runoff losses from row crops averaged a 10 per cent increase for each increase in per cent slope.
6. The length of slope made almost no difference in runoff losses.
7. On soils during periods when they had no crop (fallow), the higher the organic matter content in the soil the less the runoff.
8. A straw or gravel mulch increased infiltration of rain and decreased soil erosion (Figure 17.1).

THE HYDROLOGIC CYCLE

Water moves in a continuous cycle from ocean to clouds to earth to ocean, and from liquid to solid to vapor to liquid—always in response to

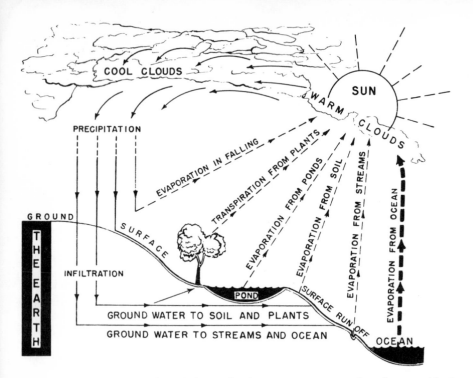

FIG. 17.2. The hydrologic cycle. The sun evaporates water from the sea and land to form clouds; then rain falls. Part of the rainfall is intercepted by plants; part is used immediately for plant growth; part is held by the soil as reserve for later use; part goes to underground reservoirs to sustain wells, springs, and streams; and part runs off to form rivers, which do their work as they return to the sea. *Source:* Soil Conservation Service.

physical and chemical laws of nature. These movements of water are known as the *hydrologic cycle.*

Figure 17.2 portrays the hydrologic cycle. Moist air from the ocean moves over land and is deflected upward. As the moist air mass rises, it cools and its ability to hold water vapor decreases. Vapor then condenses into raindrops which fall to earth. Some of the rain is intercepted by the leaves of plants, some runs off into streams, and some enters the soil. Of the water that enters the soil, some is used immediately by plants, some is stored in the soil for later use by plants, and the remainder moves downward to replenish the water table. The water table becomes the source of water for springs and wells.

Water is lost from the soil in four ways:

1. Surface runoff
2. Movement downward of drainage waters
3. Evaporation into the atmosphere from the surface of the soil
4. Transpiration through the leaves of plants into the atmosphere.

In water conservation the primary objective in humid regions is to encourage as much water as possible to enter the soil and thereby reduce runoff to a minimum. In arid regions where dryland crops are grown, the objective is the same as that in humid regions; but where water is impounded for irrigation, the main objective is to encourage as much runoff as possible without erosion.

PRECIPITATION

Precipitation in the United States averages 30 inches a year, with a variation of from under 10 to more than 80 inches (Figure 17.3).

East of the Great Plains, the precipitation is from 30 to 80 inches. The Great Plains receives from 20 to 30 inches; west of the Plains to the Rocky Mountains, 15 to 20 inches; and in the lowlands of the Intermountain Region, 10 to 15 inches. In Nevada, southeastern California, and southwestern Arizona, the precipitation is under 10 inches a year, the least in the nation. On the West Coast, precipitation is the highest in the United States, averaging more than 80 inches in places.

It is not the average annual precipitation that determines how much water is available for the growth of plants, but the amount of water which enters the soil by infiltration and is held there in a form that is available to plants at a tension between one-third and 15 atmospheres. To measure infiltration directly is difficult; for an indirect measure, the average annual runoff is subtracted from the average annual precipitation.

FIG. 17.3. Average annual precipitation in the United States. *Source: "A Water Policy for the American People," The Report of the President's Water Resources Policy Commission, Vol. 1, 1950.*

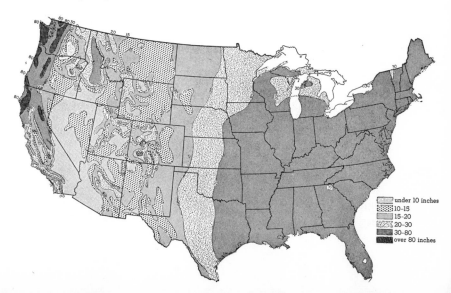

under 10 inches
10–15
15–20
20–30
30–80
over 80 inches

There are two omissions in this measure of infiltration. One is the water intercepted by the leaves, which never touches the soil, and the other is the water lost by evaporation from the surface of the soil. Intercepted water, water lost by evaporation, and runoff do not contribute to available soil water.

RUNOFF

Runoff in the United States varies from less than 1 to more than 10 inches a year. In arid and semiarid regions, loss of water by direct evaporation from the soil surface may amount to as much as half of the total precipitation. Of the three sources of water that are lost, runoff in humid regions is usually the largest in amount and the most damaging because it causes erosion. The less the average annual precipitation, the greater the percentage of runoff. From a water-conservation standpoint, it means that arid and semiarid regions have a doubly critical situation: Not only do these regions receive less precipitation, but also, a larger part of what is received is lost as runoff.

In some areas of the arid and semiarid region, runoff is encouraged as a means of obtaining larger supplies behind dams for use in irrigation. In many cases, however, the runoff is so rapid and the watershed is so poorly protected by vegetation that soil erosion is very serious.

STORAGE OF WATER IN THE SURFACE SOIL

Sand soils are capable of holding about 1 inch of rain per foot of soil depth. Clay soils often hold 1.5 inches per foot. Loams and silt loams are capable of holding 2 inches of water per foot.

Since most plants extend their roots approximately 4 feet down, the total amount of water capable of being held in the root zone is between 4 and 8 inches, depending upon soil texture, soil structure, and organic content. Only about half of this becomes available to plants; some is lost by evaporation and some is held so tightly by the soil that plants cannot get it. Water that plants can get is held with a force between one-third and 15 atmospheres.

The capacity of a soil to hold water, however, does not mean that all soils actually hold this amount. In arid and semiarid regions and in humid regions during periods of low rainfall, the actual water in the soil may be very little. The average soil in the United States may have a storage capacity

FIG. 17.4. Hardwood leaf litter is an excellent medium for storing water. Approximately one-half inch of water can be so stored. *Source:* U.S. Forest Service.

of approximately 6 inches, but because of low rainfall the actual average soil moisture stored is estimated to be 3.7 inches.

Virgin forest soils have an organic accumulation on the surface that acts as a sponge to increase the infiltration of water. Research in eastern Tennessee indicates that the leaf litter in hardwood forests serves as an excellent medium for storing water. In August, the average amount of hardwood litter on the forest floor was 4.2 tons per acre, with a field capacity of 135 per cent. This is equivalent to a water-holding capacity of 0.5 inch of rain. Following fresh leaf-fall in December, the leaf litter totalled 5.2 tons, equal to a water-holding capacity of 0.6 inch of rain. During flood periods, it is of tremendous benefit to have this much water held back on the land by the leaf litter rather than to have it contribute to the increasing of flood crests. Much of the water stored in the humus becomes available to growing plants (Figure 17.4).

STORAGE OF WATER AS GROUND WATER

Ground water is rain or melted snow that has moved downward through soil pores, plant root channels, or sand and gravel layers to meet the surface water (water table) from previous waters of infiltration. The bottom of the ground water is held in place by some impervious layer of rock or clay. Ground waters may then remain fairly stationary or they may move to form springs or water for wells.

It is estimated that the amount of ground water present in the United States consists of more fresh water than that contained in all surface reservoirs and lakes combined, including the Great Lakes. The amount is 300 inches, equal to 10 years' average rainfall or 35 years' average runoff.

Approximately one-sixth of all water used in the United States comes from the ground water. In many areas the ground water is being used faster than it is being replaced. The result is the lowering of the level of water in wells, which is a serious problem for irrigation projects, municipalities, and industries that depend upon ground water.

Some attempts have been made to recharge underground water supplies by spreading surface waters over porous beds of sand and gravel that extend to the water table. Also, some success has been obtained by discharging surface water supplies into wells, shafts, or pits.

STORAGE OF WATER IN FARM PONDS

The Soil Conservation Service has assisted farmers and ranchers in constructing 1¾ million farm ponds to serve a variety of useful purposes, including a reduction in runoff losses, an increase in the height of the water table, a source of water for livestock and human use, a site for recreation, and a source of water for irrigation and fire protection.

The storage of water in farm ponds and lakes, however, is subject to rapid losses by evaporation. To reduce such losses, the Bureau of Reclamation in the Department of the Interior began a series of research projects in 1954 in the 17 western states where the problem is most serious. One of the most promising techniques used to reduce evaporation losses was to spread on the surface of the water long-chain alcohols, such as hexadecanol and octadecanol. These techniques have proved successful and their economic feasibilities are now being tested. Research has also been conducted on the influence of these alcohols on fish and aquatic insects, and on wildlife drinking the water. After a period of 3 years, it was concluded that no pollutive or toxic effects could be detected that could not be controlled by timely rates of application.[2]

TERRACING FOR WATER CONSERVATION

For water conservation, the ridge terrace is the type that is recommended. In constructing a ridge terrace, the main object is to build a high ridge and a wide channel to impound as much water above the terrace as possible (Figure 17.5). In regions of approximately 20 inches of annual precipitation, terraces are constructed with high ridges, no grade in the

---

[2] William J. Wiltzius, *Effects of Monolayers on Insects, Fish and Wildlife,* Research Report No. 7, U.S. Department of the Interior (1967), 67 pp.

FIG. 17.5. Closed-end, level terraces in the 20-inch rainfall belt in Texas increased cotton yields an average of 60 per cent over a 26-year period. *Source:* John Deere Co.

channel (level), and with closed ends. These terraces are designed to hold *all* water that falls on a field until it soaks into the soil. Research in the 20-inch rainfall belt of Texas has shown that closed-end, level terraces impound more water and result in a 60 per cent increase in the yield of cotton as compared with cotton grown on land not terraced. These are average yields over a 26-year period.[3]

CONTOUR TILLAGE TO CONSERVE WATER

"Contour tillage" means the operation of all farm implements across the slope. By strict definition, the tillage should be exactly on the level, but in practice this condition is only approximated (Figure 17.6).

During the years 1936 to 1970, the Soil Conservation Service had assisted in the establishment of strip cropping on more than 21 million acres; and more than 47 million acres are now farmed on the contour.

Contouring is not new. In 1813 Thomas Jefferson, a good farmer as well as a good statesman, wrote:

> We now plow horizontally, following the curvature of the hills and hollows on dead level, however crooked the lines may be. Every furrow thus acts as a reservoir to receive and retain the waters; . . .

---

[3] Earl Burnett and C. E. Fisher, "The Effect of Conservation Practices on Runoff, Available Soil Moisture, and Cotton Yield," *Soil Science Society of America Proceedings* (1954).

FIG. 17.6. Contour tillage on a farm in Alabama, with winter-growing crops of crimson clover (light-colored strips) and small grain plus vetch (dark-colored strips). *Source:* Ala. Agr. Exp. Sta.

FIG. 17.7. The effect of contour tillage on runoff losses. *Source:* J. H. Stallings, *Effect of Contour Cultivation on Crop Yield, Runoff, and Erosion Losses,* Soil Conservation Service, 1945.

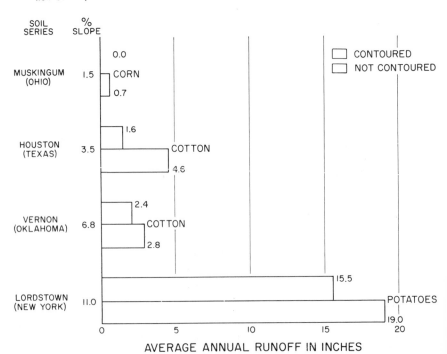

AVERAGE ANNUAL RUNOFF IN INCHES

scarcely an ounce of soil is now carried off. . . . In point of beauty nothing can exceed that of the waving lines and rows winding along the face of the hills and valleys.

The effect of contour tillage on runoff losses is given in Figure 17.7 for several crops in six states. In every instance, runoff is reduced by contour tillage. The greatest percentage reduction of runoff as a result of contouring is on Houston clay soil supporting cotton.

Benefits from contour farming may be summarized in this way:

1. Available soil moisture is increased.
2. Soil erosion is reduced.
3. Wind erosion is less.
4. Crop yields are increased.
5. Soil productivity is maintained.

FIELD CROPS AND WATER CONSERVATION

Research has shown that it requires approximately 4 inches of rainfall available in the soil to produce wheat plants from seed to mature plants; thereafter, each additional inch of rainfall available in the soil is responsible for producing 7 bushels of wheat. These estimates also depend upon adequate soil fertility.[4]

Transpiration of water by nonbeneficial vegetation, such as weeds and useless shrubs and trees, represents a loss of water in 17 western states estimated at 20 million acre-feet. A change of land use to conserve this loss of water is highly recommended.

Grasslands or any dense soil cover increase infiltration and thereby reduce runoff. Much of the water that moves into the soil later becomes available to plants. Proper land use is therefore equivalent to an increase in rainfall.

Figure 17.8 portrays the results from five erosion experiment stations comparing the runoff in inches from grasslands and corn lands. The saving in water through the use of grasslands varied from 2.1 to 3.9 inches, when compared with corn land.

Infiltration rates of water in South Carolina were studied in relation to soils under continuous cotton and soils under a cropping system of cotton-wheat-Kobe lespedeza (Figure 17.9). Water moved into the soil with con-

---

[4] David L. Guettinger, "One Hundred-Bushel Wheat," *Crops and Soils Magazine,* Vol. 18, No. 9. Madison, Wisc.: American Society of Agronomy (1966).

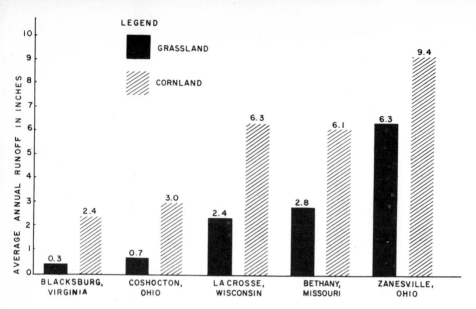

FIG. 17.8. A comparison of the runoff from grassland and cornland in the humid region. *Source:* Lloyd L. Harrold, "Effect of Increasing Grassland on Water Yields in Humid Areas," *Proceedings,* International Grassland Congress, 1952.

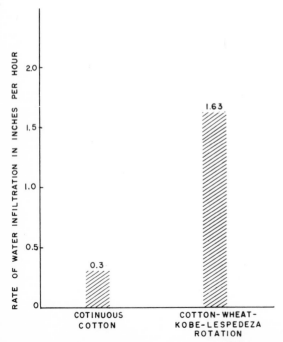

FIG. 17.9. Infiltration rates in a soil with continuous cotton as compared with cotton in a wheat-Kobe lespedeza rotation for a period of 10 years. *Notes:* (1) Soil was a Cecil sandy loam at Clemson, South Carolina. (2) The infiltration rate was measured after one hour of simulated rainfall. *Source:* T. C. Peele and O. W. Beale, "Laboratory Determination of Infiltration Rates of Disturbed Soil Samples," *Soil Science Society of America Proceedings,* 1955.

FIG. 17.10. Strips of grain sorghum stubble 3 feet wide and 50 feet apart and at right angles to the prevailing winter winds trapped more snow and thus conserved more soil moisture. Winter wind direction was from right to left (see arrow). (Colorado.) *Source:* B. W. Greg and A. L. Black, *Crop Residue Barriers for Snow Conservation in a Summer Fallow System,* Agronomy Abstracts, 1962. American Society of Agronomy.

tinuous cotton at a rate of 0.3 inch per hour, but into the soil under a rotation at the rate of 1.63 inches per hour. This is more than a fivefold increase in infiltration due to the crop rotation.

In the 25-inch precipitation belt, the Central Great Plains Field Station at Akron, Colorado, has demonstrated that parallel compact strips of grain sorghum stubble about 18 inches tall, 3 feet wide, 50 feet apart, and at right angles to the prevailing winter winds, are very effective barriers in trapping snow and thus adding more moisture to the soil (Figure 17.10). Increases in soil water as a result of the melting snow averaged 2.8 inches at a distance of 5 feet and 0.8 inches at a distance of 30 feet from the leeward side of the grain sorghum barrier.

GRASSES AND WATER CONSERVATION

Between 10 and 20 million acres of land in California lie in the foothills, receive 15 to 20 inches of annual precipitation, and mainly support worthless trees and brush. There are similar areas of land in most of the western, northwestern, and southwestern states.

The brush species are usually deep rooted and use all the available moisture in the soil. Furthermore, the brush, leaves, and stems intercept a large amount of precipitation that is evaporated directly back into the atmosphere.

Research on the conversion of such brush land to adapted grasses has demonstrated that:

1. Grasses root shallower than trees and brush.
2. Grasses become dormant earlier in the fall and thus leave more stored water in the soil for early spring growth.
3. Grasses intercept less precipitation, thus permitting more of it to enter the soil.
4. Grasses protect the soil from erosion better than brush.
5. Grasses permit more runoff water to be caught behind dams for use in irrigation. On some watersheds, as much as 2 inches more runoff have been saved.

FORESTS AND WATER CONSERVATION

Loblolly pine grows throughout the Coastal Plains and into the red hills of 11 southern states. On the Ruston soil series, the growth of loblolly pine was measured in relation to growing-season rainfall, April to September. Average height growth in feet of the dominant trees at age 50 years was the criterion used for comparative growth and was called "site index." In areas of 22 inches of growing-season rainfall, the site index of loblolly pine was 85; in regions receiving 26 inches, the site index was 89; with 30 inches, 93; and at 34 inches, 97. Even in humid areas, water thus seems to be a principal limiting factor in tree growth.[5]

In humid regions, one objective of forest management is to reduce runoff and thereby reduce flood hazards. On the other hand, in arid regions, a primary objective is to obtain as much runoff water as possible, mainly for irrigation purposes.

Forest trees, like all other plants, transpire large amounts of water as a vapor through their leaves and into the atmosphere. Forests also intercept considerable quantities of rain or snow and permit it to be evaporated back into the atmosphere before the water ever reaches the soil. Both transpiration and evaporation are often desirable losses in humid regions but are undesirable in semiarid areas.

By removing a part of the forest cover, more water will strike the soil and less per acre will be transpired. Another source of water loss appears, however, and may be a fairly large amount. The more cutting that is done in a forest, the more the sunlight strikes the ground and the more the water evaporates from the soil surface. But this loss is usually smaller than has been realized in the past.

---

[5] R. R. Covell and D. C. McClurkin, "Site Index of Loblolly Pine on Ruston Soils in the Southern Coastal Plain," *Journal of Forestry,* Vol. 65 (1967).

Note: See Chapter 6, pp. 126, 127 for a photograph and a description of the Ruston soil series.

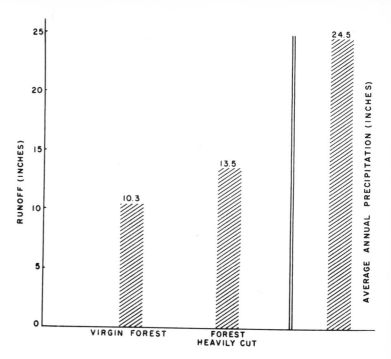

FIG. 17.11. The relationship between inches of runoff per acre in a virgin forest and in a heavily cut forest of lodgepole pine in Colorado. The average annual precipitation is shown for comparison. *Source:* H. G. Wilson, "Forests and Water," *Trees: The Yearbook of Agriculture* (1949), U.S. Dept. of Agr.

The net result of cutting a forest on the yield of water in the semiarid region of Colorado is demonstrated in Figure 17.11. This is in an area where the objective in water conservation is to obtain as much runoff water as possible for irrigation without soil erosion damage to the watershed.

Of the 24.5 inches of average annual precipitation, a virgin forest permitted a runoff of 10.3 inches per acre. On an adjoining plot, heavy cutting resulted in 13.5 inches of runoff per acre. This is a net gain of 3.2 inches of water for each acre that was heavily cut. On this basis, the increased yield of water from 10 acres of a heavily cut forest would be sufficient to irrigate an acre of sugar beets. But always the value of the increased water yield must be weighed against the possible deterioration of the watershed. Too little cutting results in more infiltration and less runoff; too much cutting, and there is insufficient protection to control soil erosion.

PLANTS AS LUXURY CONSUMERS OF WATER

Transpiration is a necessary function of plants. Under conditions of excess soil moisture, however, transpiration is a wasteful process.

Figure 17.12 illustrates the amount of water transpired by saltgrass in relation to the depth of the water table. When the water table stood at 4

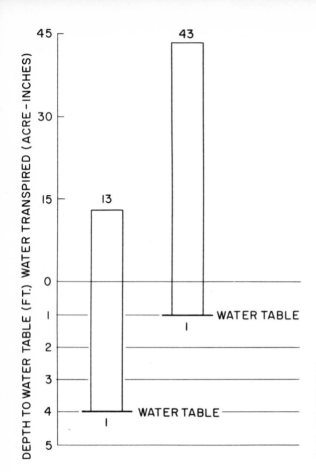

FIG. 17.12. The relationship between the amount of water transpired by saltgrass and the depth to the water table (Southern California). *Source:* Herbert C. Fletcher and Harold B. Elmendorf, "Phreatophytes—A Serious Problem in the West," *Water: The Yearbook of Agriculture* (1955), U.S. Dept. of Agr.

feet, saltgrass transpired 13 acre-inches of water a year. In contrast, 43 inches of water was transpired when the water table was only 1 foot deep.

Luxury transpiration is characteristic of nearly all plants, but especially of these plants: [6]

Saltgrass—*Distichlis spicata*
Greasewood—*Sarcobatus vermiculatus*
Salt-cedar—*Tamarix gallica*
Cottonwoods—*Populus species*
Baccharis—*Baccharis glutinosa*
Willows—*Salix species*
Mesquite—*Prosopis juliflora*

One practical application of the luxury consumption of water by certain plants was used on a ranch in Santa Clara County, California. The rancher

---

[6] Herbert C. Fletcher and Harold B. Elmendorf, "Phreatophytes—A Serious Problem in the West," *Water: The Yearbook of Agriculture* (1955), U.S.D.A., p. 424.

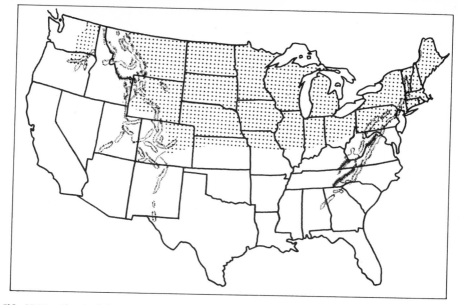

FIG. 17.13. The shaded areas represent that part of the United States where the freezing of the soil greatly reduces infiltration and increases runoff and erosion.

sprayed and killed the woody plants around a spring, and, with no additional rain, within a week the *increased* flow amounted to 28,000 gallons, or approximately 1 acre-inch of water. The plants were mostly willow, bay, white oak, and live oak.

PROBLEMS OF FROZEN SOILS

Many spring floods in the northern part of the United States are caused by the fact that water cannot move downward fast enough into the soil and must therefore flow over the surface and into streams. The cause is frozen, impermeable soils, and the result is damaging floods.

The problem of frozen soils is restricted to the northern one-third of the United States (Figure 17.13). Outside of this area, soil freezing does occur, but it is largely intermittent and of slight consequence as a factor in making soils impermeable to water.

On first thought, it may seem that the question of frozen soils is academic, that nothing can be done about it. This is not true.

Not all frozen soils are impermeable; some of them absorb water almost as fast when frozen as when not frozen. It is mainly a question of the amount of organic matter in the soil and the use of the land.

There are four principal kinds of frozen soil:

1. Concretelike
2. Honeycomb

3. Stalactite

4. Granular

Concretelike Soils: Bare soils, soils low in organic matter, and pasture soils that have been heavily grazed usually freeze into a concretelike, impermeable mass. These soils freeze to greater depths than do soils in any other type of land use.

Honeycomb: Soils high in organic matter, such as those in a well-managed forest or those that have been artificially mulched, usually freeze into a loose, porous, honeycomb type of structure.

Stalactite: Honeycomb structure that thaws on top for about 1 inch and then refreezes often forms stalactite structure. Here many icicles push upward to lift the surface crust as much as 6 inches. Stalactite structure is permeable to water. A similar type of freezing takes place when plants are heaved out of the ground.

Granular: Soils high in organic matter that freeze only to a shallow depth usually are of a granular structure. Continued freezing often changes the granular structure to a honeycomb structure. Granular structure permits water to move into it freely.

Any type of land use that adds organic matter, either living or dead, results in a soil that is open, porous, and easily penetrated by water even when frozen. Planting trees, leaving crop residues, adding manure, and planting sod crops all contribute to some type of porous structure when the soil freezes.

SUMMARY

The pressure of our growing population and its shift westward into areas of less rainfall make water one of our most critical resources. Understanding water problems will aid in their solution.

Water moves in a never-ending cycle from ocean to clouds to earth to ocean, and from a solid to a liquid to a gas, without loss. This is the hydrologic cycle.

Precipitation varies over the United States from less than 10 to more than 80 inches. Runoff can be reduced by planting trees, adopting a good crop rotation system, mulching, terracing, contour tilling, building ponds, and by grazing range and pasture land moderately.

If it is available, most plants transpire more water than they need for normal growth. One practical application of these findings is to destroy shrubs around water courses and thereby increase the amount of water available for the growth of palatable grasses and for other purposes.

A straw or gravel mulch increases infiltration of rainwater and thereby results in almost zero erosion.

If adequate organic matter is present, soils that freeze are still permeable to water during the winter.

QUESTIONS

1. Describe three ways to reduce runoff.
2. Describe a terrace system recommended to increase infiltration of water in arid and semiarid regions.
3. How can proper forest-cutting practices be used to increase water supplies for western irrigation?
4. What is meant by a "luxury consumption" of water?
5. What can be done to make soils capable of greater infiltration, even when frozen?

REFERENCES

Climate and Man: The Yearbook of Agriculture (1941). United States Department of Agriculture.

Cook, R. L., *Soil Management for Conservation and Production.* New York: John Wiley & Sons, Inc., 1962.

Donahue, Julian P., and C. R. Humphrys, *Ponds for Work and Fun.* Michigan State University Extension Bulletin E–374, 1961.

Donahue, Roy L., *Our Soils and Their Management: An Introduction to Soil and Water Conservation* (3rd ed.). Danville, Ill.: The Interstate Printers and Publishers, Inc., 1970.

Foster, Albert B., *Approved Practices in Soil Conservation.* Danville, Ill.: The Interstate Printers and Publishers, 1964.

Greenshields, Elco L., "Water Has A Key Role," in *The Farmer's World: The Yearbook of Agriculture* (1964). United States Department of Agriculture, pp. 75–96.

Hagen, R. M., H. R. Haise, and T. W. Edminster (eds.), *Irrigation of Agricultural Lands.* Madison, Wisc.: American Society of Agronomy, 1967.

Harrold, Lloyd L., *Water Intake by Soil. Experiments for High School Students.* Miscellaneous Publication No. 925, United States Department of Agriculture, 1963.

Hockensmith, Roy D. (ed.), *Water and Agriculture,* Publication No. 62, American Association for Advancement of Science. Washington: United States Government Printing Office, 1960.

The Hydrologic Cycle. Environmental Science Services Administration Pl 670003, Department of Commerce, 1967.

Jamison, V. C., D. D. Smith, and J. F. Thorton, *Soil and Water Research on a Claypan Soil.* Technical Bulletin 1378, United States Department of Agriculture and Missouri Agricultural Experiment Station, 1968, 111 pp.

Knuti, Leo L., Milton Korpi, and J. C. Hide, *Profitable Soil Management.* Englewood Cliffs, N. J.: Prentice-Hall, Inc., 1962.

McGinnies, William G., Bram J. Goldman, and Patricia Paylore (eds.), *Deserts of the World.* Tucson, Ariz.: The University of Arizona Press, 1968, 788 pp.

Patric, J. H., "Rainfall Interception by Mature Coniferous Forest of S. E. Alaska," *Journal of Soil and Water Conservation.* Vol. 21, No. 6, 1966.

Pierre, W. H., D. Kirkham, J. Pesek, and R. H. Shaw (eds.), *Plant Environment and Efficient Water Use.* Madison, Wisc.: American Society of Agronomy, 1967.

Russell, M. B. (ed.), "Water and its Relation to Soils and Crops," *Advances in Agronomy,* Vol. 11. New York: Academic Press Inc., 1959.

Schwab, Glenn O., Richard K. Frevert, Kenneth K. Barnes, and Talcott W. Edminster, *Elementary Soil and Water Engineering.* New York: John Wiley & Sons, Inc., 1957.

Science and Saving Water and Soil, Agricultural Information Bulletin 324. United States Department of Agriculture, 1967.

Soil and Water Conservation Needs—A National Inventory, Miscellaneous Publication No. 971. United States Department of Agriculture, 1965.

Soil Conservation Service, *Soil and Water Conservation Around the World.* United States Department of Agriculture, 1967.

Soil: The Yearbook of Agriculture (1957). United States Department of Agriculture.

Trees: The Yearbook of Agriculture (1949). United States Department of Agriculture.

Viets, Frank G., Jr., "Fertilizers and the Efficient Use of Water," *Advances in Agronomy* (1962), pp. 14, 223–64.

Water: The Yearbook of Agriculture (1955). United States Department of Agriculture.

Wischmeier, W. H., "Relation of Field-Plot Runoff to Management and Physical Factors," *Soil Science Society of America Proceedings.* Vol. 30, No. 2, 1966.

Woods, Lowell G., "Increasing Watershed Yield Through Management," *Journal of Soil and Water Conservation.* Vol. 21, No. 3, 1966.

Your Water Supply and Forest. Agricultural Information Bulletin 305, United States Department of Agriculture, 1966.

Note: The Environmental Science Services Administration, United States Department of Commerce, Washington, D. C., 20402, publishes bulletins and maps on the climate of each state and of the United States. These are for sale at low prices.

CHAPTER 18

SOIL CONSERVATION

Thousands of acres of land in this country are abandoned every year because the surface has been washed and gullied beyond the possibility of profitable cultivation.
—U.S.D.A. FARMERS' BULLETIN (1894)

The impact of the falling raindrop on bare soil is the major factor in starting the process of soil erosion. When plant cover, either living or dead, is present, the kinetic energy of the falling raindrop is dissipated by the springy organic carpet. As a result, the raindrop strikes the organic matter, the organic matter gives with the blow, and the raindrop slides harmlessly downward to soak into the soil for use by growing crops (Figures 18.1 and 18.2).

Untamed winds moving across unprotected soil will absorb surface moisture, break down soil granules, and carry, roll, or skip soil particles along with the wind. A plant cover, either living or dead, reduces surface wind velocities, maintains surface soil moisture, and prevents destructive wind erosion.

Soil erosion is present at higher elevations all over the United States, regardless of the amount of precipitation received.

Both water and wind erosion are serious factors tending to destroy our economy; both can be reduced to insignificance if the land is managed on the basis of scientifically proved practices.

FIG. 18.1. The impact of the falling raindrop on bare soil (*above*) beats the soil into flowing mud, and soil erosion has begun to bleed the life from the soil. This farm in Ripley, Mississippi (*below*) should have been left in trees. *Source:* Soil Conservation Service, U.S. Dept. of Agr.

FIG. 18.2. Luxuriant vegetation breaks the impact of the falling raindrop and thereby keeps erosion from starting. *Source:* Soil Conservation Service.

THE NATURE OF WATER EROSION

Soil erosion by water may be classified as (1) splash erosion, (2) erosion by surface flow, and (3) erosion by channelized flow.

Splash Erosion. One acre-inch of water weighs approximately 220,000 pounds but is capable of exerting splash erosion with a force of 3 million foot-pounds of kinetic energy, or almost 14 times its own weight.

The raindrop falls with an approximate speed of 30 feet a second (20 miles an hour). When raindrops strike bare soil, they beat it into flowing mud which splashes as much as 2 feet high and 5 feet away.

The soil textures most readily detached by raindrop erosion are fine sands and silts. Coarser particles are not shifted about much because of their greater size and weight. Most soils of finer texture, such as clays and clay loams, are not readily detached because of the strong forces of cohesion that keep them aggregated.

During a heavy rain, the soil aggregates are disrupted, splashed, shifted about, and packed together more closely. As the muddy water flows down through natural openings, the sediment plugs the channels. The result is a sealed surface which, upon drying, forms a crust slowly permeable to air and water. Water that would have moved downward during the rain now must move over the surface, carrying soil particles with it.

FIG. 18.3. Channelized flow of water results in rills, then gullies. (Montana.) *Source:* U.S. Forest Service.

Surface Flow. Runoff water is also a factor in soil erosion. Runoff water moves soil by (1) surface creep, (2) saltation (vaultation), and (3) suspension. *Surface creep* means movement of soil downhill by a rolling or dragging action of the water. *Vaultation* results when turbulent waters cause soil particles to hop or skip as they move downward. Soil particles that never touch the soil surface as they are moved along are carried by *suspension.*

Channelized Flow. As water moves over the surface of the soil, some of it concentrates in low places to form channels. Continued flow develops minor rills, and later large gullies may be formed by the scouring action of ever increasing volumes of muddy water (Figures 18.3 and 18.4).

FIG. 18.4. Unchecked rills develop into gullies. (Utah.) *Source:* U.S. Forest Service.

RAINFALL AND EROSION

A study of records for the nation reveals that there is not a good correlation between mean annual or mean monthly rainfall and erosion. More important than mean rainfall is the intensity of the rain and whether the soil has a protective cover.

In an 11-year study of erosion in the Texas Blacklands, of the 1055 rains received, only 141 of them resulted in any runoff. On the most erosive plot, 3 rainstorms caused 27 per cent of the erosion losses of soil, whereas 14 storms were responsible for 52 per cent of the erosion. More often than not, one storm a year usually caused more than half the soil loss. April and May were the months of greatest loss because the soil was bare as a result of having been prepared for planting crops at that time.

SOIL CONSERVATION DISTRICTS

Although the seriousness of soil erosion has been recognized by national and state leaders since the founding of the United States, a national movement to control erosion was not started until 1933, when the Soil Erosion Service was established. In 1935, the name of the Soil Erosion Service was changed to the Soil Conservation Service.

The Soil Conservation Service, a federal agency, offered direct assistance to farmers in controlling erosion. Starting in 1937, however, with the creation of the first Soil Conservation Districts, the Soil Conservation Service offered assistance to farmers only in organized Districts. All states, including Puerto Rico and the Virgin Islands, have passed laws authorizing the organization of Soil Conservation Districts. A total of 3026 Soil Conservation Districts had been organized by July 1, 1970, comprising more than one billion acres of land, or nearly 99 per cent of the farms and ranches in the United States.

CROPPING SYSTEMS TO REDUCE SOIL LOSSES

If a luxuriant crop covers the surface of the soil when the rains are most intense, there is almost no chance for the soil to erode. Keeping the surface protected and making a good living from the crops grown require scientific management.

An example of scientific soil management versus poor soil management in Maryland is given in Figure 18.5. A rotation of corn–soybeans–soy-

FIG. 18.5. Both soils were in a good rotation of corn-wheat-hay for 10 years. Two years before these photos were taken, the soil on the right was planted to corn-soybeans and soil structure deteriorated. The soil on the left remained in the corn-wheat-hay rotation and desirable soil structure was maintained. Source: Edward Strickling, Md. Agr. Exp. Sta.

beans resulted in a platy type of soil structure that tended to erode rapidly, whereas a rotation of corn–wheat–hay brought about a spheroidal type of soil structure that was resistant to erosion because it permitted greater infiltration of water and facilitated the preparation of a good seed bed.

Twenty years of experimental results on the relationship between cropping systems and soil erosion losses are identified for Missouri in Figure 18.6. Continuous bluegrass protected the soil and permitted only a trace of soil erosion. Next in order was a rotation of corn–wheat–red clover–timothy, which permitted 10 tons of soil loss per acre per year. The poorest

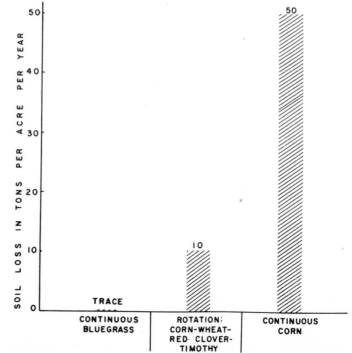

FIG. 18.6. Soil losses in relation to cropping system in Missouri. Source: Dwight D. Smith, Darnell M. Whitt, and M. F. Miller, "Cropping Systems for Soil Conservation," Mo. Agr. Exp. Sta. Bul. No. 518, 1948.

protection was given by continuous corn, which resulted in a loss of 50 tons of soil per acre each year.

When cropping systems fail to control soil erosion in the 11 southern states where loblolly pine grows naturally, planting loblolly pine has been proved a very desirable erosion control measure. Care in selection of the exact spot to plant each tree and mulching the soil around each tree planted are two practices that are necessary to assure survival.[1]

TERRACES TO CONTROL EROSION

The main reason for terracing in humid regions is to construct a ridge across a slope to guide the surplus water safely off the field. In dryland areas, terraces are constructed to increase water penetration. Dryland terraces were discussed in the preceding chapter.

In humid regions, the terracing system must be well designed and built under the supervision of a man who thoroughly understands the problems. Technicians of the Soil Conservation Service are available to carry on this type of work, in cooperation with the local soil conservation districts.

Terraces are usually recommended only on intensively used cropland. Land in a good grass cover seldom needs terracing for proper water control. Tillage parallel with the terraces is always a safe practice. When the field is plowed, the crest of the terrace should be the back furrow. Terraces need annual maintenance to keep them useful (Figure 18.7).

---

[1] Stanley J. Ursic, *Planting Loblolly Pine for Erosion Control in North Mississippi,* Southern Forest Experiment Station, Service Research Paper SO–3, United States Forest Service (1963).

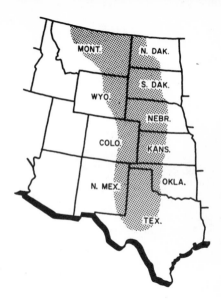

FIG. 18.8. The Great Plains region, where the most serious wind erosion takes place.

THE PROBLEM OF WIND EROSION

Some wind erosion occurs in nearly every state on unprotected sandy soils. But wind erosion is very serious on many soils in the Great Plains region (Figure 18.8). Twice during the last 40 years parts of the Great Plains have earned the unenviable name of "The Dust Bowl." On the average, approximately once in 20 years, the southern Great Plains has had a serious drought and severe wind erosion.

Most soil blowing starts on poor land that normally has a scant vegetative cover for protection. The Soil Conservation Service estimates that 14 million wind-erosion susceptible acres in the Great Plains that are now raising cultivated crops should be put in permanent grass to reduce wind erosion.

WIND EROSION A SELECTIVE PROCESS

A few years ago, soil material was blown from Texas and Oklahoma and analyzed at Clarinda, Iowa, a distance of more than 500 miles from its source. This windblown soil, as compared with the soil from whence it came, was

19 times higher in phosphorus
10 times higher in organic matter
1.5 times higher in potassium.

A mechanical analysis showed this dust to contain no sand but 97 per cent silt-plus-clay. Soil material deposited 300 miles away from its source

FIG. 18.9. Blowing sand caused severe damage to property and structures. *Source:* Soil Conservation Service.

contained 4 per cent sand, 68 per cent silt, and 25 per cent clay. This indicates that sand can be carried 300 miles but not as far as 500 miles under these conditions [2] (Figure 18.9).

CONTROLLING WIND EROSION

Foremost in wind-erosion control is to put the 14 million wind erosion-susceptible acres now in cultivation back to grass. Most of these areas cannot successfully be seeded to grass until a wet season occurs.

Stubble mulches are always recommended to control soil blowing. Crop residues on the surface help to hold winter snows, protect the soil from violent winds, improve soil structure, and increase the infiltration of rain-water.

The practice of using subsurface tillage implements that leave crop residues on or near the suface of the soil is known as *stubble mulch* farming. This method is an effective control measure of wind erosion on semiarid wheat land, and from 1936 to 1959 the Soil Conservation Service assisted farmers in establishing stubble mulch farming on approximately 84 million acres (Figures 18.10 and 18.11). The cumulative acreage of crop residue management, including stubble mulching, as of June 30, 1970, was approximately 157½ million acres.

On rangelands, the best practices to control wind erosion are as follows:

---

[2] J. H. Stallings, "Soil Fertility Losses by Erosion," *Better Crops with Plant Food Magazine* (Oct., 1951).

FIG. 18.10. This special Calkins sweep plow was designed to be used in the fall, after the harvest of grain sorghum or wheat, to provide subsurface tillage that leaves approximately 90 per cent of the stubble on the surface of the soil as a mulch to reduce wind erosion. (See Fig. 18.11.) (Texas Panhandle.) *Source:* Hugo Bryan, Soil Conservation Service.

FIG. 18.11. A stubble mulch established by the Calkins sweep plow holds winter snows, improves soil structure, increases the infiltration of water, and reduces wind erosion. *Source:* Soil Conservation Service.

1. Always leave a large percentage of the forage for reserve and residue. From 20 to 50 per cent should remain, depending on circumstances.
2. Livestock numbers (stocking) per unit area must be carefully adjusted to current rainfall and available grass.
3. Provide additional watering places for the livestock so that they will not overgaze areas adjacent to water.
4. Place salt boxes to encourage more uniform grazing. Livestock do not need salt near water.
5. Water spreading, gully control, brush eradication, and reseeding desirable grasses are other practices designed to reduce wind erosion.

The direction of the rows of crop plants in relation to the direction of the winds that prevail at the time of least surface soil moisture is very important. For example, when the rows of sorghum ran parallel to the prevailing winds, wind erosion was three times greater as compared with wind erosion when the rows of sorghum ran at right angles to the prevailing wind. This research was conducted at four locations in the Great Plains.[3]

SUMMARY

Raindrops striking bare soil are the primary cause of water erosion. Drying winds moving over unprotected soil start the process of wind erosion. The control of both water and wind erosion is achieved by providing protective cover for the soil.

Most erosion is caused by only one or two torrential rains or dust storms a year, which come at a time when the soil is least protected by crops. Both water erosion and wind erosion are selective processes, eroding the most productive parts of the surface soil. Yields of crops at any one location are greater when more topsoil is present.

The most protective cover is a healthy stand of trees or grass. Next in effectiveness is a cropping system that includes perennial grasses and legumes.

Terraces to reduce water erosion are recommended on many gently sloping cropland acres. The terraces must be laid out by a technician and carefully maintained. Terracing grasslands is usually wasted effort.

Wind erosion on the Great Plains is a very serious problem. It has been estimated that 14 million acres of cropland in this region should be planted to perennial grasses to reduce the hazard of wind erosion. The use of stubble mulches, terracing, contour furrowing, subsoiling, subsurface tillage, and row orientation are some of the practices used to reduce wind erosion.

QUESTIONS

1. What is the primary cause of water and wind erosion?
2. Describe splash erosion.
3. "Water and wind erosion are selective processes." Explain this statement.

---

[3] E. L. Skidmore, N. L. Nossaman, and N. P. Woodruff, "Wind Erosion as Influenced by Row Spacing, Row Direction, and Grain Sorghum Population," *Soil Science Society of America Proceedings,* Vol. 30, No. 4 (1966). Madison, Wisc.

Note: The four locations were Albuquerque, New Mexico; Midland, Texas; Salina, Kansas; and Great Falls, Montana.

4. Describe the ideal cropping system for erosion control.
5. In general, what can you do to control wind erosion?

REFERENCES

Beard, James B., "A Comparison of Mulches for Erosion Control and Grass Establishment on Light Soil," *Quarterly Bulletin, Michigan Agricultural Experiment Station.* Vol. 48, No. 3, 1966.

Bower, C. A., and Jesse Lunin, "Problems of Soil and Water," in *The Farmer's World: The Yearbook of Agriculture* (1964). United States Department of Agriculture, pp. 535–42.

Chepil, W. S., Earl Burnett, and F. L. Duley, *Management of Sandy Soils in the Central United States,* Farmers' Bulletin No. 2195. Washington: United States Department of Agriculture, 1963.

Cook, R. L., *Soil Management for Conservation and Production.* New York: John Wiley & Sons, Inc., 1962.

Eck, H. V., R. F. Dudley, R. H. Ford, and C. W. Gantt, Jr., "Sand Dune Stabilization Along Streams in the Southern Great Plains," *Journal of Soil and Water Conservation.* Vol. 23, No. 4 (July–Aug., 1968), pp. 131–34.

Foster, Albert B., *Approved Practices in Soil Conservation.* Danville, Ill.: The Interstate Printers and Publishers, Inc., 1964.

Foster, Albert B., and Adrian C. Fox, *Soil and Water Conservation Activities for Boy Scouts,* PA–348. United States Department of Agriculture, 1964.

Grubb, H. W., and G. S. Tolley, *Benefits and Costs of Soil Conservation in the South and its Subregions.* North Carolina Agricultural Experiment Station Technical Bulletin No. 172 (Apr., 1966).

Hafenrichter, A. L., John L. Schwendiman, Harold L. Harris, Robert S. Mac-Lauchlan, and Harold W. Miller, *Grasses and Legumes for Soil Conservation in the Pacific Northwest and Great Basin States.* Agriculture Handbook 339, Soil Conservation Service, United States Department of Agriculture (Apr., 1968), 69 pp.

Hockensmith, Roy D., and Phoebe Harrison, "Soil Conservation, A World Movement," in *The Farmers' World: The Yearbook of Agriculture* (1964). United States Department of Agriculture, pp. 69–75.

Jamison, V. C., D. D. Smith, and J. F. Thorton, *Soil and Water Research on a Claypan Soil.* Technical Bulletin 1378, United States Department of Agriculture and Missouri Agricultural Experiment Station, 1968, 111 pp.

Knuti, Leo L., Milton Korpi, and J. C. Hide, *Profitable Soil Management.* Englewood Cliffs, N. J.: Prentice-Hall, Inc., 1962.

Schwab, G. O., R. K. Frevert, T. W. Edminster, and K. K. Barnes, *Soil and Water Conservation Engineering.* New York: John Wiley & Sons, Inc., 1966.

Science and Saving Water and Soil, Agricultural Information Bulletin 324. United States Department of Agriculture, 1967.

Soil: The Yearbook of Agriculture (1957). United States Department of Agriculture.

Soil and Water Conservation Needs—A National Inventory, Miscellaneous Publication No. 971. United States Department of Agriculture, 1965.

Water: The Yearbook of Agriculture (1955). United States Department of Agriculture.

CHAPTER 19

IRRIGATION

I have made water flow in dry channels and have given an unfailing supply to the people. I have changed desert plains into well-watered land.—HAMMURABI (2067–2025 B.C.)

The amount of water used for irrigation in the United States was 20 billion gallons per day in 1900, 55 billion in 1920, 70 billion in 1940, 135 billion in 1960, and an estimated 177 billion gallons per day in 1980.[1]

The public wants to believe that atomic energy can be used economically to desalt seawater for use in irrigation. The hard facts are that water of good quality for irrigation can be so made. The only question is the *high cost*. Technology to accomplish this objective may be available at an acceptable cost at some future date but not now. What is available now is atomic energy desalting plants to produce water for use by municipalities that can afford to pay higher costs than can farmers for irrigation water.[2]

Indians in New Mexico and Arizona were using crude irrigation practices when the Spanish explorers came. But the first Anglo-Saxons to establish scientific irrigation practices were the Mormons. In 1847, the Mormons started to irrigate their crops near Salt Lake City, Utah. Today, these people are leaders in irrigation agriculture.

---

[1] *Source: Statistical Abstract of the United States,* 1965, U.S. Department of Commerce.

[2] Marion Clawson, Hans H. Landsberg, and Lyle T. Alexander, "Desalted Seawater for Agriculture: Is it Economic?" *Science,* Vol. 164 (June 6, 1969), pp. 1141–48.

When Theodore Roosevelt became President in 1901, one of his first acts was to establish an expanded program for conservation of soil and water in the arid West. As a result of this legislation, the Roosevelt Dam was started on the Salt River in Arizona in 1909. The impounded waters were used for making electricity, for providing recreation, and for irrigation.

From its beginnings in 1847, irrigation agriculture grew in the western states to nearly a 30-million-acre business. Although this represented only 3.5 per cent of the total land in farms in these western states, the crops harvested from irrigated land had a value of 35 per cent of all crops harvested in the region.

In the East, expansion in irrigation has been rapid. Irrigation is concentrated in the high plains of Texas, the rice belts of Texas, Louisiana, and Arkansas, and the vegetable areas in central and southern Florida. All other states in the humid region practice irrigation to some extent.

No one doubts the need for irrigation in the arid and semiarid West. Recent work in North Carolina may give some indication of the future need in humid regions. In an area receiving an average annual precipitation of 40 inches, as many as 50 days of drought may occur one year in five. This indicates that, even though the average precipitation seems to be adequate for crop production, irregularities in the distribution may make irrigation profitable.

A similar study was made of the "drought days" in the Lower Mississippi Valley, where the average annual precipitation varies from approximately 50 to 60 inches a year. If properly distributed throughout the growing season, this amount of rainfall would be sufficient to grow all crops except rice. Because of its distribution, however, many areas have "drought days" half the time, whereas only a few areas have "drought days" less than one-third of the time. As a consequence, irrigation is necessary to achieve maximum crop production on all but the most water-retentive soils in the Lower Mississippi Valley.

With increasing pollution of rivers, lakes, and oceans by municipal wastes, more emphasis is being placed on using city sewage effluents for fertilizing field crops in a modified irrigation system. At other locations, soil areas are being used to spray city effluent with no thought of raising farm crops. Bacteria, fungi, and other life in the soil are capable of "purifying" and decomposing large amounts of sewage in this way.

SOILS IN ARID AND SEMIARID REGIONS

Soils in arid and semiarid regions, compared with soils of humid regions, are characterized by:

1. A lower level of organic matter
2. Less acidity and more alkalinity
3. The presence of more calcium carbonate in some horizon
4. A weaker profile development
5. Coarser textures
6. Lower biological activity
7. Occasionally the presence of a desert pavement of stones, an accumulation of wind-drifted sand around clumps of vegetation, and/or moving sand dunes.

Many soils in arid and semiarid regions need only water to make them productive for most crop plants. Other soils in the West may be *saline* or *alkali* and require special treatment to make them productive.

Saline and alkali soils have been classified by the scientists at the United States Salinity Laboratory in Riverside, California, into *saline, alkali,* and *saline-alkali* soils.

Saline Soils. Soils are classified as saline if the solution extracted from a saturated soil paste has an electrical conductivity value of 4 or more milli reciprocal ohms per centimeter, usually written as 4 mmhos/cm at 25°C. This information is obtained on a special salt bridge, patterned after a common Wheatstone bridge. The amount of exchangeable sodium in saline soils is lower than 15 per cent; as a consequence, the pH is below 8.5.

For many years saline soils were called *white alkali* by soil scientists as well as by most farmers; now the term is gradually changing to *saline* soils.

Saline soils usually have a surface crust of white salts, especially in the summer, when the net movement of soil moisture is upward. Salts dissolved in the soil water move to the surface, where they are left as a crust when the water evaporates. These white salts are mostly chlorides, sulfates, and carbonates of calcium and magnesium. Owing to the small amount of sodium present, saline soils are in a flocculated condition, and consequently, when drainage will permit, the excess salts can readily be leached below the root zone with irrigation water.

Alkali Soils. The percentage of exchangeable sodium saturation in alkali soils is greater than 15; as a result, the pH is between 8.5 and 10.0. The saline content is below 4 mmhos/cm at 25°C, as measured on a salt bridge. The former name of these soils was *black alkali,* because they usually are black, owing to the effect of the high sodium content and the dispersal of organic matter. Locally, many of the areas are known as *slick spots,* because, when the soil is plowed slightly wet, it turns over in slick, rubbery furrow slices.

Because of the high exchangeable sodium content, both the clay and the organic matter are dispersed, and the result is close packing of the soil particles. The close packing of the particles reduces the size and the amount of pore spaces, and, as a consequence, water and air will not move through the soil readily. Poor aeration and high exchangeable sodium content, which is often toxic, make alkali soils difficult to reclaim.

Saline-Alkali Soils. The term "saline-alkali" applies to soils that are both saline and alkali. For example, they have

1. A conductivity of the saturated extract greater than 4 mmhos/cm at 25°C
2. Exchangeable sodium in excess of 15 per cent
3. A variable pH, depending upon the relative amounts of exchangeable sodium and soluble salts. When most of the soluble salts have leached downward, the pH will rise above 8.5 because of the remaining sodium; but when the soluble salts again accumulate near the surface, the pH may again fall below 8.5 (Figure 19.1).

RECLAMATION OF SALINE AND ALKALI SOILS

Saline soils are relatively easy to reclaim for crop production if adequate amounts of low-salt irrigation water are available. The main problem is to leach most of the salts downward and out of contact with subsequent irrigation water.

Frequently the saline soils have a high water table, a dense gypsum layer, or are fine textured. These conditions reduce the movement of irrigation water downward and therefore make it difficult to leach the salts to the desired depth. In salty soils with a high water table, artificial drainage is necessary before the excess salts can be removed. Deep chiseling or deep plowing is sometimes used on soils with impervious layers to open the soil for the desired downward movement of salt.

Reclamation of saline soils may also be hastened by the application of a surface organic mulch, as reported in the Rio Grande Valley of Texas. Cotton gin trash and chopped woody plants were equally effective when applied at the rate of 30 tons per acre. With a mulch, the surface soil salt content was less whether the area received only natural rainfall or supplemental sprinkler irrigation.[3]

The reclamation of alkali soils is another story. In alkali soils, the exchangeable sodium is so great as to make the soil almost impervious to water. But even if water *could* move downward freely in alkali soils, the

---

[3] D. L. Carter and C. D. Fanning, "Mulches Help Remove Salts," *Crops and Soils Magazine* (June–July, 1964), p. 26.

FIG. 19.1. Toxic salt accumulation (white layer on ridge at arrow No. 1) has prevented the growth of all plants on this irrigated cotton field in southwestern Texas. The accumulation of excess salt could have been caused by a slight surface depression and/or a concentration of clay in the soil profile that reduced infiltration of irrigation water, as seen in the insert (at arrow No. 2). (Scale on insert is in feet and tenths of feet.) *Source:* P. J. Lyerly, Tex. Agr. Exp. Sta., El Paso, Texas.

water alone would not leach out the excess exchangeable sodium. The sodium must be replaced by another cation and then leached downward and out of reach of plant roots.

By cationic exchange, calcium is often used to replace sodium in alkali soils. Of all calcium compounds, calcium sulfate (gypsum) is considered the best for this purpose.

Applications of 18 tons of gypsum per acre in Nevada increased water infiltration and increased the depth of water penetration. Three years after applying the gypsum, the water penetrated to a depth of 19 inches in the soil receiving the gypsum, and to 10 inches in the soil which received no gypsum. This resulted in a reduction of exchangeable sodium percentage from 42 to 18 per cent during the 3 year period. At the same time, the plot without gypsum gained in exchangeable sodium from 50 to 53 per cent. Yields of hay were increased from 0.05 to 1.02 tons per acre per year as a result of the application of gypsum.

Sulfur, which readily oxidizes to sulfuric acid, is used also in the reclamation of alkali soils and for lowering the pH of the soil. The amount of sulfur required to lower the pH from 9.0 to 6.5 for sandy soils and clay soils is given in Table 19.1. The amount of sulfur and gypsum required to reclaim alkali soils, based upon the exchangeable sodium in the soils, is shown in Table 19.2.

TABLE 19.1   AMOUNT OF SULPHUR (95%) REQUIRED TO LOWER THE SOIL pH TO 6.5 *

Soil pH	Sulfur Broadcast (lb per acre)	
	Sand Soils	Clay Soils
7.5	500	900
8.0	1250	1750
8.5	1750	1750
9.0	2500	(Not recommended)

* Fertilizer Handbook, 1963, National Plant Food Institute, Washington, p. 36.

TABLE 19.2   AMOUNTS OF GYPSUM AND SULFUR REQUIRED TO RECLAIM ALKALI SOILS *

Exchangeable Sodium in Milliequivalents per 100 Grams of Soil	Tons per Acre for Top One Foot of Soil	
	Sulfur	Gypsum
2	0.64	3.4
4	1.28	6.9
6	1.92	10.3
8	2.56	13.7
10	3.20	17.2

* L. A. Richards (ed.), Diagnosis and Improvement of Saline and Alkali Soils, Agriculture Handbook No. 60, U.S.D.A. (1969), p. 49.

SALT TOLERANCE OF FIELD CROPS

Under some circumstances it may not be feasible to reduce the salt content of soils to permit the growth of sensitive crops. The alternative is to select crops that are tolerant of salt.

A classification of field crops based upon their relative salt tolerance is shown in Table 19.3. Crops that are tolerant of salt are barley, sugar beet, rape, and upland cotton. Those intermediate in tolerance include rye, wheat, oats, sorghums, soybeans, sesbania, broadbean, corn, rice, flax, sunflower, and castor bean. Field bean is the most sensitive row crop shown here. However, during seed germination, sugar beets are sensitive to salt, and young seedlings of rice, barley, and sesbania are also sensitive.

Exchangeable sodium may be in a high enough concentration to injure crop growth, depending on how plants vary in their tolerance. For example,

TABLE 19.3   RELATIVE TOLERANCE OF FIELD CROPS TO TOTAL SALT *

High Salt Tolerance (8–12 mmhos)	Medium Salt Tolerance (4–8 mmhos)	Low Salt Tolerance (2–4 mmhos)
Barley (grain)	Rye (grain)	Bean (field)
Sugar beet	Wheat (grain)	White clover
Rape	Oats (grain)	Meadow foxtail
Cotton (upland)	Sorghum (grain)	Alsike clover
Safflower	Sorgo (sugar)	Red clover
	Soybeans	Ladino clover
	Sesbania	
	Broadbean	
	Corn (field)	
	Rice	
	Flax	
	Sunflower	
	Castor bean	

* In order of decreasing tolerance. *Source:* Leon Bernstein, *Salt Tolerance of Field Crops.* U.S.D.A., Agr. Inf. Bul. No. 217, 1960.

Figure 19.2 portrays that bean plants are severely injured at an exchangeable sodium percentage above 15, but that sugar beets can withstand higher concentrations.[4]

---

[4] George A. Pearson, *Tolerance of Crops to Exchangeable Sodium,* Agricultural Information Bulletin No. 216, U.S.D.A. (1960), p. 3.

FIG. 19.2. Beans are more sensitive to exchangeable sodium than beets. Other crops tolerant of exchangeable sodium include crested wheatgrass, tall wheatgrass, and rhodesgrass. *Source:* George A. Pearson, *Tolerance of Crops to Exchangeable Sodium,* U.S. Dept. of Agr., Agr. Inf. Bul. No. 216, 1960.

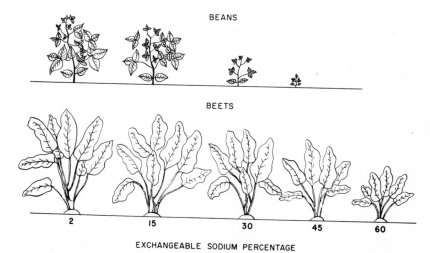

BEANS

BEETS

2   15   30   45   60

EXCHANGEABLE SODIUM PERCENTAGE

0	0.1	0.3	1.0	3.0	10

Millimhos of Calcium Sulfate

FIG. 19.3. The appearance of the bean plant after a 7-day nutrient solution experiment. All nutrient solutions had a uniformly very high concentration of sodium chloride of 50 millimhos, but each solution had varying amounts of soluble calcium as calcium sulfate, as indicated under the photograph. The calcium sulfate reduced the toxicity of sodium chloride. Source: P. A. LaHaye and Emanuel Epstein, "Salt Toleration by Plants: Enhancement with Calcium," Science, Vol. 166, No. 3903, Oct. 17, 1969.

A recent experiment in California has indicated that salt tolerance of the bean plant can be increased with additional calcium. With a uniformly very high concentration of sodium chloride of 50 mmhos in the aerated nutrient solution, increasing amounts of soluble calcium supplied as calcium sulfate enhanced the growth of the bean plant (Figure 19.3).

WATER QUALITY

Water for irrigation must not only be plentiful throughout the growing season, but it also must be of the right quality. The quality of water is determined mainly by four characteristics:

1. The total concentration of soluble salts
2. The amount of sodium in relation to calcium-plus-magnesium

3. The amount of bicarbonate

4. The presence of boron or lithium in amounts that may be toxic.

The soluble salts in irrigation waters are mainly the cations: calcium, magnesium, and sodium. Anions consist of sulfates, chlorides, and bicarbonates.

The soluble salts in eastern surface waters are seldom in sufficiently high concentration to be harmful, except in brackish waters near the ocean. In western surface waters, however, salt content is usually too high. Western waters in arid regions may vary from 70 to 3500 parts per million of soluble salts. This is a range of 0.1 to 5 tons of salt per acre-foot of water. The higher the salt concentration in irrigation waters, the greater the hazard of toxic accumulations in the soil.

Ground waters in the East are usually free of harmful salt concentrations; but in western areas, ground waters may be as high or higher in total salts than surface waters.

In irrigation language, there is a saying, "Hard water makes soft land and soft water makes hard land." The explanation of this statement is that calcium and magnesium, the two principal cations that make water hard, help in creating desirable soil structure. Sodium, the dominant cation in soft water, disperses clay and humus and creates an undesirable structure.

When there is a large amount of sodium in the irrigation water and a low percentage of calcium and magnesium, the sodium is readily absorbed on the surface of the clay and humus particles. The sodium disperses the soil and makes it less able to absorb water. If the same amount of sodium is present but there is a larger quantity of calcium and magnesium, less sodium is absorbed, and therefore a better soil structure is maintained. It is therefore the amount of sodium in relation to calcium and magnesium that is important in judging the quality of irrigation waters.

The bicarbonate ion is toxic, especially to apple trees, beans, and Dallisgrass. Specific toxicity of the bicarbonate ion may be found on other plants as more research is conducted in this field. It has been demonstrated that the bicarbonate ion accumulates in the soil and interferes with the uptake of iron in plants. This condition causes *iron chlorosis.*

Boron is present in nearly all western irrigation waters. A boron concentration as high as 3 parts per million is satisfactory only for such boron-tolerant crops as asparagus, sugar beets, alfalfa, and cabbage. This concentration of boron is toxic for the sensitive crops, such as navy beans and almost all tree fruits.

Lithium concentration in the irrigation water of 0.1 part per million may be toxic to citrus trees.

Irrigation waters may contain fairly high concentrations of selenium, molybdenum, and fluorine, which are absorbed by roots without damage to the plant; but when the plants are eaten by livestock, the concentration of any one element may be toxic to the animal.

IRRIGATION "LANGUAGE"

The common language used by the irrigator must be understood by the soil scientist before communication can be achieved.

In all large irrigation projects, the field man who makes possible the delivery of irrigation water to the farm is called a *ditchrider* in the northern states but a *zanjero* (from the Spanish word *zanja,* meaning "ditch") in the Southwest.

To the irrigator, *intake* or *infiltration* refers to the rate at which water enters the soil, and *penetration* or *permeability* is the movement of water within the soil. The surface soil immediately after irrigation is called *saturated,* and the water held by the soil after free drainage has taken place is known as *moisture-holding capacity* or *field capacity.* When plants begin to wilt, the water left in the soil is called the *wilting point.* The amount of water between the field capacity and the wilting point is termed *available water.*

On salty soils with good internal drainage, the irrigator may add an excess of water to *leach* the salts below root depth so that they will be less harmful.

Leveling and *grading* denote the movement of soil in preparing land for irrigation. *Planing, smoothing,* and *floating* are interchangeable terms and refer to the final touch-up in preparing land for irrigation or to the annual operation of eliminating small irregularities in the land.

Although *acre-inches* is a typical term among soil scientists, irrigators usually use other terms to refer to ordering or measuring water. *Second-feet* (cubic feet per second, c.f.s.), *gallons per minute* (g.p.m.), and *miner's inch* are commonly used by irrigators. Approximately 450 gallons per minute equal 1 cubic foot per second, and 1 cubic foot per second flowing for 1 hour will cover 1 acre of land 1 inch deep (1 acre-inch), assuming there is no infiltration and no evaporation. The *miner's inch* is variously defined, depending upon the state:

1. In Idaho, Kansas, Nebraska, New Mexico, North Dakota, South Dakota, Washington, and northern California, 50 miner's inches equal 1 cubic foot per second.
2. In Arizona, southern California, Montana, and Oregon, 40 miner's inches equal 1 cubic foot per second.
3. In Colorado, 38.4 miner's inches equal 1 cubic foot per second.

CONSUMPTIVE USE OF WATER

"Consumptive use" is defined as the sum of the water transpired by plants and the water evaporated from the surface of the soil.

Factors influential in determining the consumptive use for any crop are:

1. Length of growing season
2. Temperature
3. Daytime hours.

Other factors of minor importance are relative humidity and wind.

The annual consumptive use of water was determined for eight crops in South Dakota and is shown in Table 19.4. The consumptive use varies from 27 inches for alfalfa to 14 inches for dry beans. The water supplied by rain varies from 9 to 12 inches, depending on the length of time the crop is growing. Perennial crops, such as alfalfa and long-season annuals, such as sugar beets, benefit more from natural rain than do shorter-season crops. The column "Supplied by Irrigation" is an average value determined for a period of years at 32 locations in South Dakota.

A 4-year experiment was conducted to determine the amount of water used by maize. The maize used 6 acre-inches when the soil surface was covered with a plastic sheet (resulting in no evaporation), 13 acre-inches

TABLE 19.4 ANNUAL CONSUMPTIVE USE OF WATER FOR SELECTED
CROPS IN SOUTH DAKOTA *

Crop	Annual Consumptive Use of Water		Total Consumptive Use
	Supplied by Rain (in.)	Supplied by Irrigation (in.)	(in.)
Alfalfa	12	15	27
Grass, hay, or pasture	12	12	24
Sugar beets	12	10	21
Corn	9	10	19
Flax	9	10	19
Potatoes	9	10	19
Small grains	9	7	16
Dry beans	9	5	14

* Leonard J. Erie and Niel A. Dimick, *Soil Moisture Depletion by Irrigated Crops Grown in South Dakota*, South Dakota Agricultural Experiment Station Circular 104 (1954).

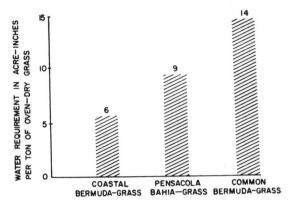

FIG. 19.4. A comparison of the water requirement (consumptive use) of three Southern grasses. *Source:* Glenn W. Burton, E. H. DeVane, and Gordon M. Prine, *Agronomy Abstracts,* 1955.

when the maize was grown under farm conditions with no irrigation, and 16 acre-inches when the maize was irrigated.[5]

The kind of grass makes a big difference in the efficient use of water. Some grasses produce a large amount of dry matter per acre-inch of water; other grasses, even though closely related, are more wasteful. Figure 19.4 illustrates this statement. Coastal Bermudagrass, a giant hybrid, produced 1 ton of dry matter for each 6 acre-inches of water. In contrast, it took 9 acre-inches of water for Pensacolagrass and Bahiagrass and 14 acre-inches for each ton of dry matter of common Bermudagrass. The common Bermudagrass, therefore, used $2\frac{1}{3}$ times as much water per unit of dry matter as did Coastal Bermudagrass.

Daily consumptive-use values have been determined for cotton and for grain sorghums on the high plains (panhandle) of Texas. The research was carried out on Pullman clay loam. Cotton uses the largest amount of water around July 15, the average date of the first bloom. At this time, the consumptive use of water for cotton is approximately 0.25 inch of water per day. Grain sorghum needs a maximum of 0.33 inch of water daily just prior to "booting." [6]

Studies have also been made of the daily consumptive use of water for forests. From June 1 to July 15, the consumptive use of water for fully stocked forest stands in Arkansas was 0.19 inch per day. Water depletion was determined weekly to a depth of 60 inches. Measurements were made on even-aged and all-aged hardwoods (southern red oak, post oak, and sweetgum), and even-aged and all-aged softwoods (loblolly pine and short-leaf pine). The consumptive use of water was the same, regardless of the composition of the forest.[7]

[5] Doyle B. Peters, "Water Use by Field Crops," *Plant Food Review,* National Plant Food Institute, Washington (Spring, 1961).

[6] E. L. Thaxton, Jr., "Irrigating Cotton and Grain Sorghum," *Texas Agricultural Progress,* Vol. 2, No. 2 (March–April, 1956), Texas A. & M. College System (Now Texas A&M University), College Station, Texas.

[7] *Soil Moisture Depletion Equal under Pines or Hardwoods,* 1955 Annual Report, Southern Forest Experiment Station, p. 29.

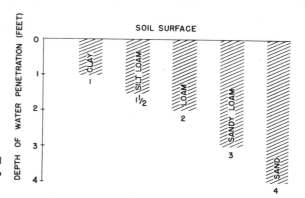

FIG. 19.5. Three inches of water will wet each soil texture approximately to the depth shown.

WATER-HOLDING CAPACITY OF SOILS

The depth of penetration in the soil of a specific amount of infiltrated water depends upon the water-holding capacity of the soil. This, in turn, is influenced mainly by soil texture.

In Figure 19.5, it may be seen that 3 inches of water, either as rain or irrigation, wets each soil texture to definite soil depths. For example, 3 inches of water wets a clay soil to a 1-foot depth, but a loam to 2 feet and a sand to a 4-foot depth.

NORMAL IRRIGATION DEPTH FOR CROPS

The depth to wet the soil for the best response of crops will vary with the depth of rooting of each crop. Probably the best rule is to irrigate to a depth where 90 per cent of plant roots are growing.

Figure 19.6 gives the normal irrigation depth for several crops. For pastures, a depth of 1.5 to 3 feet is satisfactory, whereas for alfalfa, 4 to 8 feet is the recommended depth to irrigate. Other crops should be irrigated to intermediate depths.

AMOUNT AND FREQUENCY OF IRRIGATIONS
(BY CALCULATION)

Based upon the consumptive use of the crop, the rooting habit of the crop, and the water-holding capacity of the soil, the amount of water to apply when there is no rain during the growing season can be determined for each crop in each field.

For an example of how to calculate the amount of water to apply, look at Table 19.4. These data show that corn has a consumptive use of 19 inches. Assuming that the corn is growing in a silt loam soil, Figure 19.5

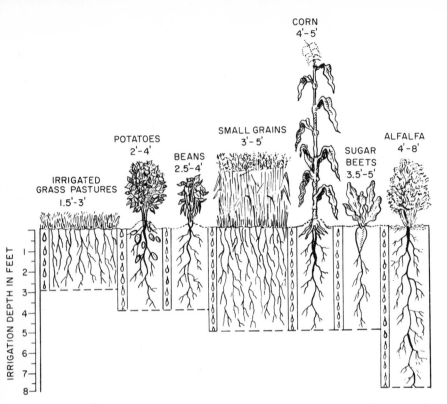

FIG. 19.6. Normal irrigation depth for common farm crops varies from one and one-half feet for some grass pastures to as much as 8 feet for alfalfa. *Source: Irrigation on Western Farms,* U.S. Dept. of Int. and U.S. Dept. of Agr., Agr. Inf. Bul. No. 199, 1959.

indicates that 3 inches of water will wet the soil to a depth of 1.5 feet. Then Figure 19.6 shows that the normal irrigation depth for corn is from 4 to 5 feet. To wet a silt loam soil to 4.5 feet requires 9 inches of water. For a dry silt loam soil, any irrigation water in excess of 9 inches would soak below 4.5 feet, and would therefore be wasted for crop use.

Since the total consumptive use for corn is 19 inches, this amount should be added as needed during the growing season. Each time water is added, a soil probe should be used to check the depth of wetting in order to stop irrigation when the soil becomes wet at a depth of 4.5 feet.

At first, it seems possible that by dividing the consumptive use (19) by the amount of water required to wet the soil to the 4.5-foot depth (6), the number of times needed to irrigate would be obtained. This calculation would be correct if the soil were allowed to become air dry each time water were applied; however, plants wilt before the soil becomes air dry. As a general rule, it is practical to irrigate when the soil is 50 per cent depleted of moisture. According to the preceding calculation, watering should be started when the soil, to a 4.5-foot depth, contains 4.5 inches of water. At

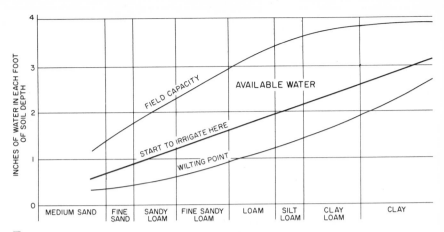

FIG. 19.7. The amount and frequency of irrigation are determined partly by soil texture. A silt loam soil has the largest available water capacity and therefore should require irrigation water less frequently than a sand or a clay soil. *Source: Water: The Yearbook of Agriculture (1955),* U.S. Dept. of Agr.

this point, 4.5 inches of water should be supplied by irrigation. On this basis, 19 divided by 4.5 equals approximately 4 irrigations, each of 4.5 inches of water, that will be required during the growing season. (These calculations assume that no rainfall is received.)

The general relationship among soil textures, field capacity, wilting points, available water, and the time to start irrigating is given in Figures 19.7 and 19.8.

AMOUNT AND FREQUENCY OF IRRIGATIONS
(BY MEASUREMENT)

Under very intensive management, certain instruments may be used in the soil to determine quite accurately the amount and frequency of irrigations. Two such instruments in common use are the tensiometer and a modified Wheatstone bridge.

A tensiometer [Figure 19.9] consists of a porous cup, a connecting tube, and a vacuum gauge. The unglazed ceramic cup is porous

FIG. 19.8. As a general rule, it is time to start irrigating when the soil pores are half depleted of their soil moisture, as depicted in this sketch. *Source:* John Box and William F. Bennett, "Irrigation and Management of Texas Soils," Tex. Agr. Ext. Serv. Bul. B-941, 1959.

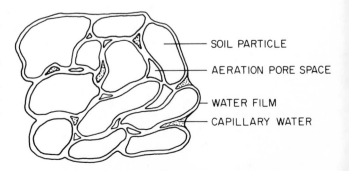

SOIL PARTICLE

AERATION PORE SPACE

WATER FILM

CAPILLARY WATER

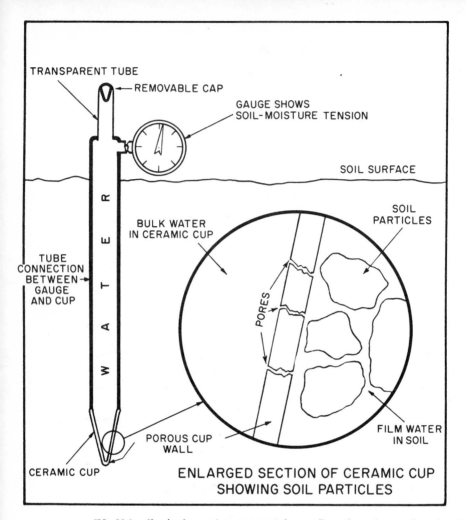

TRANSPARENT TUBE

REMOVABLE CAP

GAUGE SHOWS
SOIL-MOISTURE TENSION

SOIL SURFACE

BULK WATER
IN CERAMIC CUP

SOIL
PARTICLES

W A T E R

TUBE
CONNECTION
BETWEEN
GAUGE
AND CUP

PORES

FILM WATER
IN SOIL

CERAMIC CUP

POROUS CUP
WALL

**ENLARGED SECTION OF CERAMIC CUP
SHOWING SOIL PARTICLES**

FIG. 19.9. Sketch of a tensiometer mounted vertically in the surface of the soil, with an enlarged schematic diagram of a section of the porous ceramic cup in contact with soil water. Water in the soil should usually be kept sufficiently high in the root zone to maintain a reading of 70 or below on the vacuum gauge. *Source:* S. J. Richards and R. M. Hagan, *Soil Moisture Tensiometer,* Calif. Agr. Exp. Sta. and Ext. Serv. Leaflet No. 100, 1958.

to water, but its pores are so small that when the cup is wet they are sealed from air by the water films. The transparent upper portion of the tube enables the operator to observe changes in the water level, and the tube is provided with a plug so that the system can be kept filled with water. For convenience, the vacuum gauge has a scale of 0 to 100 that reads soil-moisture tension, or suction, in percentage of the normal atmospheric pressure.

The way a tensiometer operates can be most easily seen by referring to the schematic diagram, which shows a few enlarged soil particles adjacent to the surface of the cup. Note that water in the soil forms a film over each soil particle. The pore openings in the cup wall provide passage ways between the water in the cup and the films covering the soil particles. As the soil dries out, the water films become thinner and more tightly bound to the soil particles. The tension thus produced within these water films causes water to be pulled from the cup, and this movement produces a tension or suction within the tensiometer, which is indicated by a higher reading on the vacuum gauge. When water from irrigation or rainfall reaches the neighborhood of the cup, the tension within the water films on the soil particles is reduced, and water moves back into the tensiometer cup. The tension within the instrument is thus reduced, and the gauge shows a lower reading.

Tensiometers should be installed at depths and locations where roots are most actively absorbing water. For row crops it is suggested that the tensiometers be installed in the row at several depths within the active root zone. In orchards tensiometers should be installed at two or more depths and, in general, should be placed about one-third the distance from the drip line of the tree to the trunk.

Tensiometers at various points and depths in a field will seldom show the same readings at any one time because of the variability in topography, soils, and root activity and in the date and efficiency of the last irrigation or rain. This variability makes it desirable to base irrigations on the responses of several tensiometers placed at representative locations. Irrigations are applied when half the tensiometers in the field reach or exceed the tension values selected for the particular crop under the existing conditions.

Most established crop plants will not be limited in growth for lack of water if tensiometer readings in the active portions of the root zone are kept below 70 on the dial gauge or at corresponding readings on the manometer types. And on the other end of the scale, if tension values in the root zone remain at, or close to, zero for several days, water should be withheld, and where possible more adequate drainage should be provided.

Coarse-textured soils may need to be irrigated at tension values lower than 70 on the dial because in such soils the amount of available moisture is limited at higher tensions. Some fine-textured soils, on the other hand, may be allowed to go for some time after tensiometers have reached their maximum readings before irrigation is needed because more water is available in such soils. Irrigation at lower tension values may be desirable in saline soils. Basically, the tensiometer cannot indicate soil-moisture conditions in the dry range. If irrigation water is still withheld after the readings rise to about 80 on the dial, the readings will not continue to go up, but water will be pulled out of the tensiometer system, and the readings will cease to be valid.[8]

---

[8] S. J. Richards and R. M. Hagan, *Soil Moisture Tensiometer—How it works; How to use it.* California Agricultural Experiment Station and Extension Service, Leaflet No. 100 (1958).

FIG. 19.10. A farmer measures soil moisture in his corn field with a Bouyoucos Bridge. The wires lead from a gypsum block buried in the soil and are fastened to the Bridge to measure the electrical resistance. This method measures moisture in the soil from the wilting point to the field capacity. *Source:* Henry Corrow, N.H. Ext. Serv.

A second type of instrument used to measure soil moisture indirectly is the Bouyoucos bridge, described in Chapter 10. The Bouyoucos bridge is accurate for the determination of soil moisture throughout the range of available soil moisture—that is, from the wilting point to the field capacity. (Figure 19.10).

SPRINKLER IRRIGATION

The sprinkler irrigation system consists in pumping water through a pipe and through rotating heads to apply water to the soil in a manner similar to that received by natural rainfall. This is the most popular system in humid regions (Figure 19.11).

The advantages of the sprinkler system are as follows:

1. Land-leveling is not necessary.
2. Drainage problems are decreased.
3. Erosion is kept to a minimum.
4. Fewer special skills are required.

Some disadvantages of the sprinkler system are the following:

1. The initial investment is high.
2. Power costs are higher.

FIG. 19.11. The sprinkler irrigation system is best adapted to sloping land or sandy soils. *Source:* Sprinkler Irrigation Association.

3. More labor is required to move the pipe.
4. Wind prevents a uniform distribution, often making it necessary to irrigate at night.
5. Evaporation losses of water are higher than with other methods of irrigation.

FURROW IRRIGATION

Furrow irrigation is the oldest kind of irrigation system. In this method, water flows by gravity from a main ditch and down each furrow. On top of the furrows the crop has been planted before water is applied.

Field crops such as corn and cotton have a furrow to carry water between all rows. Crops that are planted in double rows or beds are irrigated by directing the water between the beds. Crops planted in a wide spacing, such as berries, grapes, and orchards, usually have two furrows for irrigation between each two rows of plants.

In general, soil erosion is excessive when the furrow method of irrigation is used on rows that have a slope of more than 2 per cent. An aim should be to keep the slope of the furrows less than 0.25 per cent.

The aim of the irrigator should be to obtain the maximum flow of water down each furrow without causing excessive erosion. In this way, water will soak into the soil at a fairly uniform rate all along the furrow. As a general rule, if it takes 4 hours to add sufficient water to a furrow, then the water should be turned into the furrow in such volume that it will flow to the end of the furrow in one hour (one-fourth of the elapsed time).

FIG. 19.12. Furrow irrigation of cotton, showing plastic siphon tubes in use for moving water from the main ditch to each furrow. *Source:* Drue W. Dunn, Okla. Ext. Serv.

There are several common methods of controlling the distribution of water in furrow irrigation:

1. Field lateral ditch with a small equalizing ditch leading directly to each furrow
2. Field lateral ditch with siphon tubes leading to each furrow (Figure 19.12)
3. Field lateral ditch with spiles (small straight pipes) leading directly to each row
4. Irrigation pipe with large openings (gates) emptying into each furrow (Figure 19.13)
5. Buried pipe to carry the water to the field, with risers emptying into each furrow or series of furrows

FIG. 19.13. Irrigation water is transported by pipe to the field where large openings in the pipe (gates) occur at each furrow. Note the canvas (or plastic) sleeves that lead the water into the furrows (at arrows) without causing excessive erosion. *Source:* Atto C. Wilke, Tex. Agr. Exp. Sta., Lubbock, Texas.

FIG. 19.14. Corrugation irrigation systems are adapted to sloping land, sod crops, and fine-textured soils. *Top:* Making small furrows on the contour for corrugation irrigation. *Bottom:* Water flows in a contour corrugation system at a non-erosive velocity. *Sources: Top,* U.S. Dept. of Agr.; *bottom,* E. L. Thaxton, Jr., Tex. Agr. Exp. Sta.

CORRUGATION IRRIGATION

The corrugation method of irrigation consists in running water down many small furrows for the irrigation of nonrow crops such as alfalfa, grasses, and the small grains.

This method is adapted to slopes up to 5 per cent, to fine-textured soils, and to soils that are inclined to bake when the flooding system of irrigation is used (Figure 19.14).

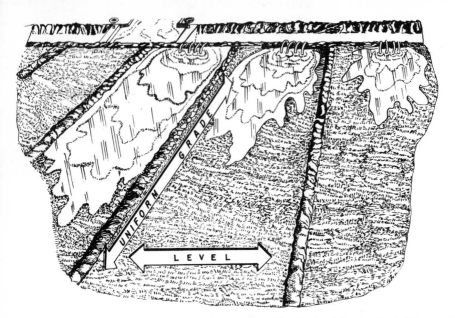

FIG. 19.15. The field has been prepared for border irrigation by building small levees around each leveled area; then the areas are flooded to irrigate them. *Source: U.S. Dept. of Agr.*

BORDER (CONTOUR LEVEE) IRRIGATION

The border method of irrigation is used on gentle slopes. Narrow strips of land are leveled, a low ridge built around them as a border to hold water, and each strip is irrigated by flooding (Figure 19.15).

Border irrigation is most satisfactory under these conditions:

1. When the surface soil is deep enough to permit the land to be leveled without leaving areas of unproductive sub-soil
2. When the infiltration rate of the soil is intermediate. Sandy soils would not permit a uniform depth of water penetration, whereas on impervious clay soils an excessive amount of water would be lost by evaporation before it soaked into the soil
3. When the slope is satisfactory to permit the construction of borders with a slope of 1 per cent or less. Land to remain in sod crops can be steeper without causing excessive erosion
4. When the land is planted to crops that are not injured by temporary flooding
5. When flooding is desirable, as with the levee system of growing rice.[9]

---

[9] "Contour-Levee Irrigation," Sec. 15, Chap. 6, *Soil Conservation Service National Engineering Handbook: Irrigation,* U.S.D.A. (July, 1969), 30 pp.

INCREASING INFILTRATION OF WATER IN SOILS

In the process of irrigation, flowing water sorts out the sand, silt, and clay. Since clay stays in suspension the longest, a thin layer of clay is deposited on the surface of the soil as the water moves into the soil. The result is a decrease of the infiltration rate of water, a ponding of irrigation water, and more loss of water by evaporation. Some success has been achieved in increasing infiltration by the use of certain green manure crops and crop residues.

Barley, cotton and corn residues, sesbania, Sudangrass, and field pea were used in several experiments on various soil textures in California to increase the infiltration of irrigation water. On a sandy loam soil that had a fairly high initial infiltration rate, no increase in infiltration was observed by using any green manure crop. When the barley on a loam soil was in the late dough stage, it caused an increase in infiltration of 93 per cent. Sudangrass incorporated with a clay loam soil increased infiltration by 45 per cent. Corn residues were more effective than cotton residues, but neither sesbania nor field pea was effective in increasing infiltration.[10]

One problem often encountered in row irrigation is soil compaction and soil crusting caused by tillage machinery and the sorting action of irrigation water flowing down the furrows. The result is slower infiltration of water into the soil, greater loss of water by evaporation and runoff, and greater hazard of salt accumulations.

The solution to this problem may be the use of wheat planted in the furrows, as was done in Riverside County, California. In this experiment, water moved into the soil 36 per cent faster where wheat had been seeded.[11]

SUMMARY

Irrigation practices are spreading rapidly in almost every state in the nation; even humid regions with frequent drought periods use irrigation.

Soils in arid and semiarid regions, where irrigation practices are most common, may be *saline, alkali* or *saline-alkali,* and the reclamation of each type of soil is different. Included in the reclamation is the selection of the

---

[10] W. A. Williams and L. D. Doneen, "Field Infiltration Studies with Green Manures and Crop Residues on Irrigated Soils," *Soil Science Society of America Proceedings,* Vol. 24 (1960), pp. 58–61.

[11] N. C. Welch and A. W. Marsh, "Water Penetration in Strawberries Aided by Seeding Grain in Furrows," *California Agriculture,* Vol. 20, No. 4 (1966).

most tolerant crops. A soil test would determine the kind of soil, and a chemical test would determine whether the water that is to be used for irrigation is satisfactory. Reclamation of saline soils may be hastened by an organic mulch.

Water for irrigation must be low in soluble salts, sodium, bicarbonates, and boron. Alfalfa has a high consumptive use, corn is intermediate, and small grains and beans have a low consumptive use of water. Three inches of irrigation water will wet a dry sandy soil to approximately 4 feet but will wet only 1 foot of a clay soil. Alfalfa should be irrigated to a depth of 4 to 8 feet; corn, 4 to 5 feet; and grasses, 1.5 to 3 feet. A crop should be irrigated when one-half the available soil water has been depleted. Sandy soils require more frequent irrigations than silt loam soils. Wheat planted in the furrows hastens infiltration of irrigation waters.

The amount and frequency of application of irrigation water can be determined by calculation or by direct measurement with a tensiometer or a modified Wheatstone bridge.

The sprinkler irrigation system is especially adapted to rolling topography and is the most popular system in the humid region. Furrow irrigation is the oldest method known and is adapted to row crops on fairly level land. Corrugation irrigation is used mostly on sod crops that grow on fairly steep slopes. The border irrigation system is well adapted to soils of intermediate infiltration capacity, such as sandy loams.

QUESTIONS

1. Describe a good quality of water for irrigation.
2. What is meant by the "consumptive use" of water?
3. On the average, how deep will 3 inches of water soak into a dry loam?
4. How deep should the soil be wetted for corn?
5. Describe briefly a method for calculating the amount and frequency of irrigation water for a small grain crop on clay loam. How could the same information be obtained by direct measurement?

REFERENCES

Bennett, O. L., B. D. Doss, and D. A. Ashley, *Cotton Irrigation in Southeastern United States,* Agricultural Information Bulletin No. 282. Washington: United States Department of Agriculture, 1964.

Climate and Man: The Yearbook of Agriculture (1941). United States Department of Agriculture.

Conservation Irrigation in Humid Areas, United States Department of Agriculture Handbook No. 107. United States Department of Agriculture, 1957.

Diagnosis and Improvement of Saline and Alkali Soils, United States Department of Agriculture Handbook No. 60. United States Salinity Laboratory Staff, United States Department of Agriculture, 1969.

Fireman, Milton, and Roy L. Branson, *Gypsum and Other Chemical Amendments for Soil Improvement.* California Extension Leaflet No. 149, 1962.

Grimes, D. W., V. T. Walhood, and W. L. Dickens, "Alternate-Row Irrigation for San Joaquin Valley Cotton," *California Agriculture.* Vol. 22, No. 5 (May, 1968).

Hagan, Robert M. (ed.), "Can Man Develop a Permanent Irrigation Agriculture?" *Proceedings, First Intersociety Conference on Irrigation and Drainage.* Denver, Colo.: United States National Committee on Irrigation and Drainage, 1959.

Hagan, Robert M., H. R. Haise, and T. W. Edminster (eds.), *Irrigation of Agricultural Lands.* Madison, Wisc.: American Society of Agronomy, 1967.

Irrigation on Western Farms, Agricultural Information Bulletin No. 199. United States Department of Interior and United States Department of Agriculture, 1959.

Maker, H. J., C. W. Keetch, and J. U. Anderson, *Soil Associations and Land Classification for Irrigation, San Juan County, New Mexico.* New Mexico State University Research Report 161 (Sept., 1969), 40 pp.

McGinnies, William G., Bram J. Goldman, and Patricia Paylore (eds.), *Deserts of the World.* Tucson, Ariz.: The University of Arizona Press, 1968, 788 pp.

Science and Saving Water and Soil. Agricultural Information Bulletin No. 324. United States Department of Agriculture, 1967.

Swenson, H. A., and H. L. Baldwin, *A Primer on Water Quality.* Washington: United States Department of the Interior, 1965.

Water: The Yearbook of Agriculture (1955). United States Department of Agriculture.

CHAPTER 20

DRAINAGE

If the land is wet it should be drained with trough-shaped ditches dug three feet wide at the surface and one foot at the bottom, and four feet deep. Bind these ditches with rock. If you have no rock, then fill them with green willow poles braced crosswise. If you have no poles, fill them with faggots. Then dig lateral trenches three feet deep and four feet wide in such a way that the water will flow from the trenches into the ditches.
—CATO (234–149 B.C.)

Artificial drainage of wet soils lowers the water table and results in these benefits:

1. Wet soils are usually the most fertile soils on the farm. Drainage permits them to be used for more productive purposes.
2. Properly drained soils warm earlier in the spring, thus permitting crops to be planted early enough to mature. It takes five times the heat to raise the temperature of water one degree as is required for dry soil.
3. Proper drainage makes the entire field more uniform in soil moisture (including the elimination of wet spots) for earlier and more efficient tractor tillage operations.
4. Drainage increases the amount of oxygen in the soil. Often an oxygen deficiency results in a chemical reduction in manganese, which may be toxic to plant growth.

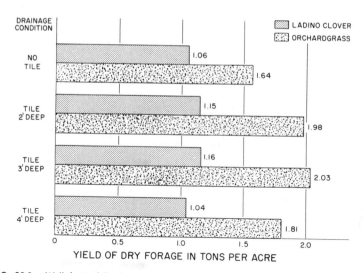

FIG. 20.1. Well-drained land is best adapted to many useful crops. Ladino clover is not responsive to tile drainage laid at varying depths. Orchardgrass is most productive if tile drainage is laid at a depth of 3 feet. *Source:* D. S. Chamblee and W. B. Gilbert, *The Influence of Drainage on the Growth of Forage Plants,* Res. Rep. No. 11, North Carolina State College, 1958. Data are averages for 1954–57 on Bladen silt loam.

5. Drainage decreases the losses of nitrogen from the soil by denitrification.

6. Drainage increases the percentage of crude protein in plants. The percentage of potassium, chlorine, and magnesium in plants is also increased by drainage.

7. Drained soils are freer from certain diseases, such as black rot of strawberries, fusarium root rot of sugar beets, and cereal root rots.

8. Drained pastures are healthier pastures because more parasites are killed by desiccation.

9. Soil structure is improved by drainage. The increase in wetting and drying, and the greater root growth, earthworm activity, and accelerated growth of bacteria and fungi aid in creating desirable soil structure.

10. Land that has been drained is adapted to a wider variety of more valuable crops (Figure 20.1).

11. Drainage permits a deeper penetration of plant roots; this increases the amount of nutrients available to growing plants and results in greater crop yields. Deeper roots also make the plants more drought resistant.

12. Drainage carries away excess surface water. This reduces the losses

FIG. 20.2. This soil requires drainage to get rid of the excess surface water. (Louisiana.) *Source:* Soil Conservation Service.

of plants due to the "heaving" action resulting from freezing and thawing of the soil (Figure 20.2).

13. Plants growing on well-drained soils utilize lime and fertilizers more efficiently.

SOILS THAT REQUIRE DRAINAGE

A soil may need artificial drainage for one of two reasons:

1. When there is a high water table that should be lowered
2. When excess surface water cannot move downward into the soil fast enough to keep from suffocating plant roots.

High water tables are common in most fibric, hemic, and sapric (peat and muck) soils and in some low-lying sandy soils (Figure 20.3). Commonly, however, level upland soils need artificial drainage because of excess surface water. Soils that permit only a slow movement of water downward may need ditching. Repeated experiments in areas of high rainfall have shown that fine-textured soils with a massive structure are most in need of artificial drainage. Much can be learned about the internal drainage of the soil by digging into the soil. Digging post holes, trench silos, or foundations will reveal some subsoil characteristics that indicate soil permeability. (Permeability is the capacity of a soil to allow movement of air and water through it.) A permeable soil seldom needs artificial drainage, except when

FIG. 20.3. Peat and muck soils require drainage because of a high water table. This is a mole drainage machine in operation. (Florida.) *Source:* Soil Conservation Service.

the soil has a high water table. Slowly permeable soils often need artificial drainage, especially when the land surface is level and rainfall is high.

Permeable soils that do not require drainage are uniform in color throughout the profile. The color may vary from brown to red. Yellow subsoils indicate intermediate permeability, especially when the texture is a clay. Subsoils that are mottled with red, yellow, and gray are more slowly permeable than are yellow subsoils. Gray clay subsoils in humid regions indicate very slow permeability and a probable need for artificial drainage. (Refer to Chapters 6 and 7 for more detail.)

DRAINAGE CAPACITY OF SOILS

It is often difficult to determine whether a soil will drain rapidly enough to permit the use of some form of subsurface drainage. With a potential investment so large, it usually pays to make some field and laboratory determinations to find what drainage system is best suited to any particular soil.

One way to obtain reliable data is to determine the field capacity, bulk density, and particle density. From the bulk density and particle density, the total pore space can be calculated. Then from this value is subtracted the per cent of pore space occupied by water at one-third atmosphere of tension. (The per cent moisture at one-third atmosphere tension is the same numerical value as the per cent pore volume at the same tension, because 1 gram of water occupies 1 cc.) The result represents the pore spaces through which water will move through the soil toward an underground drain. Such a value is known as *drainage capacity*.

To obtain total pore space, this simple equation is used:

$$\% \text{ total pore space} = 100 - \frac{\text{Bulk density}}{\text{Particle density}} \times 100$$

Pore spaces that hold water at the field capacity are not available for transmitting water through a soil in the field. These pore-space volumes may be readily obtained from field-capacity determinations. Moisture held at one-third atmosphere of tension closely approximates the moisture obtained from a field determination of field capacity.

Drainage capacity, as a percentage, is then found in this way:

$\%$ drainage capacity $= \%$ total pore space $- \%$ pore space occupied by water at the field capacity.

An example may help to further explain drainage capacity. In Mississippi, the surface soil of Memphis silt loam (developed from loess) has a total pore space of 59 per cent. The pore volume at one-third atmosphere tension (field capacity) is 20 per cent. The drainage capacity is therefore $59 - 20 = 39$ per cent.

From Table 20.1, it may be seen that Memphis silt loam, with a drainage capacity of 39 per cent, will drain readily. Bosket sandy loam will drain fairly readily, with a drainage capacity of 12 per cent. On the other hand, since its moisture at one-third atmosphere (field capacity) is greater than the total pore space, *Sharkey clay will not drain* through tile drains. This soil must be drained only by surface ditches, since water will not flow through it into tile drains.

FIG. 20.4. Some soils can best be drained by open ditches because water will not flow through the soils fast enough for a tile drainage system. The system of open ditches also permits a large volume of water to be carried off the land in a short period of time. (Louisiana.) *Source:* Soil Conservation Service.

TABLE 20.1 TOTAL PORE SPACE, PORE VOLUME AT 1/3-ATMOSPHERE TENSION, AND DRAINAGE CAPACITY OF THREE MISSISSIPPI SOILS *

Soil Type	Depth (in.)	Total Pore Space (%)	Pore Volume at 1/3 Atmosphere of Tension (%)	Drainage Capacity (%)
Memphis silt loam	0–6	59	20	39
Bosket sandy loam	0–8	51	39	12
Sharkey clay	0–6	51	81	0

* W. M. Broadfoot and W. A. Raney, *Properties Affecting Water Relations and Management of 14 Mississippi Soils,* Mississippi Agricultural Experiment Station Bulletin 521 (1954).

SURFACE DRAINAGE

Soils having a low drainage capacity must be drained by surface ditches. These open drains should be shallow enough to be crossed by machinery, if the fields are to be managed efficiently (Figure 20.4).

Drainage ditches are sometimes laid out by eye, leading from wet spot to wet spot, and finally into a protected grassy or wooded area. Too often, however, such a system of drainage becomes a maze of gullies that prevents the efficient use of the land.

Fine-textured, fairly level fields in areas of high rainfall are usually drained most advantageously by the construction of a precision land-forming system (Figure 20.5).[1] Expert information on establishing such a

---

[1] Irwin L. Savenson, *Surface Drainage of Flatlands,* U.S.D.A., in cooperation with Louisiana Agricultural Experiment Station, Miscellaneous Publication 1062 (1967), 10 pp.

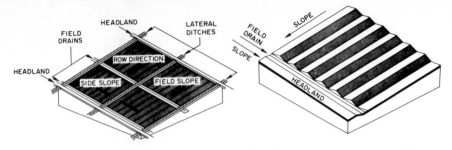

FIG. 20.5. Oblique sections of the precision land-forming system of drainage. This system is adapted to flat lands, fine-textured soils, and row crops such as cotton and corn. *Source:* U.S. Dept. of Agr. Misc. Pub. 1062, 1967, p. 6.

system may be obtained from the County Extension Agent, the Soil Conservation Service, the Agricultural Extension Service, or the Agricultural Experiment Station in each state. In some states, the Agricultural Conservation Program offers partial payment for establishing a drainage system when the work is supervised by technicians of the Soil Conservation Service.

TILE DRAINAGE

Satisfactory layout of a tile drainage system requires considerable planning and a lot of experience. Technical assistance in layout and installation of a tile drainage system can be obtained from the local Soil Conservation Service.

A tile drainage system will be satisfactory for a century or more if properly planned, adequately constructed, and carefully maintained. The depth and spacing at which to lay the lines of tile will vary with the crops grown and the type of soil. Soils with slow downward movement of water should have shallower placings of the lines of tile, and the lines should be laid closer together. Drainage to be established for alfalfa or orchards needs a depth of tile of about 4 feet. Corn needs intermediate depths, and the grasses and small grains can get along best with the tile lines placed about 2 feet deep. Spacings may vary from 40 to 300 feet between lines of tile, depending upon soil drainage capacity, which is related to soil texture and soil structure.

In clay and clay loam soils, the depth of the tile should not exceed 3 feet and the spacing no more than 70 feet. Tile lines in silt loam soils can be placed 4 feet deep and 100 feet apart. The respective maximum depth and spacing allowed in sandy soils is 4.5 and 300 feet.

Outlets for tile lines should be screened to prevent rodents from plugging them. Outlets should also be encased in cement, with a suitable apron to prevent undercutting by flowing water. Also, the last 10 feet of tile back from the outlet should be cemented at the joints. Other tile in the lines are placed end-to-end to permit water to seep between each two sections of tile (Figure 20.6).

FIG. 20.6. A tile drainage system will be satisfactory for a century or more if properly planned, adequately constructed, and carefully maintained. *Source:* U.S. Dept. of Agr.

Right: A tile ditching machine seen in operation. (West Virginia.)

Center: A tile line that has been laid but not yet covered. (Oregon.)

Bottom: A line of tile that has not yet been completed but is already draining the field into an open ditch. The outlet will be screened to keep out rodents. (Washington.)

Trees near tile lines should be cut so that the roots cannot grow into the cracks between the joints of tile. A hole in the soil above a tile line indicates that one of the sections has been broken or displaced. This should be repaired before the whole tile system is ruined. Occasionally the outlet will become plugged by "mud dauber" nests, birds' nests, or rodents living in the tile. Sometimes also, an outlet will erode and render the whole drainage system useless. All these maintenance jobs pay big dividends in extending the useful life of a tile drainage system.

Tile drains operate more effectively when a good cropping system is followed. Deep-rooted and long-season legumes and grasses are especially effective in helping to improve the drainage capacity of the soil. The land should be kept in close-growing crops as long in the rotation as possible to extend the effective life of the tile drainage system.

Heavy machinery operating on a wet soil reduces its drainage capacity by creating tillage pans. These decrease the effectiveness of the tile drainage system. The feet of grazing cattle on a wet clay soil pack the surface inches into a pasture pan and thereby lowers the efficiency of the drainage system.

SUMMARY

Drainage systems have been in use since early Roman times.

Drained soils are warmer in the spring, more efficient for the use of fertilizers, healthier for livestock, and have a more desirable structure and a wider adaptation for a greater variety of crops. Drained fields are more efficient to work with a tractor.

Soils with a fairly high drainage capacity can be tile drained; those with a low drainage capacity must be drained by open ditches. Some soils need to be drained because of a high water table; others require drainage because of excess surface water flowing onto them. Any drainage system should be laid out by a person trained to do the work and should be built by experienced machinery operators. Careful and regular maintenance extends the life of a drainage system for many decades.

QUESTIONS

1. Why do drained soils warm more rapidly in the spring?
2. Why are drained soils less droughty?
3. Explain the term "drainage capacity."
4. How does the drainage capacity of a soil determine what type of a drainage system should be installed?
5. How should a tile drainage system be maintained?

REFERENCES

Agricultural Statistics 1969. Washington: United States Department of Agriculture.

Cook, R. L., *Soil Management for Conservation and Production.* New York: John Wiley & Sons, Inc., 1962.

De Mooy, C. J., and John Pesek, "Nodulation Response of Soybeans to Added Phosphorus, Potassium, and Calcium Salts," *Agronomy Journal.* Vol. 58, No. 3, 1966.

Doll, E. C., *Lime for Michigan Soils.* Extension Bulletin 471. Cooperative Extension Service, Michigan State University, 1966.

Hagan, Robert M., Howard R. Haise, and Talcott W. Edminster (eds.), *Irrigation of Agricultural Lands,* Agronomy Monograph No. 11. Madison, Wisc.: American Society of Agronomy, 1967.

Knuti, Leo L., Milton Korpi, and J. C. Hide, *Profitable Soil Management.* Englewood Cliffs, N. J.: Prentice-Hall, Inc., 1962.

Schwab, Glenn O., Richard K. Frevert, Kenneth K. Barnes, and Talcott W. Edminster, *Elementary Soil and Water Engineering.* New York: John Wiley & Sons, Inc., 1957.

Science and Saving Water and Soil, Agricultural Information Bulletin No. 324. United States Department of Agriculture, 1967.

Shaw, Earle J., *Western Fertilizer Handbook* (4th ed.). Sacramento, Calif.: Soil Improvement Committee, California Fertilizer Association, 1965.

CHAPTER 21

ORGANIC AMENDMENTS

The earth neither grows old nor wears out if it is dunged [manured].—COLUMELLA (*ca.* A.D. 45)

With the relatively low cost of commercial fertilizers and the high cost of farm machinery and farm labor, farmers are tempted to consider all organic amendments, such as animal manures, crop residues, and green-manure crops, as waste products to be discarded at the least possible expense. The problem of utilization of animal manures is further aggravated by the fact that poultry are now being raised by the hundreds of thousands, hogs by the thousands, and dairy and feeder beef cattle by the hundreds. Under these conditions of high concentration of numbers of animals, in many instances it may not pay to haul all of the manure and spread it on the fields. This statement does not mean that the animal manure on all farms should be considered a waste product.

The situation on each farm must be studied and the decision regarding the disposition of animal manure must be made, using the following facts on the technical aspects of the problem, especially the short- and long-term effects of animal manures and other organic soil amendments on soil productivity.

There is nothing sacred about maintaining a particular level of organic matter in the soil except as this relates to the long-term economy of plant production. Experiments, demonstrations, and farmers' experiences have reinforced this statement: Management of animal manures, crop residues, and green-manure crops may make or break a farmer; management of animal manures, peat, compost, sewage sludge, and sawdust may separate successful gardeners from those who live on excuses for failures, such as, "I don't have a green thumb."

Both successes and "green thumbs" are achieved only by the timely application of valid information.

COMPOSITON OF COW MANURE

An average ton of cow manure plus bedding contains 500 pounds of organic matter. This organic matter contains 10 pounds of N, 5 pounds of P_2O_5, and 10 pounds of K_2O. Water makes up the remaining 1500 pounds, or three-fourths of the ton of manure. Dairy cow manure also contains (in pounds per ton): calcium, 5.6; magnesium, 2.2; sulfur, 1.0; iron, 0.08; boron, 0.03; zinc, 0.03; manganese, 0.02; copper, 0.01; and molybdenum, 0.002.

Approximately 50 per cent of the nitrogen is in the solid portion and the other 50 per cent is in the liquid part of the manure. By contrast, nearly all of the phosphorus (99 per cent) is in the solid portion. Eighty-four per cent of the potash is in the liquid and only 16 per cent is in the solid part (Figure 21.1).

THE CARE OF MANURE

Manure is difficult to use without waste because it is bulky and perishable. Watertight gutters, adequate bedding, reinforcement with superphosphate and the spreading of manue on the fields each day all aid in reducing

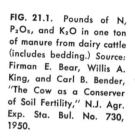

FIG. 21.1. Pounds of N, P_2O_5, and K_2O in one ton of manure from dairy cattle (includes bedding.) Source: Firman E. Bear, Willis A. King, and Carl B. Bender, "The Cow as a Conserver of Soil Fertility," N.J. Agr. Exp. Sta. Bul. No. 730, 1950.

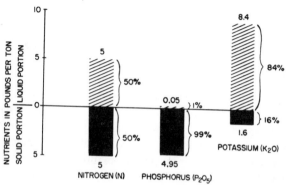

FIG. 21.2. To conserve as many plant nutrients as possible, manure should be spread daily. However, it should not be spread on frozen soil. *Source:* New Idea Farm Equipment Co.

losses of nutrients from manure. Of special importance in the care of manure is keeping it protected from the weather until it is spread on the field (Figure 21.2).

POULTRY MANURE

The average laying hen produces approximately 33 pounds of manure each year, calculated on an oven-dry basis. Without litter, the manure analyzes as follows:

Nitrogen (%N)	Phosphorus (%P)	Potassium (%K)	Calcium (%CaCO₃)	Magnesium (%MgCO₃)	Boron (p.p.m.B)
2.00	1.88	1.85	2.5	0.4	5

A hen in a year therefore produces manure containing approximately 0.66 pound of N, 0.62 pound of P, 0.61 pound of K_2O, and appreciable amounts of calcium, magnesium, and boron.

Most laying houses in winter have a strong odor of ammonia. This is evidence that nitrogen in the manure is being lost. Applications on the litter of 2 pounds per day of superphosphate per 100 birds will reduce nitrogen losses approximately by half. Materials which have been used for reducing losses of ammonia from poultry manure, in order of effectiveness are superphosphate, quicklime, gypsum, and peat moss.

On an oven-dry basis, fresh turkey manure has an average chemical composition as follows:

Nitrogen (%N)	Phosphorus (%P)	Potassium (%K)
1.31	0.31	0.41

A mature turkey each year produces approximately 40 pounds of manure. On the basis of the analysis given here, each year a turkey produces manure containing 0.5 pound of N, 0.1 pound of P, and 0.2 pound of K. Broiler (poultry) manure, on an oven-dry basis, contains

Nitrogen (%N)	Phosphorus (%P)	Potassium (%K)
2.30	1.08	1.69

RABBIT MANURE

Rabbit manure is a valuable fertilizer, analyzing higher in percentage of N, P, and K than any other common manure.

In a year, a doe with her four litters produces approximately 90 pounds of manure, calculated on an oven-dry basis. Without straw or hay refuse, the manure will analyze as follows: [1]

Nitrogen (%N)	Phosphorus (%P)	Potassium (%K)
2.40	0.62	0.05

On this basis, manure produced by a doe and her young during a year contains 2.2 pounds of N, 0.6 pound of P, and 0.04 pound of K. A dry doe or herd buck produces approximately half these amounts.

---

[1] George S. Templeton, *Value and Use of Rabbit Manure*, Agriculture Handbook No. 89, U.S.D.A. (1946).

MANURE AS A MULCH ON CORN

In a 3-year test in Ohio, 10 tons of manure per acre as a surface mulch increased the yield of corn an average of 10 bushels more per acre than did the same amount of manure plowed under. The increased yield resulting from the mulch was attributed to: (1) a protection of the soil from beating raindrops; (2) more water entering the soil for use by the corn; (3) a better structure that permitted corn roots to obtain more oxygen; and (4) a cooling effect of the mulch.

MANURE AND AVAILABLE WATER

At the Rothamsted Experiment Station near London, England, 14 tons of barnyard manure per acre per year was applied for 100 years. A similar plot received no manure. The results show that the plow layer of the manured plot is capable of supplying 0.7 more inches of available water for crops than the plot that received no manure.

MANURE AND CROP YIELDS

Such crops as corn and alfalfa respond readily to applications of manure, both in humid regions and in arid regions under irrigation.

Because of high labor costs and the bulkiness of animal manures, many producers of livestock consider that the use of manure on fields is not economical. Especially dairy operators and broiler producers often have more manure available than the fields and pastures need; therefore, they often use large quantities per acre.

Field experiments on the use of broiler manure on corn, oats, cotton, cabbage, and forage crops were conducted in Georgia and some results are summarized here.[2]

1. Significant increases in yield were obtained on all crops with applications of broiler manure, ranging from 2.5 to 16 tons per acre.

---

[2] *Source:* H. F. Perkins, M. B. Parker, and M. L. Walker, *Chicken Manure—Its Production, Composition, and Use as a Fertilizer,* Georgia Agricultural Experiment Station Bulletin N.S. 123 (1964), 24 pp.

2. Spring weed infestation was increased in the *first cutting only* by broiler manure on Coastal Bermudagrass.
3. Broiler manure as well as commercial N fertilizer decreased the percentage of legumes in a legume–grass pasture.
4. The soil pH was increased for a few days following the application of broiler manure, but the long-time (42-month) effect of 16 tons per acre was to make the soil more acid.
5. An annual application of 12 tons of broiler manure per acre gave comparable corn yields to an annual application of commercial fertilizer per acre containing:

$$N \qquad 120\,lb$$
$$P_2O_5 \qquad 100\,lb$$
$$K_2O \qquad 100\,lb$$

For comparison, the 12 tons of manure contained approximately

$$N \qquad 550\,lb$$
$$P_2O_5 \qquad 590\,lb$$
$$K_2O \qquad 480\,lb$$

It must be added, however, that residual effects following one application of broiler manure were significant for 3 years following *one* application. By contrast, commercial fertilizers were effective only the first year of application.

CROP RESIDUES

The newest trend in maintaining a satisfactory level of soil organic matter on crop land is to:

1. Fertilize liberally and according to the results of a periodic soil test.
2. Plant all crops in narrow rows with a very high plant population.
3. Practice minimum tillage. This usually means that crop stubble is left standing in the field until time for planting; then manipulating land preparation, seeding, fertilizing, weediciding, and pesticiding all in one operation over the field.

Crop residues are therefore more dense and are used to maintain soil organic matter instead of green-manure crops or dense-growing hay or pasture crops in rotation.

GREEN-MANURE CROPS

Green-manure crops are usually short-season legumes grown in rotation between two field crops for the purpose of restoring soil organic matter. The most common green-manure leguminous crops are vetch, annual sweetclover, Austrian winter peas, crotalaria, and rye (a non-legume).

Many research stations have shown that green-manure crops are not to be recommended as widely as they were at one time because of their low effectiveness in maintaining soil organic matter and crop yields. In special situations, however, such as the use of a winter legume to reduce cotton root rot, green-manure crops are very effective.

PEAT

Peat is a naturally occurring organic material that has accumulated in wet and cool places. Commercial producers of peat are in 25 states in all regions of the United States, with Michigan accounting for 40 per cent of the total production. Following in order of production are Colorado, Ohio, Florida, Washington, Pennsylvania, Illinois, and Minnesota. Total reserves of useful peat are estimated at 14 billion tons on an air-dry basis. Minnesota accounts for half the total reserves.

Many kinds of peats are valuable for use as mulches, soil conditioners, and as acidifying agents. There are at least five types of highly organic materials that are sold as peat [3] (Figures 21.3 and 21.4).

---

[3] Robert E. Lucas, Paul E. Rieke, and Rouse S. Fornham, *Peats for Soil Improvement and Soil Mixes,* Extension Bulletin 516 (1965). Michigan State University.

FIG. 21.3. Examples of peat sold as soil conditioners. (1) *Peat humus,* is of intermediate to low value because of its advanced stage of decomposition. (2) *Reed-sedge peat,* is intermediate in value as a soil conditioner. *Source:* R. E. Lucas, Crop and Soil Sciences Department, Mich. State U.

Sedimentary peat originates at the bottom of certain ponds from partially decomposed algae, plankton, water lilies, and pond weeds. The color is usually dark gray to black and the particles are so finely divided that they swell upon being wetted and contract upon being dried. (When moist they are "rubbery.") Furthermore, some sedimentary peat contains marl (lime), silt, and clay. Sedimentary peat has *no value* as a soil conditioner, a top-dressing, or as a soil mix (classified as *Sapric* soil material).

Muck soil is a product of decomposed peats that is dark gray to black, finely divided, variable in composition, and of *very little value* for use as a soil conditioner, a top-dressing, or as a soil mix (classified as *Hemic* soil material).

Peat humus is a product of advanced decomposition of hypnum moss peat or reed-sedge peat. It is dark brown to black in color and is slightly effective when used as a soil conditioner, top-dressing for lawns, and as a golf green soil mix (classified as *Hemic* soil material).

Reed-sedge peat originates from residues of reeds, sedges, marsh grasses, and cattails. It is brown to reddish brown in color and has a moderate water-holding capacity. It is reasonably effective when used as a soil conditioner, a top-dressing for lawns, and as a potting soil mix (classified as *Fibric* soil material).

Moss peat originates mainly from sphagnum moss or hypnum moss. The characteristics are a brown color, a high acidity, and a very high water-holding capacity. It is very effective when used for soil conditioning, top-dressing on lawns, for surface mulching, for rooting cuttings, for acidifying the soil, and for use in mixing a potting soil (classified as *Fibric* soil material).

FIG. 21.4. Examples of peat sold as soil conditioners. (1) *Sphagnum moss peat,* is one of the best for use as a soil conditioner, top-dressing for lawns, for surface mulching, and for acidifying the soil. (2) *Hypnum moss peat,* ranks along with sphagnum moss peat as a soil conditioner. *Source:* R. E. Lucas, Crop and Soil Sciences Department, Mich. State U.

COMPOST

A desirable, weedfree substitute for well-rotted manure can be made and will be ready for use within a year. Grass clippings, garden weeds, hay, biodegradable garbage, tree leaves, sawdust, and peat, together with soil, sod, lime, and fertilizers, can be used in making a compost pile.

Alternate layers of various organic materials should be piled in 6-inch depths to a height of 5 feet. The best width is 4 feet. As the various layers of organic materials, soil, fertilizers, and lime are applied, the sides of the pile should be as vertical as possible and the top depressed to absorb the rain.

As alternate layers of organic materials are put in the pile, lime and fertilizers are added to hasten decomposition and to reinforce the compost. The recommended amounts per ton of dry organic material are:

Dolomitic limestone	60 lb
Ammonium nitrate	40 lb
Superphosphate	30 lb

SEWAGE SLUDGE

Most countries consider human wastes to be an essential part of their agriculture; in the United States even the idea seems repulsive. Especially in the Orient, human excreta (night soil) is used as a fertilizer; without it, starvation would be much more serious than at present.

Sewage in America has found some use as a fertilizer. After its value as a fertilizer becomes fully realized, sewage will probably be in much greater demand.

Sludge produced from city sewage plants are of two general types, *activated* and *digested*. Activated sewage sludge is made by bubbling a large volume of air for several hours through raw sewage in the presence of aerobic bacteria. Digested sewage sludge consists of anaerobic decomposition of the raw sewage in large open vats for at least 2 weeks.

Activated sludge, although not commonly used as a fertilizer, is richer than digested sludge in all essential elements except manganese. Both kinds of sludge are especially high in percentage of zinc. There is almost no potassium in either kind of sludge, because it is leached out by the large quantities of water used in processing.

Near many cities sewage is used for fertilizing crops by irrigating directly with the effluent. Care must be taken, however, to avoid the spread of certain human diseases.

SAWDUST

There are millions of tons of sawdust that could be used as a mulch, bedding material, compost, or for direct application to the soil. Although some sawdust is now being used for these purposes, a greater knowledge of its properties will result in its expanded use.

As a fertilizing material, sawdust ranks very low. A comparison of its nutrient content with that of wheat straw and alfalfa is given in Table 21.1. Sawdust contains 4 pounds of N, 2 pounds of P, and 4 pounds of K per ton of material, on an oven-dry basis. Sawdust is richer than wheat straw only in its calcium content. Alfalfa is from 5 to 15 times as plentifully supplied with the essential elements as is sawdust.

Sawdust as a mulch is usually a good practice if certain precautions are used. Blueberries and strawberries are especially benefited by a sawdust mulch. To decrease the tieup of available soil nitrogen, approximately 25 pounds of N should be mixed with each ton of dry sawdust used.

As a bedding material to absorb liquids, sawdust is a satisfactory product. Depending on its fineness, sawdust is capable of absorbing from 2 to 5 pounds of water per pound. This compares with 3 pounds of water absorbed for each pound of chopped hay.

Sawdust is often used in a garden to make fine-textured soils more easily worked. This is a desirable practice if weathered sawdust is used and if at least 25 pounds of N per ton of sawdust is first added. Sometimes it is desirable to use small amounts of a phosphorus fertilizer to overcome the possible tieup of phosphorus by the decomposing bacteria. Sawdust has very little influence on soil acidity.

Wood chips are available in some areas and can be used in a similar way.

TABLE 21.1 THE PRINCIPAL PLANT NUTRIENTS IN SAWDUST, WHEAT STRAW, ALFALFA HAY, AND SEWAGE SLUDGE IN POUNDS PER TON OF DRY MATERIAL *

Organic Material	Nitrogen (N)	Phosphorus (P)	Potassium (K)	Calcium ($CaCO_3$)	Magnesium ($MgCO_3$)
Sawdust	4	2	4	11	1
Wheat straw	10	3	12	7	2
Alfalfa hay	48	10	28	50	15
Digested sludge	48	54	Trace	—	—
Activated sludge	112	114	Trace	—	—

* Sources: F. E. Allison and M. S. Anderson, *The Use of Sawdust for Mulches and Soil Improvement*, U.S.D.A. Circular 891 (1951). M. S. Anderson, "Composition of Sewage Sludge as Influenced by Type of Disposal System," a paper presented before the Soil Science Society of America, Davis, Calif. (1955).

The ammoniation of sawdust is being used to some extent to supply a fertilizer that contains more slowly available nitrogen. Sometimes phosphoric acid is also used to treat sawdust, thus making the material a carrier of phosphorus as well as nitrogen.

Sawdust, shavings, and wood chips from eastern hardwood trees usually decompose at a moderate rate and therefore require nitrogen fertilizer to be added so that crop plants will not be starved from lack of nitrogen. This is not true for western softwoods, however. Wood products from such western species as California incense cedar, cypress, redwood, western larch, and red fir decompose so slowly in the soil that bacteria do not tie up the nitrogen and thus starve crop plants. Sawdust, shavings, and wood chips from these western forest trees can be used advantageously on the soil as a mulch without adding any nitrogen fertilizer.[4]

SUMMARY

The high cost of labor and the relatively low cost of chemical fertilizers has forced every producer of organic materials to reexamine the question of the use of organic soil amendments.

Manure contains the greatest fertilizing value when it is fresh; this means that manure should be applied to the land as soon as possible after it is produced. When it cannot be spread immediately, manure should be stored in a compact, moist pile under shelter or in a manure pit until it is used.

Losses in storage are higher in the summer because of increased bacterial activity. It is therefore more urgent to spread manure daily in warm weather than it is in cool weather.

Greater efficiency per ton of manure is obtained when manure is applied in small amounts and more often. Approximately 10 tons per acre on every acre each year will result in greater increases in crop growth than will 20 tons per acre every other year.

The returns per ton of manure are higher when manure is used on infertile and eroded soils than when used on the best soils. By contrast, commercial fertilizers usually respond best on productive soils.

The use of manure and other organic materials improves soil structure, increases the available water capacity of soils, and increases crop yields.

---

[4] F. E. Allison and C. J. Klein, "Comparative Rates of Decomposition in Soil of Wood and Bark Particles of Several Softwood Species," *Soil Science Society of America Proceedings*, Vol. 25, No. 3 (1961).

Fibric, Hemic and Sapric soil materials (peats and mucks) have high value when used as mulches, soil conditioners, acidifying agents, and as a mixture in preparing potting soil. Best results can be obtained, however, when the right kind of peat or muck is selected.

Compost, sewage sludge, and sawdust are good sources of organic matter, and their use will increase in proportion to the information available concerning their value.

QUESTIONS

1. Compare the chemical composition of hen, turkey, and rabbit manures.
2. Explain the use of manure as a mulch.
3. Why do organic materials improve soil structure?
4. Describe the construction of a compost pile.
5. Why is it necessary to add nitrogen and sometimes phosphorus fertilizer to sawdust from eastern hardwoods before it is used?

REFERENCES

Benne, E. J., C. R. Hoglund, E. D. Longnecker, and R. L. Cook, *Animal Manures: What Are They Worth Today?* Michigan Agricultural Experiment Station Circular Bulletin 231, 1961.

Donahue, Roy L., *Our Soils and Their Management: An Introduction to Soil and Water Conservation* (3rd ed.). Danville, Ill.: The Interstate Printers and Publishers, Inc., 1970.

Eno, Charles F., *Chicken Manure.* Florida Agricultural Experiment Station Circular S–140, 1962.

Lucas, Robert E., Paul E. Rieke, and Rouse S. Farnham, *Peats for Soil Improvement and Soil Mixes.* Extension Bulletin 516. Michigan State University, 1965.

Perkins, H. F., M. B. Parker, and M. L. Walker, *Chicken Manure—Its Production, Composition and Use as a Fertilizer.* Georgia Agricultural Experiment Station Bulletin N. S. 123, 1964, 24 pp.

CHAPTER 22

SOIL DIAGNOSIS

*. . . A field examination should precede laboratory
studies.**

When nutrient imbalances arise in the soil, the plant is subject to
nutrient stresses that result in reduced growth and quality. A constant,
balanced supply of elements to the plant is essential, otherwise nutrient
deficiencies or toxicities will result.

When soil analyses are used to determine the kind and the amount of
the various nutrients to be applied to the soil, the likelihood of nutrient
imbalances resulting from their application will be minimal. However, if
a shortage or excess of a particular nutrient does occur, soil analysis is an
important diagnostic tool for evaluating the problem.

Soil analysis is no panacea. It will *not* supply answers to unsatisfactory
plant growth when the cause is dry weather, compacted soils, critically
low or high temperatures, inadequate soil drainage, improper placement of
fertilizer, insect damage, competition from weeds, or untimeliness of operations. These probable limiting factors must be inventoried before the
soil tests can be interpreted into a fertilizer recommendation (Figure 22.1).

In many instances, the cause of poor growth may not be due to a
single factor, but may be associated with several interacting and interrelated factors. Corn roots require a high content of soil oxygen before they

---

* George Nelson Coffey, *A Study of Soils of the U.S.,* Bureau of Soils Bulletin
85, Washington (1912), p. 6.

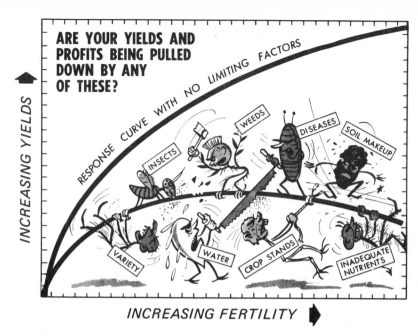

FIG. 22.1. Growth response from an increase in soil fertility will not reach its maximum until all other growth-limiting factors, such as drainage, diseases, insects, weeds, and plant populations, have been corrected. *Source: American Potash Institute.*

can absorb potassium normally. In a wet soil, the water displaces a large amount of oxygen and corn roots are not able to absorb sufficient potassium even though the soil tests "high" in available potassium. For example, a potassium deficiency was observed on corn growing on a Clyde silt loam from a *poorly drained* field in northeastern Iowa. Chemical analysis of Clyde silt loam indicated a *high* content of available potassium, but additional potassium is generally required to produce high yields of corn. Even though there is a good supply of soil potassium, some factor, such as excess soil moisture, is limiting potassium absorption by the plant. Compacted soils have a similar effect, due to restricted oxygen.

FIG. 22.2. All soils in the photograph tested *high* in phosphorus. During a *wet* and *cold* spring, additional phosphorus was applied at the time of planting to the corn in the background (at B), but none to the corn in the foreground (at A), where the plants show phosphorus deficiency (leaves turn purple). When the season became drier and warmer, the phosphorus deficiency disappeared, and yields were the same on all plots. *Source: R. E. Lucas, Crop and Soil Sciences Department, Mich. State U.*

The young corn plants shown in the foreground of Figure 22.2 growing in a soil testing *high* (71 pounds per acre) in available phosphorus (Bray P_1) but receiving no planting-time phosphorus, are suffering from inadequate phosphorus nutrition (at A). Low soil temperatures reduced the uptake of phosphorus by the young plants, which was later corrected when the soil temperature, with the advance of the season, increased to a more favorable level. No significant yield difference resulted between the plants receiving phosphorus, shown in the background of Figure 22.2 (at B), and those receiving no phosphorus.

An experiment in Oklahoma on the effect of soil temperature in relation to the response of wheat to phosphorus fertilizer indicated that when the soil temperature was 50°F, even on soil *low* in phosphorus, there was no response to applications of phosphorus fertilizer. On the same *low*-phosphorus soil, dry matter yields of wheat were doubled with the application of 4.4 pounds of phosphorus per acre when soil temperatures were maintained at 65 and 90°F.[1]

Chemical soil analysis is a valuable diagnostic tool for preventing the occurrence of, and providing answers to, complex nutritional problems. It should be supplemented with a thorough inventory of soil and cropping information. Obtaining a soil sample that adequately represents the area is a key step in soil chemical diagnosis.

SAMPLING THE SOIL

One of the major objectives of carrying out a chemical inventory of the soil is to determine its ability to supply the essential elements in the right proportions and in adequate amounts throughout the growing season. The soil sample to be analyzed for the nutrients should be representative of the area. Lime and fertilizer recommendations cannot be made accurately when the soil sample does not represent the field.

In general, for most field crops, soil should be sampled once every 2 to 3 years. For soils under intensive use, however, as in gardens and greenhouses or for high-value-per-acre field crops, soils should be tested before planting *each* crop.

Soil samples may be taken at any time during the year when the soil is not frozen and the moisture conditions permit. It is suggested, due to the variation in nutrient availability that may be associated with time of sampling, that any given area be sampled about the same month each year.

---

[1] Joe R. Gingrich, "Effect of Soil Temperature on the Response of Winter Wheat to Phosphorus Fertilization," *Agronomy Journal*, Vol. 57, No. 1 (1965), pp. 41–44.

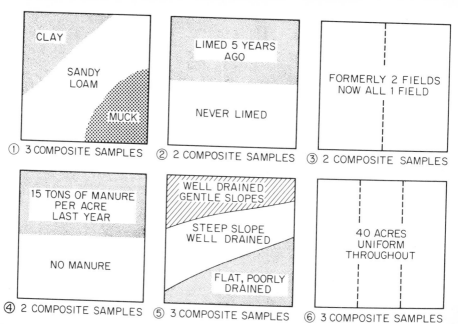

FIG. 22.3. Conditions should be studied before sampling a field. Variations in soils, productivity, drainage, and past management will determine the number of samples that must be taken to obtain a representative soil sample. The two textures and muck in Field 1 require three-composite samples. Due to the differences in management of Fields 2, 3, and 4, two-composite samples need to be taken in each. Differences in topography and drainage represented in Field 5 necessitate three composite samples. Field 6 is a uniform 40-acre field, but requires three-composite samples to insure a representative soil analysis of the entire field. Source: J. C. Shickluna, Crop and Soil Sciences Department, Mich. State U.

UNIFORM SAMPLING AREAS [2]

Before sampling the field, it should be examined for differences in soil characteristics. Consideration should be given to soil productivity, topography, texture, drainage, color of topsoil, and past management. If these features are uniform throughout the field, each composite sample of the topsoil can represent 10 to 15 acres. If there is a great variation in these features, the field should be divided for taking a composite sample from each predetermined area. A composite sample made up of samplings from two distinctly different areas is not representative of either area (Figure 22.3).

From each predetermined area, prepare a composite sample by taking not less than 20 samplings consisting of vertical columns or cores of soil

[2] J. C. Shickluna, *Sampling Soils for Fertilizer and Lime Recommendations,* Extension Bulletin E–498, Cooperative Extension Service, East Lansing, Mich.: Michigan State University (1971).

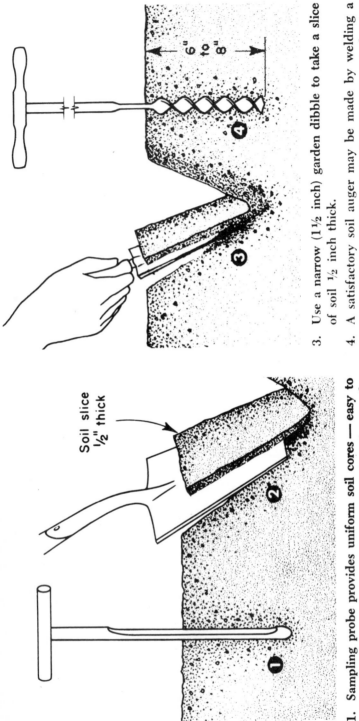

1. Sampling probe provides uniform soil cores — easy to use — saves time — best tool for sampling farm soils.

2. A spade or shovel can do the job for the home gardner.

3. Use a narrow (1½ inch) garden dibble to take a slice of soil ½ inch thick.

4. A satisfactory soil auger may be made by welding a 1¼ inch or 1½ inch wood bit into a ½ inch pipe equipped with T-handle.

FIG. 22.4. Suitable tools to obtain a representative soil sample. The soil probe provides uniform soil cores. A spade, garden dibble, or auger may also be used satisfactorily. Take 20 samplings per composite sample from a given area to plow depth. One subsoil sample per field, taken at a depth of 18 to 24 inches, will aid in evaluating subsoil fertility level. Source: J. C. Shickluna, Crop and Soil Sciences Department, Mich. State U.

approximately 0.5 inch in cross section and to plow depth, as shown in Figure 22.4.

Generally, 20 well-taken samplings or soil cores for a composite sample from a given area will result in laboratory tests that can be duplicated much more frequently than where only 5 or 10 samplings comprise the composite sample. Furthermore, 20 samplings per area appear to give as good results as 40 or even 100 samplings.

Avoid sampling unusual areas unless such locations are sampled and packaged separately. Areas close to gravel roads, previous locations of brush, lime or manure piles, or burned-over areas should be avoided.

Subsoil samples taken at a depth of from 18 to 24 inches, especially with organic soil, will often aid in making lime recommendations. A subsoil sample need not be a composite.

The individual samplings should be placed in a clean pail and mixed thoroughly. The composite sample should then be transferred to a clean container for shipment to the soil testing laboratory. (Container is usually supplied by the laboratory.)

PROVIDE COMPLETE INFORMATION

The more complete the information provided, the better will be the fertilizer recommendation. The following information should accompany the sample: (1) previous crop grown; (2) crop or crops to be grown; (3) yield goal; (4) when the field was last limed and the rate of application; (5) whether the field will be manured for the crop being grown; (6) depth of plowing; (7) soil type, series or soil management group; (8) whether drainage is good, intermediate, or poor; (9) other special problems or conditions that may affect plant growth.

UNDERSTANDING SOIL TESTING

The general impression many people have of soil testing is that it is a rather simple procedure to determine which nutrients are in short supply in the soil, so that the needed fertilizer can be added and an abundant harvest will be assured.[3]

Recommendations made to a farmer are based on the relationship between the soil test and the outcome of closely controlled field fertility ex-

---

[3] A. Bauer, L. Hanson, and J. Grava, *Understanding Soil Testing. Better Crops with Plant Food,* American Potash Institute, Inc., Washington (1960).

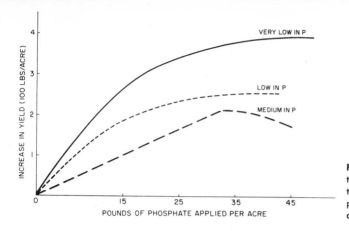

periments. A detailed study of soil testing reveals a rather complex set of relationships and procedures, involving precise chemical analytical methods based upon a vast amount of soil and plant research.

Several studies have been made that indicate that the lower the soil test for a particular element, the greater the plant response when this element is added as fertilizer. For example, Figure 22.5 demonstrates graphically that the lower the soil test for phosphorus, the higher the response of wheat to phosphate fertilizer. When the soil phosphorus tested *very low,* 35 pounds of P_2O_5 gave an increase of 375 pounds of wheat; when the soil phosphorus tested *low,* the increase was 250 pounds per acre of wheat; and when the test for soil phosphorus was *medium,* the increase in wheat from the application of 35 pounds of P_2O_5 was 200 pounds per acre.

Chemically extracted nutrients can never reflect the exact amounts of nutrients that the plant roots will take up during the entire growing season. Neither is plant nutrient uptake the same from year to year. Too, a soil sample usually represents only the surface 6–8 inches, whereas plant roots may penetrate 3–5 or more feet into the soil.

THE FERTILIZER RESPONSE CURVE

Plants respond to fertilizer in a nonlinear (curvilinear) fashion, as illustrated by the yield response curve in Figure 22.6. The maximum yield is regulated by climate, water, and insect damage.

The fertilizer cost curve based on yield is shown at the bottom of Figure 22.6. What point on the yield response curve represents the most profitable investment from fertilizer? The maximum dollar return per dollar spent on fertilizer will be realized at the low rates where, for example, 4 dollars is returned for each 1 dollar invested in fertilizer. However, profit per acre is greatest at the higher fertilizer rates as shown in Figure 22.6 by the dotted line. If a farmer could always predict the dollar response curve, he should stop adding fertilizer just before he is "trading dollars."

TECHNOLOGY HAS OUTDATED SOIL TESTING KIT

The rapid advance in instrumental methods of analysis has made obsolete the methods that were popular several years ago. Portable soil testing kits that were once relied upon by the diagnostician to evaluate nutrient status of soils have now been replaced with more expensive, sophisticated and reliable testing techniques. It now costs at least 10,000 dollars to purchase the basic equipment required for a modern soil testing laboratory.

Where portable soil testing kits are used for diagnosing soil problems and determining fertilizer needs, it is imperative that the operator be well trained in carrying out the tests, have an appreciation of the shortcomings of the technique, and be knowledgeable about the interpretation of the test results and subsequent recommendations.

Soil testing kits have been widely used by crop and soil specialists and fertilizer dealers for diagnosing on-the-spot field problems. Caution should be exercised in handling, storing, and systematically replacing the chemical reagents used in the kit. The reagents are subject to contamination. Should this happen, the soil test will be misleading.

SOIL CHEMICAL DETERMINANTS FOR INSURING OPTIMUM PLANT GROWTH

To obtain the maximum growth response from an added nutrient element, all other essential elements must be in adequate, but not injurious, amounts. Justus von Liebig (1803–1873), postulated the law of the minimum, which states that the growth of any plant is regulated by the plant nutrient element present in the least amount. (See Chapter 1 for further details about Liebig.) Soil chemical analyses are widely used for probing and evaluating the nutrient status of the plant root environment, thus un-

FIG. 22.6. Each additional increment of fertilizer does not produce equal increases in plant yield. The most profitable application rate is just before the point where the last dollar invested in fertilizer returns one dollar in yield increase. *Source:* S. A. Barber, Purdue U., taken from American Potash Institute.

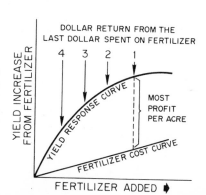

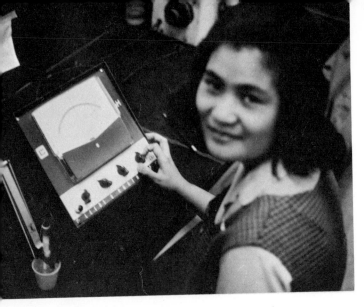

FIG. 22.7. The pH meter is used for rapid and accurate measurements of soil acidity. The availability of many of the micronutrients is influenced by the soil pH. This instrument is also used for carrying out the lime requirement test to determine lime needs. Source: J. C. Shickluna, Crop and Soil Sciences Department, Mich. State U.

covering possible deficiencies and toxicities; they should be performed on soils at regular intervals to avoid nutrient imbalances.

SOIL ACIDITY

Some plants are more tolerant of soil acidity than others. Excess acidity may create toxic plant conditions due to the increase in solubility of such elements as aluminum and manganese, and subsequently cause a nutrient imbalance in the plant. On the other hand, soils that are too alkaline may cause a deficiency of several essential nutrient elements, such as boron, iron, zinc, and manganese. From a pH measurement (Figure 22.7), one will readily ascertain the degree of soil acidity or alkalinity. Since the availability of many of the micronutrients is influenced by the acidity of the soil, it is imperative that accurate pH measurements be made. The pH meter is also used for the lime requirement test. The pH of the soil does not, by itself, indicate the amount of limestone required to effect a given change in the pH of the soil. An additional test referred to as a *lime requirement determination* is required to evaluate accurately the amount of liming material necessary to bring about the desired change in pH. Permissible soil pH ranges for various crops are represented in Chapter 3.

POTASSIUM AND SODIUM

The use of the flame photometer in modern soil testing laboratories for the determination of such elements as potassium and sodium, as shown in Figure 22.8, has facilitated rapid analysis without sacrificing accuracy.

FIG. 22.8. The flame photometer is used in modern soil testing laboratories for rapidly and accurately determining potassium and sodium. *Source:* J. C. Shickluna, Crop and Soil Sciences Department, Mich. State U.

Potassium tests usually involve acid or salt solution extractions, which, when reacting with the soil, replace some of the potassium held electrically by the clay and/or organic matter particles. This form of potassium is referred to as *exchangeable* potassium and is considered a reliable index of the ability of the soil to supply potassium to the plant throughout the growing season.

PHOSPHORUS

Many extracting solutions are employed for the determination of available soil phosphorus. These tests usually involve treating the soil with either a strong or a weak (dilute) acid or a special salt solution to extract the more soluble phosphorus fractions. A quantitative determination of the soil phosphorus may be carried out on a photoelectric colorimeter. Solutions used for extracting soil phosphorus are usually similar within a soil region.[4] It does not necessarily mean that the method in use is the only acceptable one, but it emphasizes the point that adequate calibration, relating soil phosphorus level to plant response, must be carried out before

---

[4] Frank T. Bingham, "Chemical Soil Tests for Available Phosphorus," *Soil Science,* Vol. 94 (1962), pp. 87–95.

a new soil extractant is used. Phosphorus present in acid soil probably can be extracted by a variety of solutions with equally good results, but neutral and alkaline soils can be better evaluated with milder extractants.

CALCIUM AND MAGNESIUM

The secondary nutrients, calcium and magnesium, are commonly extracted from the soil with neutral normal ammonium acetate, and their concentration in the soil extract is determined on an atomic absorption spectrophotometer. This technique offers a high degree of speed, precision, and accuracy that cannot be duplicated with the portable soil testing kit. Like potassium, these cations are absorbed on the surface of negatively charged clay and organic matter particles and are replaced from these exchange sites by the positively charged ammonium (NH_4^+) ion. The concentration of these nutrients in the soil may be classified as adequate or inadequate by the use of soil test calibration data; corrective measures, if needed, may be taken.

MICRONUTRIENTS

Determinations of such nutrients as manganese, iron, copper, and zinc, are readily carried out on the atomic absorption spectrophotometer. Since the amounts of these elements in the soil are generally small, it was formerly difficult to determine their availability with any degree of accuracy. New techniques, however, have made this a much simpler task.

As discussed in Chapter 11, the availability of these nutrients is closely associated with soil pH. Recent research information relating plant needs to the concentrations of the micronutrients in the soil and in the plant are given in Chapter 23.

INTERPRETATION OF SOIL TEST INFORMATION

The interpretation of soil tests involves determining how much of a particular nutrient will be needed throughout the growing season to provide a sufficient supply of this element to the plant for a predicted yield. In the same soil test range, soils with a low supplying power of available nutrients will require larger amounts of fertilizer to supply the needs of the plant.

In determining the quantity of nitrogen, phosphorus, and potassium that will be needed to produce a certain yield of a crop, it is important to consider the carryover of these nutrients from previously applied fertilizer and manure, and the inclusion of legumes in the rotation.

On mineral soils, the immediate nitrogen needs of a crop depend more on the system of management than on the soil type or soil test at the time of planting. Too, there is no reliable quick soil test for plant-available soil nitrogen. Nitrogen fertilizer recommendations, therefore, are determined largely by the previous crop grown and the yield goal for the immediate crop.

The percentage of plant nutrient elements in soil, fertilizer, and manure which become available to the first crop after their application has been estimated as 40 per cent of the available N, P_2O_5, and K_2O in soil, as determined by soil tests; 60 per cent of the N, 30 per cent of P_2O_5, and 50 per cent of K_2O in fertilizer; and 30 per cent of N and P and 50 per cent of K in manure (Table 22.1).

TABLE 22.1   FERTILIZER PRESCRIPTION FOR 100-BUSHEL-PER-ACRE CORN YIELD *

	N (lb)		P_2O_5 (lb)		K_2O (lb)	
Source of Nutrient	Present in Soil	Crop Gets	Present in Soil	Crop Gets	Present in Soil	Crop Gets
Soil (by test)	200	40% or 80	80	40% or 32	133	40% or 53
10 tons manure	100	30% or 30	50	30% or 15	100	50% or 50
200 lb 6–24–24 (starter)	12	60% or 7	48	30% or 15	48	50% or 24
200 lb 30–0–0	60	60% or 40	—	—	—	—
Totals for crop	—	157	—	62	—	127
Totals needed for 100 bushels	—	150	—	60	—	120

*Source: Fertilizer Handbook. (Washington: National Plant Food Institute, 1963), p. 57.
Note: Data applies especially in Wisconsin where the work originated, but can be adapted to other soil and climatic zones.

USE OF COMPUTERS TO INTERPRET SOIL TEST RESULTS AND MAKE FERTILIZER RECOMMENDATIONS

Many laboratories are now using electronic computers to interpret the soil test results and other input data. Generally, computer-programmed recommendations can be more accurate and more rapid than handwritten recommendations because a greater number of relevant factors can be taken into consideration. With the computer, factors such as soil type or soil management group, yield potential, subsoil fertility, cropping sequence,

FARMER: JOHN DOE
R 3
COLUMBUS O 43210
COUNTY FRANKLIN

THE OHIO STATE UNIVERSITY
SOIL TESTING LABORATORY-COOPERATIVE EXTENSION SERVICE
1885 NEIL AVENUE · COLUMBUS, OHIO 43210

INDUSTRY: LIME FERTILIZER CO
22 GREEN THUMB ST
COLUMBUS O 43210

RECEIVED SAMPLE: 6-05 DATE PRINTED: 6- 9-69 REPORT NO 46524 PLAN NO 3

LIME AND FERTILIZER RECOMMENDATIONS

PLAN A – IMMEDIATE BUILD-UP OF SOIL FERTILITY AND MAINTENANCE
PLAN B – GRADUAL BUILD-UP OF SOIL FERTILITY AND/OR MAINTENANCE

FOLLOW PLAN A when corrective fertilizer is used.
FOLLOW PLAN B when corrective fertilizer is not used.

YOUR SAMPLE NUMBER	CORRECTIVE LIME (Ag-Ground in T/A to raise soil to pH indicated) pH	T/A	CORRECTIVE FERTILIZER PHOSPHATE P_2O_5 lb/A	POTASH K_2O lb/A	YEAR	CROPS	YIELD GOALS	PLAN A ANNUAL CROP FERTILIZER NITROGEN N lb/A	PHOSPHATE P_2O_5 lb/A	POTASH K_2O lb/A	PLAN B ANNUAL CROP FERTILIZER NITROGEN N lb/A	PHOSPHATE P_2O_5 lb/A	POTASH K_2O lb/A	COMMENTS SEE BELOW
					LAST	CONTINUOUS CORN								
1A	6.5	2.5	200	0	1970	CONTINUOUS CORN	120BU/A	140	40	20	140	60	20	
1					1971	CONTINUOUS CORN	145BU/A	180	50	30	180	70	30	
					1972	CONTINUOUS CORN	150BU/A	220	60	40	220	80	40	
	pH	T/A			LAST	CORN								
2A	6.5	0.0	0	200	1970	CORN	150BU/A	220	25	70	220	25	120	
2	7.0	0.0			1971	WHEAT (SEEDED)	50BU/A	20	20	35	20	20	55	
					1972	ALFALFA	5T/A	***	10	140	***	10	195	
	pH	T/A			LAST	CORN								
3A	6.5	4.5	0	0	1970	MEADOW SEEDING		***	***	***	20	40	40	
3	7.0	5.5			1971	MEADOW	5T/A	***	***	***	125	10	110	1
					1972	MEADOW	5T/A	***	***	***	125	10	110	2

COMMENTS:
1, 2 IF 50 PERCENT OR MORE LEGUME OMIT N, IF 20 PERCENT OR LESS APPLY 50 LBS MORE N.
1, 2 SPLIT NITROGEN INTO SPRING AND FALL APPLICATIONS.

RESULTS OF LABORATORY ANALYSIS FOR EACH SAMPLE (Standard Test; Standard Test Plus Calcium and Magnesium; Minor Elements)

	SAMPLE IDENTIFICATION: YOUR SAMPLE NUMBER	CODE NUMBER	LAB. NUMBER	SAMPLE INFORMATION: SOIL NUMBER	SIZE OF AREA ACRES	DEPTH PLOWED	LAST LIMED YEAR	LAST LIMED T/A	SOIL pH	* LIME TEST INDEX	AVAILABLE PHOSPHORUS P lb/A	EXCHANGEABLE POTASSIUM K lb/A	EXCHANGEABLE CALCIUM Ca lb/A	EXCHANGEABLE MAGNESIUM Mg lb/A	CATION EXCHANGE CAPACITY meq/100g	PERCENT BASE SATURATION (Computed) % Ca	% Mg	% K	CALCIUM MAGNESIUM Ca/Mg RATIO	AVAILABLE Manganese Mn lb A	BORON B lb A	ZINC Zn lb A
1	1A	00010	0110	608	20	9	1966	2	6.3	6.6	17	390	3780	317	17.0	55.5	7.8	2.9	7.2	50	0.5	5.0
2	2A	00011	0111	602	12	8	1969	5	6.4	6.7	132	145	2875	428	13.5	53.3	13.2	1.4	4.0	32	1.5	16
3	3A	00012	0112	608	5	8			5.5	6.4	200	298	2575	142	16.1	40.1	3.7	2.4	10.9		0.5	

Lime Test Index · Use when adjusting lime recommendation. (See Agronomy Guide)

FARMER COPY

FIG. 22.9. Computers are now being used to speed the interpretation of soil test results and to make fertilizer recommendations. This computer-processed soil test report with lime and fertilizer recommendations is now being used by the Ohio State University Soil Testing Laboratory and by many other laboratories throughout the United States. *Source:* Ray Linville, Agronomy Department, Ohio State U.

crop variety, subsoil moisture, weather probabilities, and the managerial ability of the farmer can all be considered simultaneously, and the fertilizer recommendations can be made available in a few seconds in the computer print-out.[5]

An example of a computer-processed soil test report with lime and fertilizer recommendations, as used by the Ohio State University Soil Testing Laboratory, is shown in Figure 22.9. The sample identification, sample information, and the results of laboratory analyses for each sample are shown across the bottom of the print-out. The lime and fertilizer recommendations occur directly above. Fertilizer recommendations are given for both an immediate buildup of soil fertility [Plan A] and for a gradual buildup of soil fertility [Plan B]. Plan A consists of a corrective fertilizer application applied the first crop year, coupled with annual applications, including the first crop year. Annual fertilizer applications alone (no corrective applications) are suggested in Plan B.

For example, if Plan A is to be followed for Sample 1–A (circled), where corn is to be grown with an anticipated yield goal of 120 bushels per acre, it is suggested that a corrective application of 200 pounds of P_2O_5 per acre be applied the first crop year. In addition, 140 pounds per acre of N, 40 pounds of P_2O_5, and 20 pounds of K_2O are to be applied annually.

The lime recommendations are based on raising the soil to the desired pH level of either 6.5 or 7.0. For Sample 1–A (soil pH 6.3), a lime requirement of 2.5 tons per acre is suggested to raise the pH to 6.5.

In the case of Sample 3–A (circled), 4.5 and 5.5 tons of limestone per acre is required to raise the soil pH to 6.5 and 7.0, respectively. Under the column designated "Comments," there are two notes for Sample 3–A. This soil is to be used for establishing a meadow, and it is suggested that if the meadow is to consist of 50 per cent or more legume, that the 125 pounds of recommended N be omitted. If, on the other hand, the meadow consists of 20 per cent or less legume, it is suggested that the N application rate be increased by 50 pounds per acre. It is also suggested that the N be split into spring and fall applications.

The laboratory analyses for magnesium, manganese, boron, and zinc are recorded in the lower right side of the report. The absence of fertilizer recommendations for these nutrients indicates that the available soil levels for these elements are satisfactory for good plant growth.

---

[5] J. C. Shickluna and L. M. Walsh, *Application of Electronic Data Processing Equipment to Soil Testing,* Soil Testing and Plant Analysis, Part 1, Soil Science Society of America, Special Publication No. 2 (1967).

GREENHOUSE SOILS

The production of greenhouse crops requires an intensive fertility program to insure maximum growth and quality in as short a time period as possible. A monitored fertilizer program using chemical soil tests, therefore, is basic to insure that the essential nutrients are not limiting plant growth and, conversely, are not in excess to create an imbalanced or a toxic soil condition.

TABLE 22.2 STANDARD SOIL NUTRIENT TEST FOR GREENHOUSE FLORICULTURAL CROP PRODUCTION FOR SOILS HAVING A DENSITY OF 0.8–1.2 Gm PER CC

Nutrient	Deficient (p.p.m. in Soil)	Optimum (p.p.m. in Soil)	Excess (p.p.m. in Soil)
Nitrate nitrogen (NO₃) *	Below 8	20–40	Above 100
Phosphorus (P) †	Below 6	20–60	Above 200
Potassium (K) ‡	Below 60	180–300	Above 500
Calcium (Ca) ‡	Below 500	—	—
Magnesium (Mg) ‡	Below 60	—	—
Iron (Fe) †	—	—	100
Manganese (Mn) †	—	—	50
Sulfates (SO₄) †	—	—	1800
Chlorides (Cl) †	—	—	200
Sodium (Na) †	—	—	100

* Soils extracted with sodium acetate—acetic acid solution (Morgan's extract).
† Soils extracted with 0.018 normal acetic acid (soil–to–solution ratio of 1:8).
‡ Soils extracted with neutral normal ammonium acetate (soil–to–solution ratio of 1:8).
Source: P. E. Rieke and R. E. Lucas, *Greenhouse Soils Notes*, Mimeographed Report, Crop and Soil Sciences Department, East Lansing, Michigan State University (1969).

NUTRIENT LEVELS FOR GREENHOUSE FLORICULTURAL CROPS

Table 22.2 shows nutrient soil levels for greenhouse floricultural crops considered deficient, optimum, or toxic for nitrate nitrogen, phosphorus, and potassium; deficient levels of calcium or magnesium; or excessive levels of iron, manganese, sulfates, chlorides, or sodium. The nutrient values are expressed in parts per million (p.p.m.) of the soil, and the standard values for these nutrients are suggested for soils with densities of 0.8 to 1.2 grams per cubic centimeter.[6] Light-density synthetic soils, in which vermiculite,

---

[6] P. E. Rieke and R. E. Lucas, *Greenhouse Soils Notes*, Mimeographed Report, Crop and Soil Sciences Department, East Lansing, Michigan State University (1969).

perlite, or peats are used, will require correspondingly higher levels of nutrients for optimum growth.

NITRATE NITROGEN

If the recommended range for nitrates for a specific crop is 20–40 p.p.m. in the soil, then nitrogen should be applied when the test drops below 20 p.p.m. A proper application of nitrogen will raise the test to 40 p.p.m. or slightly higher. If the nitrate level approaches 100 p.p.m., growth may be affected, depending upon the kind and age of plants.

PHOSPHORUS

There is little danger from high phosphorus levels, although micronutrients may be less available if soil levels exceed 200 p.p.m. of phosphorus.

POTASSIUM

When chloride carriers are being used to supply potassium, soil tests for this element showing 300 p.p.m. of K or greater may indicate soluble salt problems. If soluble salts are low, soils may test as high as 500 p.p.m. potassium without difficulty, depending upon the crop. On very sandy or porous soils (except those containing vermiculite in the soil mix), more frequent potassium application is suggested.

CALCIUM

Calcium is very closely related to soil pH. Because of high organic matter content, if soil pH is above 5.5, calcium will be adequate in most soils. A soil test level of 500 p.p.m. or more should be adequate for most crops; some soils, with high cation exchange capacities, may test below this and have sufficient calcium.

MAGNESIUM

Available amounts of magnesium are usually closely associated with calcium, but of lower magnitude. To prevent possible magnesium deficiency from developing, the potassium–to–magnesium ratio should not exceed 3:1.

TABLE 22.3 EXAMPLES OF SOME FERTILIZERS AND THEIR APPLICATION RATES FOR GREENHOUSE SOILS

Fertilizer	N (%)	P_2O_5 (%)	K_2O (%)	For Use as Soluble Fertilizers	In Soil Mixes (Pounds per Cubic Yard)	On Bench Crops (Pounds per 100 sq ft)	General Salt Level	Characteristics
NITROGEN								
Ammonium nitrate	33	0	0	Excellent	¼	½	Very high	Rapid availability, acidifying
Calcium nitrate *	15	—	—	Excellent	½	1–2	Moderate	Rapid availability, alkalizing
PHOSPHORUS								
Treble superphosphate	0	46	0	No	1–2	1–3	Low	Moderate availability
Diammonium phosphate *	21	53	0	Excellent	½	1	Moderate	Rapid availability, acidifying
POTASSIUM								
Potassium chloride	0	0	60	Acceptable	¼	½	Very high	Rapid availability
Potassium nitrate *	13	0	44	Excellent	¼	½–1	High	Very rapid availability
ORGANIC								
Activated sewage sludge *	6	3	0	No	1–2	2–4	Low	Medium availability
COMPLETE FERTILIZERS								
20–20–20 *	20	20	20	Excellent †	¼	½–1	—	Some sources water-soluble
5–20–20 *	5	20	20	No †	½–1	½–1	—	Phosphorus not very soluble

* Provides more than one nutrient.
† Suitability for use in soluble fertilization depends upon components. Look for solubility information on fertilizer bags. Source: P. E. Rieke and R. E. Lucas, Greenhouse Soils Notes, Mimeographed Report, Crop and Soil Sciences Department, Michigan State University, East Lansing (1969).

442

IRON AND MANGANESE

If toxic levels of iron and manganese are present, raising the soil pH by the addition of lime is the easiest means of reducing their solubility and toxicity.

SULFATES, CHLORIDES, AND SODIUM

If toxic levels exist, the soil should be leached, amended, or replaced. The cause of the high salts should be determined and corrected. A guide to greenhouse fertilizer and general application rates is shown in Table 22.3.

SOLUBLE SALTS

Soluble salts resulting from too much fertilizer are frequently a problem in greenhouse crop production. Pennsylvania State University researchers [7] reported that approximately 20 per cent of 300 greenhouse samples tested had accumulated excessive amounts of soluble salts. Other researchers in New York [8] and Ohio [9] have reported similar soluble salt problems. A survey of 73 greenhouse soils tested over a period of one year by the Michigan State University Soil Testing Laboratory indicated that 65 per cent of these samples contained excessive amounts of soluble salts and an additional 8 per cent contained threshold levels of salt.[10]

Soluble salt levels of soil increase rapidly with the addition of fertilizers, especially those containing sodium, chlorides, and sulfates. The application of fertilizer materials to greenhouse soils often exceeds 3000 pounds per acre.[11] Water used for irrigation purposes can also be an important source

---

[7] E. C. Dunkle and F. G. Merkle, "The Conductivity of Soil Extracts in Relation to Germination and Growth of Certain Plants," *Soil Science Society of America Proceedings,* Vol. 8 (1943), pp. 185–88.

[8] W. C. Kelly, *Soil Testing for Greenhouse Crops,* Annual Report, Vegetable Growers Association of America (1954), pp. 83–87.

[9] J. B. Page, *Physical and Chemical Analysis of Ohio Greenhouse Soils,* Annual Report, Vegetable Growers Association of America (1945), pp. 290–93.

[10] W. W. McCall, R. F. Stinson, and R. S. Lindstrom, *Soluble Salt Studies with Greenhouse Floriculture Plants,* Quarterly Bulletin of the Michigan Agricultural Experiment Station, East Lansing, Michigan State University, Vol. 41 (1959), pp. 798–804.

[11] See footnote 10 above.

of soluble salts. Occasionally, nitrates may be too high if there is a source of nitrogen leaching into the water supply.

A guideline for soluble salt tests in soils for greenhouse plants is shown in Table 22.4. Exceptions to these levels may result from the variation in soil, moisture, age, kind of plant, or source of salt.

TABLE 22.4 STANDARD SOLUBLE SALT LEVELS (CONDUCTIVITY) IN SOILS FOR GREENHOUSE FLORICULTURAL CROPS

Conductivity Readings †		
2:1 *(Weight Basis)* * 5:1 *(Volume Basis)* *		
Instrument Reading (EC $\times$ 10⁵)		**Relation to Plant Growth**
0–25	0–12	Very low salt levels present indicate very low nutrient levels
25–75	12–35	Desirable for susceptible plants, particularly bedding plants
75–125	35–60	Desirable range for most established plants; reduced growth may occur for some
125–175	60–85	Reduced growth and vigor common; toxic symptoms on leaf often appear
175–225	85–110	Toxic salt symptoms common, often severe
225+	110+	Toxic to most plants

† Values determined on an AC line-operated Solubridge, Model RD, manufactured by Industrial Instruments. *Source:* P. E. Rieke and R. E. Lucas, *Greenhouse Soils Notes*, Mimeographed Report, Crop and Soil Sciences Department, East Lansing, Michigan State University (1969).

* The 2:1 water-to-soil ratio is used for most greenhouse soils; on highly organic soils, a 5:1 ratio is employed.

SUMMARY

Soil chemical analysis is a valuable deterrent in preventing plant nutrient deficiencies and toxicities from occurring. Before the analysis can be properly interpreted, however, a thorough study is essential of specific soil management and cropping systems.

Soil sampling is an important link in the diagnostic chain. Lime and fertilizer recommendations can be no better than the quality of the samples submitted to the laboratory for testing. It is imperative that the samples represent the field being sampled. At least 20 samplings are required for a composite sample in a given area. Twenty well-taken samples have been shown to be as good as 40 or 100. The number of composite samples required to represent a field depends upon its uniformity. Variations in productivity, topography, texture, drainage, color of topsoil, and past management will necessitate additional composite samples.

The interpretations of soil analyses are based on the relationship between the soil tests and the plant response that can be expected from added plant food. Such relationships are the result of a vast amount of soil and plant research. Although such relationships are not always perfect, research has shown that the lower the soil test for a particular element, the greater the plant response when this element is added as fertilizer.

The increase in growth of a plant from fertilizer is not a straight line relationship. Each successive increment of plant food results in a lower plant response than the former increment. If a farmer could always predict the dollar response curve, he would stop adding fertilizer just before he is trading dollars.

Modern soil testing techniques for determining nutrient needs have outdated the soil testing kit. The basic equipment in the modern soil testing laboratory costs around 10,000 dollars. The pH meter has replaced the color dyes that were once relied upon for soil acidity measurements. The flame photometer and atomic absorption spectrophotometer now make it possible to rapidly and accurately determine potassium, sodium, calcium, magnesium, manganese, iron, copper, and zinc in soils.

Computers are being used by many laboratories as aids in interpreting soil test results, and fertilizer and lime recommendations become available in a few seconds in a print-out.

Maintaining nutrient balance and low salt levels in greenhouse soils is a continual problem and requires a monitored fertilizer program, based upon periodic soil tests, to insure the growth of healthy and vigorous plants. Greenhouse soils should be sampled and tested as often as once per crop for short-term crops or three to four times annually for longer duration crops.

QUESTIONS

1. Under what conditions will soil analysis *not* supply satisfactory answers?
2. Briefly describe the criteria that should be used in determining a uniform sampling area, and how to take a sample of soil for chemical analysis.
3. Briefly describe the advantages of modern soil testing techniques over the tests that can be carried out using a soil testing kit.
4. What advantages do computer-processed soil test reports offer over the conventional handwritten recommendations?
5. List the major problems encountered in the fertilization of greenhouse soils.

REFERENCES

Bartholomew, W. V., and Francis E. Clark, *Soil Nitrogen*. Madison, Wisc.: American Society of Agronomy, 1965.

Black, C. A., *Soil-Plant Relationships*. New York: John Wiley and Sons, Inc., 1968.

Cook, R. L., *Soil Management for Conservation and Production*. New York: John Wiley and Sons, Inc., 1962.

Donahue, Roy L., *Our Soils and Their Management*. Danville, Ill.: The Interstate Printers and Publishers, Inc., 1970.

Fried, M., and Hans Broeshart, *The Soil-Plant System*. New York: Academic Press Inc., 1967.

Kilmer, V. J., S. E. Younts, and N. C. Brady, *The Role of Potassium in Agriculture*. Madison, Wisc.: American Society of Agronomy, 1968.

Muhr, Gilbert R., N. P. Datta, H. Sankarasubramoney, V. K. Leley, and Roy L. Donahue, *Soil Testing in India* (2nd ed.). New Delhi, India: United States Agency for International Development, 1965, 120 pp.

Pearson, Robert W., and Fred Adams, *Soil Acidity and Liming*. Madison, Wisc.: American Society of Agronomy, 1967.

Soil Testing and Plant Analyses. Soil Testing: Part 1. Madison, Wisc.: Special Publication No. 2, Soil Science Society of America, 1967.

Sprague, Howard B., *et al.*, *Hunger Signs in Crops*. The National Fertilizer Association and the American Society of Agronomy, Washington: David McKay Co., Inc., 1964.

Tisdale, Samuel L., and Werner L. Nelson, *Soil Fertility and Fertilizers*. New York: The Macmillan Company, 1966.

CHAPTER 23

PLANT DIAGNOSIS

The plant is able, for better production, to utilize the growth factor occuring in minimum amount, in proportion as the other growth factors occur at the optimum value for the plant.—PROFESSOR G. LIEBSCHER (1895)

The importance of total plant analysis as a diagnostic tool for evaluating plant nutrient needs was recognized early in the nineteenth century (1840) by a German scientist, Justus von Liebig.

A popular theory in Liebig's day was that the mineral elements determined in the plant ash were a valid and accurate indication of the elements necessary for normal plant growth. Liebig, however, expressed doubt as to the accuracy of this hypothesis and believed that some elements taken up by the plant served no metabolic function.[1] Many of the elements absorbed by plants have not, to this day, been shown as essential for plant growth.

The number of elements designated essential for plant growth is dependent upon the criteria employed for essentiality. Arnon and Stout in 1939 proposed the following three criteria that must be met by an element before it can be classified as essential: [2]

---

[1] *Liebig And After Liebig,* Publication of the American Association for the Advancement of Science, No. 16, Washington: Smithsonian Institution Building (1942).

[2] D. I. Arnon and P. R. Stout, "The Essentiality Of Certain Elements in Minute Quantity For Plants With Special Reference To Copper," *Plant Physiology,* Vol. 14 (1939), pp. 371–75.

1. A deficiency of the element makes it impossible for the plant to complete the vegetative or reproductive stage of its life cycle.
2. The symptoms resulting from a deficiency of the specific element can be prevented or corrected *only* by supplying this element.
3. The element is directly involved in the nutrition of the plant, quite distinct from its possible effect in correcting some microbiological or chemical condition of the external medium.

Using these criteria, 16 elements are considered as essential (see Chapter 11). More recently, it has been suggested that as many as 20 elements may be considered as essential for some plants, and this number will no doubt increase in the future. Nicholas [3] has suggested that the "specific element criterion" (No. 2 above) is probably too specific, and he has illustrated the argument with two examples. First, molybdenum is required for nitrogen fixation by azotobacter; however, vanadium may substitute for molybdenum in certain plant species. A second example is the substitution of bromine, and other halides, for chlorine. It has been proposed, therefore, that the term "functional" or "metabolism nutrient" be employed to cover any mineral element that functions in plant metabolism, regardless of its specificity.[4] Using this addendum to Arnon's criterion of essentiality, vanadium, silicon, cobalt, and sodium may be added to the 16 essential nutrient elements discussed in Chapter 11.

A combination of methods is frequently necessary to diagnose accurately a specific plant growth problem. For example, if a plant is shown by green tissue tests to be low in phosphorus or potassium, a soil test for these elements will support the diagnosis that the soil is low in these nutrients and is unable to supply the needs of the plant. Other factors, such as low soil temperature, low soil moisture, poor drainage, or root damage, could result in a low level of these elements in the plant, even though the levels of available soil phosphorus and potassium are adequate. The results of a study conducted in California (Table 23.1) show how an *increase in available water increased the yield* of grapes and the uptake of nitrates, phosphate, and potassium. Soil levels of these nutrients appeared to be adequate but their availability to the plant was *limited by low soil moisture*.

---

[3] D. J. D. Nicholas, "Role of Trace Elements in the Nitrogen Metabolism of Plants with Special Reference to Micro-organisms," *Journal of The Science of Food and Agriculture*, 8,S15 (1957), and D. J. D. Nicholas, "The Function of Trace Metals in the Nitrogen Metabolism of Plants," *Annals of Botany* (*London*), Vol. 21 (1957), pp. 587–98.

[4] D. J. D. Nicholas, "Minor Mineral Nutrients," *Annual Review of Plant Physiology*, Vol. 12 (1961), pp. 63–90.

Additional support in the identification of the plant growth problem may be supplied through the use of plant nutrient deficiency symptoms. The answer to plant nutrient problems will be more readily discernible when a combination of diagnostic techniques are employed; and the chance of diagnosing a problem wrongly is considerably less when all the available diagnostic techniques have been "pooled" to solve the problem.

TABLE 23.1   ADEQUATE SOIL MOISTURE INCREASES NUTRIENT UPTAKE BY GRAPE PLANT

	Nutrient Uptake				
Irrigation Water (Inches)	Nitrate (NO₃) (p.p.m.) *	Phosphate (P) (%)	Potassium (K) (%)	Average Yield † (lb per Vine) 1959	1960
0	780	0.27	0.29	22.7	15
6	1020	0.33	0.84	26.0	31
18	1210	0.50	1.50	37.3	48

* Petioles of leaves opposite cluster taken on Sept. 5.
† Harvested at 21 per cent sugar. *Note:* All differences are significant. *Source:* National Plant Food Institute, University of California (1961).

TOTAL PLANT ANALYSIS

Obtaining the plant sample is a critical step in plant diagnosis. Not only is the physical position of the plant part to be selected for analysis important, but its stage of development is important in its ability to reflect the nutrient needs of the plant. It has been suggested that the tissue for analysis should be selected on the basis of physiological age rather than chronological (calendar) age. Ideally, a sample from a normal, healthy plant should be obtained at the same time the problem plants are sampled. Then, the diagnostician can base his conclusions on comparative tests. Simultaneously, a soil sample can be taken from the area in question. California workers have shown that factors other than those associated with the soil have a pronounced influence on the chemical composition of leaf tissue (Table 23.2).

Differences in chemical composition of the orange leaves shown in Table 23.2 will result from differences in the age of the leaf. Standardization of sampling procedures has been emphasized to maximize the usefulness derived from foliar analysis. Kenworthy has stated that one of the primary benefits from foliar analysis is the possibility of comparing data

| Method of Sampling | Per Cent of Dry Weight † | | | |
Leaves	N	P	K	Mg
From nonfruiting terminals	2.36	0.128	0.86	0.250
From behind young fruits	1.53	0.083	0.38	0.349

* From a field experiment near Pala, Calif.

† Each value represents the mean of 24 determinations. Paired samples were obtained from each of 24 trees.

Note: All differences had statistical significance at the 1 per cent level.

Source: Walter Reuther, W. W. Jones, T. W. Embleton, and C. K. Labanauskas, "Leaf Analysis As a Guide to Orange Nutrition," *Plant Analyses—A Special Issue,* American Potash Institute, Inc., Washington (1962).

on a species or crop grown in various environments throughout the world; this can be done only if sampling methods are standardized.[5] When the analyses are to be used for nutritional studies or for diagnostic purposes, it is important that the plant samples be free from disease, insect damage, and physical or chemical injury.

Sampling procedures suggested for several herbaceous plants are recorded in Table 23.3, and for corn in Figure 23.1.

The results from a 2-year plant analysis study showed that about half the samples classed as normal by farmers actually tested low or deficient in one or more elements, and that the farmers submitting the plant samples were completely unaware that nutrient deficiencies existed. Obviously, plant analysis is valuable in detecting hidden hunger (a condition of marginal deficiency) before the symptoms of a particular nutrient deficiency are manifested. Plants containing below-normal levels of nutrients were most frequently from farms not engaged in a soil testing program and/or which failed to follow the soil test recommendations.[6]

After completing the analyses of 17,000 plant samples for 13 elements, the Ohio workers concluded that the major fertility problems of corn and soybeans are macronutrients (N, P, and K) and not micronutrients. Micro-

5 A. L. Kenworthy, *Fruit, Nut, and Plantation Crops—Deciduous and Evergreen. A Guide for Collecting Foliar Samples for Nutrient-Element Analyses,* Horticultural Report No. 11, Horticulture Department, Michigan State University, East Lansing (Sept., 1969).

6 J. B. Jones, Jr., "A Useful Tool for Farmers—Plant Analyses," *Ohio Report,* Vol. 55, No. 2 (March–April, 1967). Ohio Agricultural Research and Development Center, Wooster, Ohio.

SAMPLE FIRST
MATURE LEAF
BELOW WHORL

FIG. 23.1. For plant analysis, corn should be sampled just prior to tasseling by taking the first fully-developed leaf below the whorl (at arrow). One leaf from each of fifteen to twenty stalks should be taken and composited to make one sample. *Source:* Ohio Agricultural Research and Development Center, Wooster, Ohio.

TABLE 23.3 SUGGESTED SAMPLING INSTRUCTIONS FOR SOME HERBACEOUS PLANTS

Plant	Age, Stage, Position or Condition of Growth	Plant Part	Number of Sub-samples for Each Composite Sample
Alfalfa * (Medicago sativa)	Prior to flowering	Top 6 inches	20–25
Corn * (Zea mays)	(a) Prior to tasseling (b) At silk initiation	(a) First fully developed leaf below whorl (b) Ear leaf	(a) 15–20 (b) 15–20
Rice † (Oryza sativa)	First 2 leaf blades from top of plant at time of flowering	Recently mature leaf blades	200 or more leaves
Sorghum † (Sorghum vulgare)	Second leaf from top of plant when head has fully emerged	Leaves	No less than 50 leaves
Soybeans * (Glycine soja)	Prior to or during initial flowering	Upper fully open leaves	20–25
Sugar beets * (Beta saccharifera)	8 to 10 weeks after seedling emergence	Center leaves and petioles	15–20
Sugar cane † (Saccharum officinarum)	Leaves from top of 4-month old plants; cut an 8-inch section from middle of leaf and remove midrib	Leaf sections (midribs removed). Punch samples of lamina from leaves; or cut sections of third, fourth, fifth, and sixth leaves	No less than 50

* J. B. Jones, Jr., "Plant Analysis," *Ohio Report*, Vol. 52, No. 2 (March–April, 1967), pp. 24–26.

† H. D. Chapman, "Foliar Sampling for Determining the Nutrient Status of Crops," *World Crops*, Vol. 16, No. 3 (1964), pp. 36–46.

nutrients were frequently found to be low or in excess, but not to the extent of the macronutrients.[7] The sampling procedures to be followed for several important woody plants are given in Table 23.4.

TABLE 23.4 SUGGESTED SAMPLING INSTRUCTIONS FOR
SELECTED WOODY PLANTS *

Plant	Age, Stage, Position or Condition of Growth	Plant Part
Apple (Malus sylvestris)	8–12 weeks after full bloom; or 2–4 weeks after formation of terminal buds on bearing trees	Leaves from middle of terminal shoot growth
Banana (Musa sapientum)	6–8 months of age	Third fully expanded leaf
Coffee (Coffea arabica)	6 weeks after bloom and before hardening of pit	Third or fourth pair of leaves
Grape (Vitis vinifera)	Near peak of bloom period	Leaves on nodes having primary flower clusters; keep petiole and discard blade of leaf
Lemon (Citrus limon)	4–5 months after bloom flush (spring flush) of growth	Nonfruiting shoots; one leaf from middle of each selected shoot
Olive (Olea europaea)	About 6 months after main bloom period or in Jan.–Feb.	Pairs of leaves in middle of current terminal growth
Peach (Prunus persica)	Same as apple	Leaves from middle of shoot
Pear (Pyrus communis)	Same as apple	Same as apple
Sour cherry (Prunus cerasus)	Same as apple	Same as apple
Sweet cherry (Prunus avium)	Same as apple	Same as apple
Tea (Thea sinensis)	8–12 weeks after peak harvesting period	First normal leaf left on shoot after harvesting

* A. L. Kenworthy, *Fruit, Nut, and Plantation Crops—Deciduous and Evergreen. A Guide for Collecting Foliar Samples for Nutrient-Element Analysis,* Mimeograph Report, Horticulture Department, East Lansing, Michigan State University (June, 1964).

PREPARATION OF PLANT SAMPLES FOR ANALYSIS

To prevent spoilage of the plant material, it is important that the sample be air dried for at least 1 day prior to mailing in the container

---

[7] J. B. Jones, Jr., and M. H. Warner, "Plant Analysis—What We Are Learning From It," *Crops and Soils* (April–May, 1968).

supplied by the laboratory. Samples should not be mailed in polyethylene bags or any air-tight containers or the sample will spoil.

Where the plant sample has been contaminated from soils, sprays, or residues of other materials, it is imperative that it be cleaned before drying, especially when micronutrient analyses are to be performed. It has been suggested that the washing be carried out as rapidly as possibly on the green tissue (1 minute or less), since such nutrients as potassium, sodium, and chlorine may be readily leached from the plant material, particularly if the sample is somewhat dry or contains necrotic (dead) tissue. Washing with water is acceptable if excess water is removed by blotting and the washing is not prolonged or harsh. Detergents are not generally recommended since they contain appreciable amounts of phosphorus, which may contaminate the sample.

PLANT ANALYSES AND THEIR INTERPRETATION

Total plant analyses usually include tests for nitrogen, phosphorus, potassium, calcium, magnesium, manganese, iron, boron, copper, molybdenum, aluminum and zinc.

The success of plant analysis as a diagnostic tool depends upon accurate and practical interpretation of the test results. Before such an interpretation can be made, it is essential that the diagnostician have a complete history of the management factors for the plant being considered. The form used by the St. Louis Testing Laboratories, Inc., to provide this information is shown in Figure 23.2.

The following categories and their descriptions are commonly used for interpreting plant tissue analysis.

1. *Deficient:* Plants are generally showing clear, visible symptoms of a nutritional deficiency.
2. *Low:* Plants are commonly normal in appearance, but will likely respond to an application of the low-testing elements.
3. *Sufficient:* Plants are normal in appearance and have adequate concentrations of the element for good growth and maximum yield.
4. *High:* Plants appear normal with anticipated optimum yields, but concentration of the particular element is higher than normal.
5. *Excess:* Plants may appear normal or manifest symptoms of nutritional disorder. Reduced yields may result.

When an essential element is classified as deficient on the basis of plant analyses, the diagnostician must determine the cause before a treatment

PLANT ANALYSIS FOR NUTRIENT EVALUATION

St. Louis Testing Laboratories, Inc.
2810 CLARK AVENUE · ST. LOUIS, MO. 63103
531-8080 Code 314
Chemical · Metallurgical · Physical · Non-Destructive · Spectrographic
Agricultural Testing and Analysis
Investigations, Research and Development, Inspection, Field Services
ST LOUIS TEST

SEE ENCLOSED SAMPLING INSTRUCTION BEFORE PROCEEDING

GROWER'S NAME _____

ADDRESS _____

CITY, STATE, ZIP CODE _____

Place Number here which appears on your sample envelope _____

Mail results to above, or

IDENTIFICATION of area or field represented by this information and sample

_____

_____

_____

_____

─────── GROWER COMPLETE THIS SECTION ───────

Date Sampled _____ Crop _____

Variety or hybrid _____ Plant Part Sampled _____

Position on plant _____ Stage of Growth _____

Appearance of plants sampled _____

Date Planted _____ Previous Crop _____

Fertilizer Used _____ KIND _____ at rate of _____ lbs/A

Method of Application _____ ROW, BROADCAST, SIDE DRESSED, FOLIAR

Fertilizer Used _____ KIND _____ at rate of _____ lbs/A

Method of Application _____ ROW, BROADCAST, SIDE DRESSED, FOLIAR

Fertilizer Used _____ KIND _____ at rate of _____ lbs/A

Method of Application _____ ROW, BROADCAST, SIDE DRESSED, FOLIAR

Lime Applied Date _____ Rate _____ tons/A

Sprays, dust applied: What _____ How Much _____ When _____

Soil Type _____ Drainage good ☐   poor ☐   excessive ☐

Previous Soil Test pH _____ , P _____ lbs/A., K _____ lbs/A.

Rainfall Last 30 Days: normal ☐   above normal ☐   below normal ☐

Air Temperature Last 10 Days: normal ☐   above normal ☐   below normal ☐

Describe Area and character of plant growth _____

Results of Analysis

MAJOR ELEMENTS IN %							MICRO-NUTRIENTS IN PARTS PER MILLION							
N	P	K	Na	Ca	Mg	S	Cu	Fe	Zn	B	Mn	Al	Ba	Mo

INTERPRETATIONS AND RECOMMENDATIONS:

LEVEL* OF MAJOR ELEMENTS							LEVEL* OF MICRO-NUTRIENTS							
N	P	K	Na	Ca	Mg	S	Cu	Fe	Zn	B	Mn	Al	Ba	Mo

*Level: D-deficient; L-low, S-sufficient, H-high, E-excess, C-dust or soil contaminated.

Recommendations: _____

_____

_____

By _____
CONSULTING AGRONOMIST

N-Nitrogen; P-Phosphorus; K-Potassium; Na-Sodium; Ca-Calcium; Mg-Magnesium; Cu-Copper; Fe-Iron; Zn-Zinc; B-Boron; Mn-Manganese; Al-Aluminum; Ba-Barium; Mo-Molybdenum; S-Sulfur.

can be prescribed. This involves a thorough review of the history accompanying the plant sample. For example, before the deficiency of the element can be associated with a low available supply of that element in the soil, one must know the level of the nutrient in the soil, climatic conditions, fertilizer treatment, soil pH, drainage, and soil type. The primary cause for the deficiency must be determined before a corrective measure is initiated. The optimum plant nutrient levels for the crops shown in Table 23.5 have been reported by several workers and they serve as a guide for interpreting plant nutrient analyses.

TABLE 23.5 OPTIMUM PLANT NUTRIENT LEVELS FOR SELECTED CROPS *

Nutrient	Valencia Orange (Leaf)	Celery (Petiole)	Onion (Leaf)	Potato (Petiole)	Corn (Ear Leaf)	Soybean (Fully-Open Top Leaves)
N%	2.4–2.6	2.7–4.2	2.5–4.0	—	2.7–3.3	4.5–5.5
P%	0.12–0.16	0.3–0.8	0.25–0.40	0.18–0.22	0.25–0.35	0.3–0.5
K%	1.2–1.7	3.0–9.0	4.0–7.0	6.0–9.0	1.7–2.3	1.5–3.0
Ca%	3.0–5.5	1.5–3.0	1.5–2.5	0.36–0.50	0.25–0.40	0.4–2.0
Mg%	0.26–0.6	0.25–0.45	0.25–0.40	0.17–0.22	0.25–0.40	0.3–1.5
Cl%	Less than 0.3	—	—	—	—	—
S%	0.2–0.3	—	—	—	0.1–0.2	—
Na%	Less than 0.16	—	—	—	—	—
Fe p.p.m.	60–120	75–300	75–200	30–200 **	20–250	30–200
Mn p.p.m.	25–200?	40–200	60–300	30–200	20–150	15–50
Zn p.p.m.	25–100?	30–100	30–100	30–100	20–75	20–70
Cu p.p.m.	5–16?	7–15	8–15	7–30	6–20	—
B p.p.m.	31–100	20–50	15–30	15–40	4–20	20–40
Mo p.p.m.	0.10–0.29?	0.6–4.0	0.8–3.0	0.5–4.0	0.2–2.0	0.5–2.0

* *Source:* Vegetable and field crop nutrient levels supplied by the Soil Science Department, Michigan State University, East Lansing, 1969. Nutrient levels for orange obtained from W. Reuther, *et al.* (see footnote to Table 23.2).
** The upper limits for iron in potato have not been firmly established.

THE USE OF COMPUTERS FOR INTERPRETING

PLANT ANALYSIS

Plant analysis reports, as with soil test reports, are now being prepared by computers. An example of such a report, prepared by the Horticulture

FIG. 23.2. (Opposite.) The diagnostician must have reliable background information on the environment under which the plant to be analyzed was growing before the plant analysis can be properly interpreted, as indicated in this form. *Source:* St. Louis Testing Laboratories, Inc.

NUTRIENT ELEMENT BALANCE CHART -- 1968

MCLACHLIN
KEWADIN, MICH
LABORATORY NO. 180

VARIETY - TART
AGE OF TREE - 05
LOCATION -
GROWER SAMPLE NO. 180

SHORTAGE	I	BELOW NORMAL	I	NORMAL	I	ABOVE NORMAL	I	EXCESS

NITROGEN   XXXXXXXXXXXXXXXXXXXXXXXXXXXXXXXXXXXXXXXXXXXXXXXX

POTASSIUM XXXXXXXXXXXXXXXXXXXXXXXXXXXXXXXXXXXXXXX

PHOSPHORUS XXXXXXXXXXXXXXXXXXXXXXXXXXXXXXXXXXXXXXXXXXXXXXXXXX

CALCIUM   XXXXXXXXXXXXXXXXXXXXXXXXXXXXXXXXXXXXXXX

MAGNESIUM XXXXXXXXXXXXXXXXXXXXXXXXXXXXXXXXXXXXXXX

MANGANESE XXXXXXXXXXXXXXXXXXXXXXXXXXXXXXXXXXXXXXX

IRON   XXXXXXXXXXXXXXXXXXXXXXXXXXXXXXXXXXX

COPPER   XXXXXXXXXXXXXXXXXXXXXXXXXXXXXXXXXXXXXXX

BORON   XXXXXXXXXXXXXXXXXXXXXXXXXXXXXXXXXXXXXXX

ZINC   XXXXXXXXXXXXXXXXXXXXXXXXXXXXXXXXXXX

NUTRIENT ELEMENT	PERCENT IN LEAVES	CHART INDEX
NITROGEN	3.08	104
POTASSIUM	1.12	77
PHOSPHORUS	0.28	106
CALCIUM	1.65	86
MAGNESIUM	0.57	88

NUTRIENT ELEMENT	PPM IN LEAVES	CHART INDEX
MANGANESE	111	89
IRON	130	83
COPPER	15	83
BORON	43	92
ZINC	21	79

DEPARTMENT OF HORTICULTURE, MICHIGAN STATE UNIVERSITY

Department of Michigan State University, is shown in Figure 23.3. Four copies of each report are prepared by the computer. One copy is kept at the university laboratory and three are forwarded to the district horticulture agent. The latter retains a copy and forwards the two remaining copies to the county agricultural agent, who also keeps a copy and forwards the final copy of the report to the grower. This system keeps all interested parties informed. The report is divided into two parts. Part 1 is the nutrient element balance chart (Figure 23.3). Part 2 contains the diagnostics for the interpretations and fertilizer suggestions.

As implied in Figure 23.3, the results reported by the Horticulture Department of Michigan State University are expressed as "balance indices," i.e., relative to a predetermined standard. The composition for a sample is converted into per cent of standard values, and the standard values are adjusted for the normal variation that can be expected for each of the elements. This permits a graphical representation of balance between nutrients having widely different composition values.[8]

The relative balance among the 10 elements for each plant sample is shown in Figure 23.3. The following five categories are used to express these relationships: shortage, below normal, normal, above normal, and excess.

Most frequently, however, the report will specify that a particular nutrient element is either above or below normal, but no corrective practice is necessary. Nitrogen applications are based on the amount applied the previous year, whereas a definite soil or foliage application is suggested for the other elements. The grower uses this information to formulate a fertilizer program.

GREEN TISSUE ANALYSIS

Green tissue analysis differs from total plant analysis in that the green succulent plant tissue is analyzed semiquantitatively for the concentration of soluble nutrients—particularly nitrogen, phosphorus, and potassium—in the plant sap. Plants growing under adequate fertility conditions absorb

---

[8] A. L. Kenworthy, *Plant Analyses and Interpretation of Analyses for Horticulture Crops, Soil Testing and Plant Analysis; Part 2, Plant Analyses*, Special Publication No. 2. Soil Science Society of America, Madison, Wisc. (1967).

FIG. 23.3. (Opposite.) The relationships among the 10 elements are shown on this computer print-out, and are classified according to one of the following five categories: shortage, below normal, above normal, and excess. Crop: tart cherry. Source: A. L. Kenworthy, Horticulture Department, Mich. State U.

more nutrients than they can actually assimilate at any one time, thus making a surplus of nutrients present in the tissue. The tests can be performed rather rapidly, either in the laboratory or in the field where the plants are growing. For this reason, green tissue tests are commonly referred to as "quick tissue tests."

In selecting the part of the plant to be used for green tissue analysis, it is imperative to remember that the three major nutrient elements, nitrogen, phosphorus, and potassium, move from the old to the new tissue when a deficiency of these elements develops. It may be advantageous, therefore, to sample both the old and the new tissue, but, generally, a test of the old tissue is sufficient. Suggestions for sampling a few plants are shown in Table 23.6.

TABLE 23.6 PARTS OF PLANTS MOST SUITABLE FOR GREEN TISSUE TESTING *

Plant	Plant Part
Beets	Leaf petioles
Corn	Leaf sheaths †
Grains	Stems
Alfalfa	Stems
Beans	Leaf petioles
Potato	Leaf petioles or stems
Tomato	Lower leaf petioles
Geranium	Leaf petioles

† Use the stalk on very young plants.
* Source: R. L. Cook and C. E. Millar, Plant Nutrient Deficiencies, Special Bulletin 353, Michigan Agricultural Experiment Station, Michigan State University (1953).

To obtain an accurate evaluation of the availability of plant nutrients, it is best to sample the tissue at regular intervals, perhaps six times throughout the growing season. If, on the other hand, only one test will be made during the growing season, it is best to make the test when the plant is under greatest stress. This will likely occur about midseason when flowering and seed setting is initiated. The value of green tissue analyses generally diminishes as the plant reaches maturity, since the plant parts used for this purpose are usually low in available nutrients.

Researchers in Arkansas compared green tissue analysis with total plant analysis on cotton at various sampling dates. The correlation between these two methods of analysis for nitrate nitrogen (NO_3) and potassium (K) was high. Phosphorus (P) correlations, however, were lower with the advance of the growing season.

To evaluate adequately by green tissue analysis the nutrient status of the plants in the field, test at least 10 to 15 samples from average plants within the area being sampled. The tests should be run on the plant materials immediately. After the tests have been completed, the values should be averaged.

As with total plant analysis, the results obtained from the green tissue analysis of deficient plants are most meaningful when they are compared with tests obtained from normal plant tissue. Plants may vary in the amount of the element they contain when in the deficiency syndrome, and, consequently, it is good practice to make tests simultaneously on normal and deficient plant tissues for comparative purposes.

There are two methods that are commonly employed for carrying out green tissue analysis: the Paper Test for NPK, and the Glass Vial Test for NPK.[9]

Essentially, the materials and equipment required for making tests for nitrogen, phosphorus, and potassium by the Paper Test Method are potassium test papers (three spots on the paper contain dipicrylamine); nitrate powder; P–K Reagent No. 1; P Reagent No. 2; sharp knife; and needle-nosed pliers.[10]

The nitrate nitrogen test can be carried out by placing the cut portion of green plant tissue (petiole or stem) on a clear portion of the folded test paper. Add nitrate powder to the tissue and squeeze the paper and tissue together with the pliers (Figure 23.4). In the presence of nitrate nitrogen, the powder turns red. A faint pink color indicates a low or deficient level of

---

[9] Information on these kits may be obtained from: Denham Laboratory, Route 1, Wilmer, Alabama; Urbana Laboratories, Urbana, Illinois; Lee Lab, 1412 Russell Boulevard, Columbia, Missouri; Edwards Laboratory, Norwalk, Ohio.

[10] N. D. Morgan, and G. A. Wickstrom, "Give Your Plants A Blood Test," *Better Crops with Plant Food,* American Potash Institute, Inc. (1956).

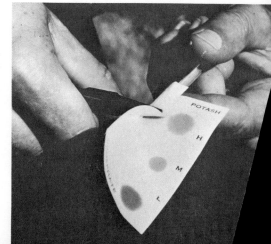

FIG. 23.4. A rapid test for determining the nitrate nitrogen content of green-plant tissue can be carried out in the field where the plant is growing. The paper is folded over the sample of plant tissue, to which nitrate powder has been added, and the paper and tissue are squeezed together with pliers. In the presence of nitrate nitrogen, the powder turns red. Nitrogen, phosphorus, and potassium can be determined on similar strips of test paper. *Source:* American Potash Institute.

FIG. 23.5. For young corn plants, the nitrate test can be determined directly on the plant tissue by cutting the stalk with a clean knife and adding nitrate powder directly on green plant tissue. The presence of nitrates is indicated by the development of a red color. *Source:* American Potash Institute.

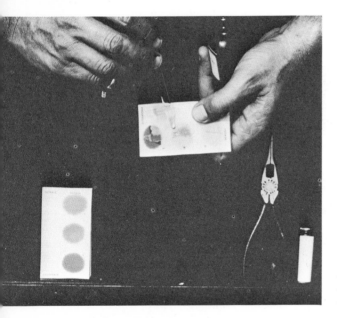

FIG. 23.6. The phosphorus is determined in the sap from the green plant tissue by adding chemical reagents PK-1 and PK-2. The presence of phosphorus is indicated by the development of a blue color. The more intense the blue color formed, the higher the phosphorus content of the plant tissue. *Source:* American Potash Institute.

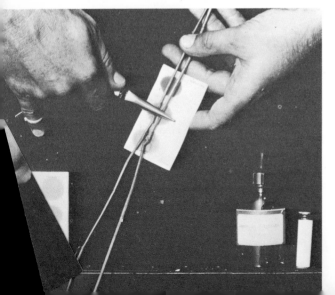

FIG. 23.7. Potassium is determined in the plant by squeezing the green tissue on each of the orange dots imbedded in the paper (the dots contain dipicrylamine) and adding reagent PK-1. The disappearance of orange color from all the dots indicates a very low test for potassium; a high level of potassium for most crops is indicated when the orange color persists in all three dots. *Source:* American Potash Institute.

nitrate nitrogen. A red color is indicative of a high test or adequate nitrogen.[11] The nitrate nitrogen test can be made directly on the green tissue, on a cut corn stalk, for example, as shown in Figure 23.5.

Some plants, such as beets and cotton, contain red pigments, thereby making the Paper Test invalid. On such plants, the Glass Vial Method or the Spot Plate Technique is more satisfactory.

Tests for phosphate-phosphorus can be readily carried out by squeezing the sap from the freshly cut plant tissue onto the paper strips. [Reagents No. 1, stannous oxalate, and No. 2 ammonium molybdate (Figure 23.6).] An adequate supply of phosphorus is indicated by a medium blue to a dark blue color. A light blue color denotes a deficiency of phosphorus.

For the potassium tests, the plant sap is squeezed on the three test spots containing dipicrylamine, and Reagent No. 1 is added (Figure 23.7) to give an orange color. Variations in the brightness of the color signify the amount of potassium present. The disappearance of orange color from all three dots on the test paper signifies a *very low* test for potassium; orange in the middle and bottom dots indicates a *medium* level; and when the orange persists in all three dots, a *high* level of potassium for most crops is indicated.

INTERPRETATION OF GREEN TISSUE ANALYSIS

As with soil analysis and total plant analysis, an accurate interpretation of green tissue tests is dependent upon adequate soil cropping information. The following factors must be considered: pH, phosphorus and potassium levels of the soil, drainage, rate and method of fertilizer placement, moisture supply, insect damage, diseases, unsatisfactory plant populations, variety, and other management factors that may affect the environment and growth of the plant. If one of these is limiting, the plant tissue test will probably not reflect the nutrient status. For example, if a low potassium test is obtained on the plant tissue and the soil test for this element is high, one should investigate other causative factors, such as root damage, poor drainage conditions, and inadequate soil moisture. Too, if one element is in low supply, other nutrients may accumulate in the plant. It cannot be assumed, therefore, that other elements are in adequate supply. It is possible to evaluate only the elements showing a low or deficient level.

---

[11] G. A. Wickstrom, N. D. Morgan, and A. N. Plant, *Ask the Plant, Fight Hidden Hunger with Chemistry*, American Potash Institute, Inc. (1964).

Some precautions should be mentioned. It is not a good practice to analyze the plants early in the morning or on very cloudy days, because results of the nitrate test may be misleading. Nitrate nitrogen is that form of nitrogen which has not yet been converted to plant proteins. In the absence of adequate sunlight, nitrate nitrogen accumulates in the plant, and, under these conditions, it may give higher results than normally expected. It is not advisable to test for nitrate nitrogen during a drought or immediately after a rain because of the accumulation of nitrate nitrogen in the soil surface during the drought and subsequent rapid movement to the leaves following the rain. Plants will likely show a satisfactory supply of nitrogen under these conditions even though a deficiency exists.

PLANT SYMPTOMS OF NUTRIENT DEFICIENCY

Plants exhibit external symptoms of starvation as a result of a nutrient deficiency or as a result of nutrient imbalance. Nutrient deficiencies are frequently the result of the soil's inability to supply a sufficient quantity of that element in time for normal growth and metabolism of the plant. If the soil is found to contain sufficient quantities of that element, a search must be made for other physical and chemical factors that are affecting its uptake by the plant. Nutrient imbalances may create a deficiency of a certain element due to an excess supply of one or more elements relative to the amount of another. A common example of nutrient imbalances frequently results in coarse-textured soils when overfertilization with potassium may induce a magnesium deficiency. Soil and plant analyses are important diagnostic tools for finding the causes and in preventing these conditions from developing. Nutrient deficiencies cause a reduction in the formation of the green pigment, chlorophyll. The pattern of development that the deficiency of various elements shows within a certain plant family are generally quite similar. This is true because members of each plant family have similar nutrient requirements. Plants of the goosefoot and mustard families, for example, have a high requirement for boron, whereas members of pea family are very sensitive to a deficiency of potassium.[12] Similar sensitivity to a lack of zinc is shown by certain pea bean varieties.

Recognizing plant nutrient deficiencies can be complicated when more than one nutrient is limiting growth or when the abnormal appearance of

---

[12] R. L. Cook and C. E. Millar, *Plant Nutrient Deficiencies*, Special Bulletin 353, Michigan Agricultural Experiment Station, East Lansing (1953) Mich. State U.

the plant is the result of disease or injury. In such cases, it is necessary for the diagnostician to make use of *all* testing methods to determine the cause.

DEFICIENCY SYMPTOMS IN CORN

Phosphorus. Young phosphorus-deficient corn plants become stunted and, generally, are dark green in color because of their high nitrogen content. The deficiency of phosphorus promotes the accumulation of sugars in the corn plant and increases anthocyanin (reddish purple) pigment. With a decrease in chlorophyll (green pigment) content, the anthocyanin pigment predominates and gives a purple color to the leaf. The reddening or purpling starts at the tip of the leaf and proceeds along the edges. This coloring may disappear when the plant develops a more extensive root system, and the only symptom remaining is the stunted condition of the plant. Not all corn varieties possess the same genetic factors for the formation of anthocyanin pigment and in such cases a bronze discoloration will replace the purpling effect.

Potassium. The symptoms of potassium deficiency are generally indicated by a stunting of the plant and shortening of the internodes.[13] During later stages of growth, plants may bend over (lodge). Leaf symptoms are characterized by a yellowing that starts at the tip of the older leaves and proceeds along the edges to the base of the leaf. Eventually, the edges become brown and dry (necrotic). This symptom is incorrectly called "leaf scorch."

Nitrogen. As the soil is depleted of its available supply of nitrogen or reaches a level too low for the demands of the corn plant, the entire plant may become light green or yellow (chlorotic), and the oldest leaves will turn yellow at the tip. The chlorosis progresses down the midrib of the leaf toward the stalk, with the leaf margins remaining green. As the deficiency becomes more acute, the second and third oldest leaves develop the same deficiency pattern and the oldest leaf by this time will have turned completely brown (necrotic). When detected during the early stages of growth, nitrogen deficiencies can be corrected with a side-dress application of nitrogen fertilizer.

Magnesium. Magnesium deficiency most commonly occurs on acid sand soils with low levels of magnesium, and on soils that do not contain much clay in the subsoil. The symptoms of deficiency of this element in corn first become obvious as a general chlorosis of the older leaves. Later,

[13] Howard B. Sprague (ed.), *Hunger Signs in Crops* (New York: David McKay Co., Inc., 1964).

whitish stripes appear between the veins, and in advanced cases, the edges and tips of the leaves turn a reddish-purple color.

Zinc. Corn is very sensitive to a shortage of zinc. A deficiency of zinc commonly occurs on soils that have been land leveled, or over tile lines that have been back filled. Such disturbed soils have subsoil on the surface which is low in available zinc.

Although the deficiency is associated with a great variety of soils that vary in texture and chemical properties, it is most frequently associated with coarse-textured (sand) soils or soils that are calcareous. The deficiency is also more prevalent under cool, wet conditions, but the plant may later outgrow the deficiency symptoms as the soils become drier and warmer. Zinc deficiency of corn is characterized by a whitish area on either side of the midrib toward the base of the new leaf. It is frequently evident as the leaf unfolds from the whorl. The remainder of the leaf—midrib, and margins—is green and the plant becomes stunted from a shortening of the internodes.

Sulfur. The deficiency of sulfur in corn results in a general chlorosis of the leaves, poor growth, and delayed maturity. An interveinal chlorosis of the leaves may develop, showing symptoms similar to iron or zinc deficiency. Sulfur deficiency, however, is most likely to occur on acid soils, whereas zinc deficiency is associated with alkaline soil conditions.[14] Note: For comparison, water deficiency on corn is manifest as a wilting (rolling upward) and later death of both upper and lower leaves.

DEFICIENCY SYMPTOMS IN RICE

Nitrogen. A deficiency of nitrogen is first characterized by a general yellowing, especially on the older leaves. The young leaves assume an erect position. As the deficiency becomes more advanced, the chlorotic areas on the older leaves turn brown. The chlorosis in the young leaves develops more slowly and the plant takes on a yellowish-green color. The leaf discoloration is accompanied by a decrease in growth of the plant, and the newly developed leaves are narrower than the normal healthy leaves. The chlorosis first develops at the tip of the older leaves with irregular streaks of yellow on either side of the main vein. Later, a general yellowing near the base of the leaf blade occurs.

---

[14] Howard B. Sprague (ed.), *Hunger Signs in Crops* (New York: David McKay Co., Inc., 1964).

Phosphorus. One of the first symptoms of phosphorus deficiency is the erect position assumed by the younger leaves, similar to nitrogen deficiency, but the older leaves take on a dark green color. Later, the tips of the older leaves become chlorotic between the veins and a speckled orange-brown color results near the tip of the leaf. The discoloration progresses down the leaf; the leaf blade rolls shut and turns an orange-brown color within 2 or 3 days.

Potassium. Potassium deficiency is manifested by stunted growth and early chlorosis of the older leaves. The discoloration typically starts at the tips of the leaves and progresses along the margins of the leaf blade. The leaf initially takes on a rusty appearance and later the tip and margin of the leaves turn yellow and become dry.

Magnesium. Deficiency symptoms of this element are similar to those of potassium deficiency in that the chlorosis starts at the tip of the older leaves and then moves along the edges. The discoloration resulting from a deficiency of magnesium, however, is grayish black as compared with a rust color for potassium deficiency. Later, yellow streaks appear near the edges of the leaf blade and a general chlorosis results in the older leaves.

Calcium. The early foliar symptoms of calcium deficiency appear as a diffused yellow-to-white area about one-third of the distance from the tip of the youngest leaf blade. The next leaf formed may be chlorotic in the upper half or the entire leaf may be chlorotic and rolled. New leaves may fail to form unless calcium is supplied to the plant. The new leaves of the tillers respond in a similar manner followed by death, and although new tillers may continue to form, they may fail to produce a panicle.[15]

DEFICIENCY SYMPTOMS IN COTTON

Nitrogen. A shortage of nitrogen is possibly the most commonly recognized nutrient deficiency symptom of cotton. The deficiency is most likely to be encountered when cotton is grown on sandy soils of low organic matter content and on soils with steep slopes. A light green or yellow chlorosis is general over the entire leaf. The nitrogen-deficient leaf is smaller than leaves from normal healthy plants. The yellowing starts with the older leaves and progresses up the plant. In advanced stages of nitrogen deficiency, the leaves turn brown and eventually fall from the plant. Nitrogen

---

[15] K. L. Olsen, *Mineral Deficiency Symptoms in Rice,* Agricultural Experiment Station, University of Arkansas, Bulletin 605 (1958).

FIG. 23.8. On soil with pH of 7.5 and deficient in available manganese, a normal green color of bean leaves was exhibited only after 120 pounds of manganese (as manganese sulfate) per acre was applied (*right*); 40 pounds of manganese per acre did not supply sufficient manganese for normal color (*center*); and where no manganese was applied to the soil (*left*), the leaves were nearly white except for the veins. *Source:* J. Rumpel, B. G. Ellis, and J. F. Davis, Mich. State U.

deficiency symptoms are frequently mistaken for dry weather damage (moisture stress).

Potassium. The potassium deficiency symptoms on cotton are usually associated with soils that contain large amounts of calcium. The deficiency is particularly noticeable if the vegetative growth of the cotton plant is restricted and fruiting is very heavy.[16] Potassium-deficient cotton leaves are commonly called cotton rust because of the rusty appearance. Initial symptoms are characterized by a chlorosis of the leaf and yellow spots occurring between the veins. The center of the spots eventually dies and brown spots occur between the veins and on the tips and margins of the leaf. The leaf tends to curl downward.[17]

DEFICIENCY SYMPTOMS IN OTHER CROPS

Manganese Deficiency in Beans. The deficiency of manganese is similar on pea beans, soybeans, and garden beans. It is most likely to occur on coarse-textured rather than on fine-textured soils, especially when the soil pH is above 6.5. The deficiency is generally uniform over the leaf with the veins remaining green. It appears first as a mottled effect on the new leaves (Figure 23.8). If the symptoms develop early in the season

---

[16] Howard B. Sprague (ed.), *Hunger Signs in Crops* (New York: David McKay Co., Inc., 1964).

[17] J. B. Jones, Jr., University of Georgia; personal communication.

FIG. 23.9. Alfalfa plant on the left is normal, while the two on the right exhibit boron deficiency, a whitening of the terminal leaves. *Source:* Pacific Coast Borax Co.

when the plants are small, a side-dress application of manganese sulfate or a spray application of manganese will generally correct the deficiency.

Boron Deficiency in Alfalfa. Boron deficiency in alfalfa first appears as a whitening of the terminal leaves. The deficiency retards the terminal growth of the plant and shortens the internodes; stunted growth is due to shortening of the internodes (Figure 23.9). Boron deficiency should not be confused with leaf hopper damage. Leaf hoppers cause a similar yellowing and bronzing, but not the characteristic shortening of the internodes and the rosetting effect of boron deficiency.

Magnesium Deficiency in Apples. Magnesium deficiency on apple leaves shown in Figure 23.10 is characterized by a spotty chlorosis that develops between the lateral leaf veins of the older leaves. The deficiency

FIG. 23.10. Apple leaves on the left are deficient in magnesium; the normal leaf is shown on the right. *Source:* R. A. Cline, Department of Agriculture and Food, Horticultural Research Institute of Ontario, Vineland Station, Ontario, Canada.

FIG. 23.11. Molybdenum deficiency is common in the *Brassica* group, as shown here on brussels sprouts (at arrow). *Source:* R. P. White, Canada Department of Agriculture, Charlottetown, P.E.I., Canada.

initially appears as a slight yellowing, and eventually the chlorotic areas become necrotic (dead) and holes may appear in the leaf tissue.

Molybdenum Deficiency in Brussels Sprouts. Molybdenum deficiency commonly occurs in plants of the *Brassica* group (cauliflower, broccoli, brussels sprouts, and cabbage). The deficiency is shown in Figure 23.11 on brussels sprouts. Note the cupping of the marginal leaf areas with some interveinal chlorosis. The midrib may become twisted as the leaf develops.

Iron Deficiency in Soybeans. "Iron chlorosis," as this deficiency is commonly called, most frequently occurs on calcareous soils. The young leaf becomes very yellow, with veins remaining green. In advanced stages, the leaf may appear almost white (Figure 23.12).

SUMMARY

Total plant analysis and green tissue analysis are valuable diagnostic techniques for determining deficient, sufficient, or excessive amounts of elements in plant tissue. A combination of methods involving soil tests, plant tissue tests, and plant nutrient deficiency symptoms are frequently necessary to diagnose a plant growth problem. It is equally important that the plants selected for analysis represent the plants growing in the entire field.

FIG. 23.12. Soybean plants on the left are iron deficient, while the plants on the right received a spray application of iron sulphate. The soil is calcareous. *Source:* Department of Agronomy, U. of Nebraska, Lincoln.

Both the botanical part of the plant and the stage of development of the particular plant part are critical in their ability to reflect the nutrient needs of the plant. Tissue for analysis should be selected on the basis of physiological rather than chronological (calendar) age.

Before an accurate and practical interpretation of the test results can be made, it is essential that the diagnostician have access to a complete history of the management factors for the plants being considered, as well as knowledge concerning disease, insect damage, and physical or chemical injury.

Green tissue tests differ from total plant analyses in that the former method uses semiquantitative tests for measuring the concentration of soluble nutrients in the plant sap. In total plant analyses, sophisticated techniques and instrumentation are employed for a quantitative evaluation of the total amount of the nutrients in the plant sample.

Nutrient deficiencies cause a reduction in the formation of chlorophyll by the plant and the characteristics of chlorosis (yellowing) are exhibited externally. Although the subsequent symptoms arising from specific nutrient deficiencies are valuable guides in evaluating the deficiency of a particular element, it becomes increasingly difficult to discern the problem on the basis of deficiency symptoms alone when more than one nutrient is limiting growth, or when the abnormal appearance of the plant is a result of disease or injury. In such cases, it is necessary that the diagnostician make use of all the diagnostic tools to determine the cause of the problem.

QUESTIONS

1. What is the difference between total plant analysis and green tissue testing?
2. Under what conditions will a tissue test greatly supplement soil test information?
3. What considerations must be made when sampling plant tissue for chemical analysis?
4. List the information that is essential in the interpretation of green tissue tests and total plant analysis.
5. In corn and rice, compare the plant symptoms of nutrient deficiency for nitrogen, phosphorus, and potassium.

REFERENCES

Ask The Plant With Chemistry. Washington: American Potash Institute, 1964.

Chapman, H. D., "Foliar Sampling For Determining The Nutrient Status of Crops," *World Crops.* Vol. 16, No. 3, 1964, pp. 36–46.

Kenworthy, A. L., *Fruit, Nut, and Plantation Crops—Deciduous and Evergreen. A Guide For Collecting Foliar Samples For Nutrient-Element Analysis.* East Lansing, Mich.: Horticultural Report No. 11, Horticulture Department, Michigan State University, 1969.

Pierre, W. H., Don Kirkham, John Pesek, and Robert Shaw, *Plant Environment and Efficient Water Use.* Madison, Wisc.: American Society of Agronomy, 1966.

Plant Testing. Washington: American Potash Institute, 1962.

Reuther, Walter (ed.), *Plant Analyses and Fertilizer Problems.* Washington: American Society for Horticultural Science, 1964.

Soil Testing and Plant Analysis. Part 2, Plant Analysis. Madison, Wisc.: Special Publication No. 2, Soil Science Society of America, 1967.

Sprague, Howard B. (ed.), *Hunger Signs in Crops.* New York: David McKay Co., Inc., 1964.

Tanaka, Akira and Shouichi Yoshida, *Nutritional Disorders of the Rice Plant in Asia.* Technical Bulletin 10, The International Rice Research Institute, 1970, 51 pp.

Tisdale, Samuel L., and Werner L. Nelson, *Soil Fertility and Fertilizers.* New York: The Macmillan Company, 1966.

CHAPTER 24

SOIL ECOLOGY AND ENVIRONMENTAL POLLUTION

My confidence in my countrymen generally leaves me without much fear for the future.—THOMAS JEFFERSON

Soil and life in the soil (ecology) are a part of the environment; comprising land, water, air, and plants and animals. Pollution is any change in the character of any of the components of the environment which adversely affects man's enjoyment of his environment.

Waste products are one source of materials that pollute the environment. Experts estimate that the nation is producing 1.3 billion tons of agricultural manure and refuse per year.[1] Some of this represents health hazards. Other wastes include 1 billion tons of mining wastes, 350 million tons of residential and industrial rubbish and sewage, 15 million tons of scrapped autos, and 142 million tons of toxic materials, which are spewn into the air from automobiles, homes, power plants, and factories. The quantities of wastes produced each year by all segments of society is increasing at a tremendous rate.

Wastes are everywhere—on the farm as well as in the factory, in the high rise apartment as well as in the state camping grounds, in the city streets, along freeways, near mines, and even in the virgin forests. The organization Keep America Beautiful estimates that American motorists drop 16,000 pieces of litter per mile per year along the primary state and national highways. The National Space Agency counted 1745 items that

---

[1] *U. S. News and World Report* (June 9, 1969).

FIG. 24.1. An American ecology that is fast disappearing because of pollution from barnyards, septic tanks, over-fertilized fields, pesticides, erosion sediments, and factory wastes. (West Virginia.) *Source:* Hermann Postlethwaite, Soil Conservation Service, U.S. Dept. of Agr.

are circling the earth, discarded by space explorers; and the astronauts left certain materials on the moon. Thus, some are saying, "Even the moon is polluted."

Pollution has not always been a problem of such magnitude. Historically man accepted his environment and did not seriously try to change it. In fact, his activities had little effect upon the environment.

Man's first pollution problem was associated with soil erosion. Overgrazing resulted in the removal of adequate vegetative cover, and soil particles were washed into both quiet and flowing waters. Although accelerated erosion affected the productivity of the soil and the quality of the water, man was not greatly concerned because he could always move to another site with greener pastures and cleaner waters.

Times have changed. Space in this world is now at a premium. People are becoming increasingly concerned about both *where* they live and *how* they live. They are vitally interested in their environment: the land, the air, and the water, which are essential to their well-being and esthetic pleasures. As a result, the pollution problem is now receiving increasing attention from both individuals and private organizations, as well as from state and national governments (Figures 24.1 and 24.2).

FIG. 24.2. Man can control his water environment if he wants to concentrate his energies in this direction. Water in the rocky river (*above*) can be (or can not be) controlled all of the way on its journey to the sea. When not controlled (*below*), sediment-laden flood waters pollute the environment by contaminating drinking water, spreading diseases such as typhoid fever, silting the highways and roads, and flooding the homes; in addition, flood waters often drown animals and people. (*Above*, North Carolina; *below*, Oklahoma.) *Source:* Soil Conservation Service, U.S. Dept. of Agr.

The words "ecology," "environment" and "pollution" are now frequently used to emphasize the significance of pollution. Environmental pollution is now defined as the unfavorable alteration of man's surroundings, including the plant and animal life (ecology), through direct or indirect effects of changes in energy patterns, radiation levels, chemical and physical constitution and the abundance of organisms.[2] One of the major considera-

---

[2] *A National Program of Research for Environmental Quality,* U.S.D.A. (Sept., 1968), Room 318-E, Administration Building, Washington, D. C. 20250.

tions in this definition is the undesirable or unfavorable components of the environment. Thus, it is possible to add something to the environment and not create a pollution problem. The definition of environmental pollution recognizes that some naturally occuring components of the environment may be harmful to some people but not to others. For example, pollen in the air is a natural pollutant harmful to many but not to all individuals. The definition also does not establish standards on levels of pollution, because what might be a safe level for one individual may not be a safe or comfortable level for another. For example, all natural waters contain some nitrogen in the nitrate form; frequently, too little to measure. However, the level may be high enough to be toxic to human beings. The critical level is considerably lower for babies and small children than for adults.

A growing number of eminent scientists maintain that *waste* is simply some substance that man has not yet found useful. These scientists ask this question: If man has the genius to mass assemble and to mass distribute, why can he not mass disassemble and massively reuse these wastes? [3]

TYPES OF POLLUTANTS

In considering environmental problems, man himself is sometimes classed as a pollutant. With the population on the verge of explosion in most parts of the world, this concept may be valid.

All industries contribute to the pollution problem. It is untrue to say that agriculture is not guilty of polluting the environment. As with other industries, the exact contribution of agriculture is not well known. On the average, the total environmental pollution caused by agriculture is considered to be relatively small, even though in specific instances the pollution may be substantial.

To list the probabilities for pollution from agriculture to occur is not the same as pointing a finger at the guilty. The formulation of a list of possibilities for pollution represents an honest attempt to inventory the extent of pollution (Figure 24.3). Every human activity is, in fact, a probable source of pollution. Solutions to the environmental pollution problems would become more evident if all industries would inventory their own contributions. [4]

---

[3] Athelston Spilhaus, "The Next Industrial Revolution," *American Association for the Advancement of Science,* Vol. 167, No. 3926 (Mar., 27, 1970).

[4] For an excellent review, see C. H. Wadleigh, *Wastes in Relation to Agriculture and Forestry,* Miscellaneous Publication No. 1065, U.S.D.A. (Mar., 1968).

There are at least 14 types of pollutants, all of which can be either indirectly or directly related to agriculture. The 14 types, many of which are interrelated, are listed as follows:

1. Animal wastes
2. Domestic wastes
3. Processing wastes
4. Infectious agents
5. Plant residues
6. Sediments
7. Plant nutrients
8. Minerals
9. Pesticides
10. Radioactive materials
11. Airborne substances
12. Heat
13. Noise
14. Light

ANIMAL WASTES

The domestic animals in the United States produce over 1 billion tons per year of fecal wastes. Large cattle operations result in the development of special problems. For example, 10,000 head of cattle on a feedlot produce 260 tons of manure per day. If the manure accumulates, it produces offensive odors, may be a source of infectious agents, provides a breeding

FIG. 24.3. A power soil sampler is being used to inventory the extent of pollution caused by the downward movement of plant nutrients derived from fertilizers, manures, and crop residues. With this device, soil samples can easily be collected to a depth of 10 feet. The samples can then be analyzed by both chemical and biological methods to determine possible pollution to underground waters. Source: Larry S. Murphy, Department of Agronomy, Kansas State U.

ground for flies, becomes an unsavory dust on drying, and during heavy rains produces a runoff high in biochemical oxygen demands.[5]

Manure is primarily organic material and can be incorporated into the soil to serve not only as a source of essential plant nutrients but also as a source of organic matter. Limited research suggests that such use of manure is not now economically feasible. However, in the future this may change when society acknowledges that the cost of handling manure should be borne as much by the consumer as by the farmer.

DOMESTIC WASTES

The term "domestic wastes" refers to the human excreta (night soil), waste arising from food preparation, and other discarded products both mineral and organic which come from our concentrated and complex society. Domestic wastes pollute the land, the water, and the air.

Many of the domestic wastes directly enter surface waters. Consider, for example, detergents, which are used in all homes. Most detergents contain phosphate chemicals which may be biodegradable.[6] Such phosphorus, when added to either rivers or lakes, is likely to act as a fertilizer and stimulate growth of undesirable aquatic plants.

The use of human sewage as a source of plant nutrients for crop production needs to be scientifically investigated in more detail. It is estimated that human sewage in the United States alone could now provide 300,000 tons of nitrogen and 200,000 tons of phosphorus each year for enhancing crop production. In certain parts of the world, human sewage has been used for centuries to maintain the fertility level of the soil.

Soil and soil materials can effectively serve not only as a filtering agent, but also as a fixing agent for domestic wastes. Spraying treated domestic wastes over land surfaces may become a common practice for many communities in the future.[7]

The city of Braunschweig, Germany, is reported to have in operation a 7500-acre sprinkler irrigation project designed to efficiently use sewage effluents. Melbourne, Australia has a 20,000-acre sewage farm. In the United States, the city of State College, Pennsylvania, is using sewage in an experiment to spray on nearby soils.

---

[5] Oxygen required for decomposition of organic matter.

[6] Biodegradable means capable of being broken down by life in the soil (ecology).

[7] Herman Bouwer, "Returning Wastes to the Land, A New Role for Agriculture," *Journal of Soil and Water Conservation* (Sept.–Oct., 1968).

PROCESSING WASTES

Processing wastes include wastes from canneries, the cotton textiles industry, meat processing, the poultry industry, and soybean processing. At present, many of these wastes are flushed directly into streams. If the present trends continue, organic processing and industrial wastes will become a greater source of river pollution than municipal sewage.

Soil and materials beneath the soil again offer an alternative method and location for disposing of such materials. The details and economics of methodology have not been fully evaluated. The use of soil represents a potential for maintaining a quality environment with little opportunity for the pollution of either water or the atmosphere.

INFECTIOUS AGENTS

The total impact of the infectious agents, toxins, and allergens apparently has not been fully assessed. More than 10 million workdays are estimated as lost each year in the United States by people suffering from effects of allergens (hay fever). The annual cost of medical treatment for desensitization to only ragweed pollen is estimated at 10 million dollars.

PLANT RESIDUES

Plant residues include those materials from both farms and forest operations. Disposing of such residues is developing into a major problem in certain areas. In the past, burning was considered to be a suitable means for disposing of the residues from forestry and orchard operations. Burning of such materials serves as a source of not only smoke but also noxious gases. Again, one should not overlook the potentials of microorganisms and other life in the soil (soil ecology) as powerful decomposing agents for disposing of organic materials by decomposition.

SEDIMENTS

Most sediment consists of soil and rock particles, plus some organic matter, that have been eroded from land disturbed by animals or abused by man.

Sediments represent a depletion of land resources, in addition to a reduction in the quality of water resources. The sediment becomes a pollutant when it fills lakes, rivers, and water storage reservoirs, and when it kills fish or wild life or otherwise upsets the ecological balance.

The turbidity of water laden with sediment detracts from the use of water for both consumptive and recreational uses. Sediment-laden water is more expensive to treat for human use. Sediment shortens the life of water distribution systems. Colloidal clay sediments act as carriers for other pollutants, such as pesticides, phosphorus and other fertilizers, lead, and mercury.

Construction projects are an ever-increasing source of sediment. Highway construction and suburban housing developments too frequently have resulted in excessive erosion and sedimentation. Planners and developers need to become aware of the potential for soil erosion and adopt methods of control. The U.S.D.A. Soil Conservation Service is in a good position to inventory and to plan methods for the prevention of erosion sediments in construction projects.

The significance of soil erosion in metropolitan areas is often overlooked.[8] For example, in Detroit, Michigan, where the annual precipitation is about 30 inches, as much erosion as 24 tons of soil per acre per year was measured near a new construction project.

In the United States, sediment from agricultural areas is a major problem on 179 million acres and a secondary problem on an additional 50 million acres.

The erosion problem is serious but not as critical as it would have been without the early state research in Missouri, Alabama, and Texas, and the later research and action programs of the U.S. Soil Conservation Service[9] (Figure 24.4).

PLANT NUTRIENTS

Plant nutrients are elements or groups of elements used by plants which are essential for growth and reproduction. The term "plant nutrients" is not interchangeable with the word "fertilizer." Fertilizer is a material used to supply elements essential for plant growth. Detergents contain phosphorus but are not fertilizers.

---

[8] J. R. Thompson, "Soil Erosion in the Detroit Metropolitan Area," *Journal of Soil and Water Conservation* (Jan.–Feb., 1970).

[9] C. H. Wadleigh, *Wastes in Relation to Agriculture and Forestry,* Miscellaneous Publication No. 1065, U.S.D.A. (Mar., 1968).

FIG. 24.4. Sheet and gully erosion is often serious on farmers' fields (A); and on bare soils during the construction of highways, sheet erosion may be serious (B). Soil erosion losses on farm land and methods for their control are being studied in Missouri (C). The first research station to study soil erosion was established in 1917 by M. F. Miller at the Missouri Agricultural Experiment Station. *Source:* Lynn S. Robertson, Crop and Soil Sciences Department, Mich. State U.

Under certain circumstances, any of the fertilizer elements may pollute water. The negatively charged particles (anions), such as nitrates and phosphates, are most likely to create problems. The fertilizer element that today is creating the most concern is nitrogen, especially as nitrates. Nitrogen in the ammonium form moves only very short distances in the soil, but nitrogen in the nitrate form is free to move in any direction with soil water. If water containing nitrate nitrogen moves into ditches or tile drains, so will the nitrate nitrogen; and if the nitrate level of the water is sufficiently high, pollution occurs.

Plant nutrients cause problems when they are present at a sufficiently high level to be toxic to plant and animal life. In water, plant nutrients may cause eutrophication. Eutrophication is a biological process that increases the rate of aging of a lake. In eutrophication, aquatic plants become plentiful and finally deplete the oxygen in the water, which is essential for other higher forms of plant and animal life (ecology). A lake depleted of oxygen may be considered *ecologically dead.*

The role that commercial fertilizers play in water pollution is not well known. In 1967, George E. Smith [10] attempted to summarize what was known about fertilizer and water pollution. He concluded that nutrients in underground aquifers or surface waters could originate from livestock feeding operations; sewage disposal systems; mineralization of natural soil

---

[10] George E. Smith, *Fertilizer Nutrients as Contaminants in Water Supplies,* American Association for the Advancement of Science Publication 85 (1967), pp. 173–186.

humus, legume residues, and manures; or the excessive use of commercial fertilizer. The quantity of nutrients in water originating from natural sources varies greatly. Calcium is the nutrient most common in underground waters. In humid areas, or under irrigation, nitrate nitrogen is likely to be found in greatest concentration in waters of percolation, but very little phosphorus is present. The nitrates in rural water supplies seem to be derived from waste disposal systems or from livestock feeding operations. Soils fix phosphorus and little is soluble; phosphorus enters water supplies primarily as a part of soil sediment. More fertilizer nutrients are lost in erosion than in leaching, and more fertilizer nutrients are removed in crops than are added as fertilizers. Under superior management practices, nutrient losses from properly fertilized soils *can be less* than from soils under poor management where no fertilizer has been added.

Summarizing, it seems that commercial fertilizers, when properly used, probably do not contribute significantly to pollution. However, when used at excessively high levels, well above those currently recommended, nitrogen as nitrates may move by deep percolation in the underground water reservoir and appear in streams and lakes as a pollutant harmful to man.

MINERALS

Minerals and inorganic materials from some chemical and metallurgical plants, including salts, metals, acids, and bases, are grouped into this class of contaminants. These may be lethal to both plants and animals if present in sufficient quantities. Some of the materials, such as lead, are classed as *cumulative* poisons; thus, the concentration at any one time does not need to be high in order to be harmful over a period of years.

The concentration of lead in the atmosphere is increased by exhausts from automobiles that use leaded gasoline. In the Los Angeles, California, area where smog is often a health hazard, the lead content of the atmosphere along major highways is estimated to be approximately 10 times the concentration of that away from the highways. The highest level of concentration, however, is only one-tenth the threshold level currently established for human safety.

PESTICIDES

Pesticides are major pollutants of both the land and water. The use of such materials has increased tremendously in recent years, and this trend may continue because pesticides are so effective. The side effects create concern about the life cycles of many species of plants and animals. Some pesticides are very persistent and are cumulative in nature. In some orchards, for example, 20 years after eliminating arsenical sprays, arsenic residues continue to injure sensitive plants. The accumulation of arsenic in such soils is now well understood and contaminated soils can eventually be returned to production, even though the process may be slow.

Simple chemicals are no longer extensively used for pest control. Complex organic compounds have proven to be more effective and in some instances more persistent and therefore are greater potential pollutants.

DDT is acknowledged as an effective insect control material. Such material, however, is persistent and cumulates in soils and in the bodies of animals, including man. In some orchard soils, it is possible to measure in excess of 100 pounds per acre of DDT.[11] Such accumulations have a profound impact upon the ecology of a region, because many species of fish and birds are sensitive to DDT. The use of long-lasting pesticides such as DDT are now being prohibited by state and federal laws.

Pesticides on the market today should be handled according to directions on the package. In this way, neither the user nor his neighbors are injured and neither soil ecology nor the environment are polluted.

RADIOACTIVE MATERIALS

Soils are a receptor for radioactive materials. The opportunity for damage to agriculture seems to be present in the form of nuclear fallout because of war, the testing of atomic weapons, and the accidental spill from power plants using nuclear reactors.

Limited research shows that radiation hazards from polluted soils can be minimized with deep plowing and the maintenance of a very high soil fertility level. Strontium-90 uptake by plants is minimized in soils with a pH of 7 or above and where calcium represents a very high percentage of the exchangeable cations (Figure 24.5).

AIRBORNE SUBSTANCES

The major chemical contaminants of the atmosphere include certain hydrocarbons, ethylene, ammonia, sulfur dioxide, chlorides, carbon mon-

---

[11] T. J. Sheets, *Pesticide Build-up in Soils*, American Association for the Advancement of Science Publication 85 (1967), pp. 311–30.

FIG. 24.5. The soil and the crops grown at this experiment station have been monitored for radioactive fall-out. Such research also serves as a basis for many of the recommendations made by the Cooperative Extension Service. Fertilizer recommendations based upon field research results are not likely to create water pollution problems when the recommendations are based on carefully planned work such as this. (Michigan.) *Source:* Lynn S. Robertson, Crop and Soil Sciences Department, Mich. State U.

oxide, ozone, and nitrogen oxides, all of which are toxic to plants and animals if present in sufficient quantities. Ozone in smog in California has killed more than 1 million pine trees.

It has been estimated that 13 million tons of natural dusts annually enter the atmosphere in the United States. Damage from dust to vegetation occurs when dust makes contact with the leaves. Dusts also contribute to respiratory problems of both man and animals. Dusts affect both the air and highway vision in addition to causing more work and repairs when dusts permeate buildings and machinery.

Sulfur dioxide is unique in that it may be a contaminant of the air, but at the same time it is a source of sulfur in crop production.

HEAT

Heat acts primarily as a *water* pollutant, although, in some instances *air* may be contaminated by heat. The hazards of heat pollution are reflected in increased rates of both chemical and biological reactions, including respiration. Only slight increases in water temperature can change the environment sufficiently in streams and lakes, making them unsuitable for certain species of fish, or for use as coolants in industry.

NOISE

Noise is defined as unwanted sound. In extreme cases noise may be inaudible to humans. Most scientists are inclined to consider noise as a pollutant of the air or the atmosphere. Very productive soils are "quieter" because they support luxuriant vegetation that "deadens" sound waves. Dense forests are especially quiet (Figure 24.6).

FIG. 24.6. No pollution is evident in this *quiet* California Redwood campground. Ecology-minded campers can drive into a campsite like this, stay over night, eat their food, and dispose of trash, as directed. In this way the campsite will remain a pleasant and relaxing location for others to use and enjoy. *Source:* Redwood Empire Association.

LIGHT

Unwanted light is a pollutant in some environments. Excessive light in man's environment can be made more tolerable by the substitution of luxuriant grasses and trees on productive soil for glaring pavements.

THE ROLE OF SOILS IN POLLUTION

When mismanaged, soils of the world contribute to pollution. Silt-laden waters originating from agricultural land, recreational areas, highway construction, and urban developments, are significant pollutants. Dust-laden winds originating from overgrazed ranges, excessively tilled fields, and unprotected areas are also major contributors to pollution of man's environment (Figure 24.7).

Somewhat less conspicuous is the role soil plays when it serves as a donor of many soluble materials, both mineral and organic, to surface and

FIG. 24.7. *Air Pollution.* The ground-hugging, cloud-like mass shown between the arrows consists of dust particles that are being swept into the air by turbulent winds in Colorado. *Source:* James A. Porter.

FIG. 24.8. Soils respond differently when used for the control of pollution. For example, soil (A) in Virginia is not suitable for any use in pollution control because the
limestone bedrock is at or near the soil surface. Soils (B) and (C) are both in New
York State, but the Hudson silt loam (B) would not be as desirable for use as a drainage field for a septic tank as the Colonie loamy fine sand (C), because the Colonie
soil has a permeability rate for water which is approximately 10 times greater than
that of the Hudson soil. For growing crops, however, the Hudson soil would be superior, because it is less acid, contains more available nutrients, and has an available
moisture capacity almost 3 times greater than the Colonie soil. *Sources:* (A), Lynn S.
Robertson, Crop and Soil Sciences Department, Mich. State U.; (B) and (C), Soil Conservation Service, U.S. Dept. of Agr.

underground waters. Under natural conditions, the quantities of materials moving into waters are usually small and insignificant. However, when the soil is used intensively by man for crop or livestock production, the quantities are likely to be greatly increased. Soils managed by an unconcerned individual, or one with little foresight, will supply more pollutants to the environment than those soils managed by a concerned, creative, and knowledgeable individual.

In some instances, the timing of a management practice is important. The farmer who applies even low rates of livestock manure or commercial fertilizer to *frozen* soil may be polluting the environment. Also the farmer who uses *excessive* rates of fertilizer, manure, or pesticides is gambling with the quality of his environment—as is the individual or the government that uses an airplane to apply chemicals which are slow to decompose, such as DDT.

In contrast, soils can be used to reduce, if not actually eliminate, some of the pollutants. Soils in some areas have been used for years to dispose of waste products. In the past, the "out of sight, out of mind" concept prevailed. Even today, this is evident. Some municipalities continuously seek new landfill areas in which to dispose of trash and waste materials, sometimes without regard to the movement of drainage water and without recognizing that many wastes are biodegradable. It seems that many city administrators and an increasing number of farmers have forgotten that some wastes, such as slaughter house refuse, can actually be used to improve the productivity of the soil.

Soils have certain characteristics and compositions that make them eminently suited for use in waste disposal programs (Figure 24.8):

1. Wide distribution
2. Depth
3. Filtering capacity
4. Exchange property
5. Biological activity (soil ecology)
6. Chemical precipitation.

WIDE DISTRIBUTION

One of the reasons soils have been referred to as "the universal decontaminator" or as "nature's decontaminator" is that they are widespread. In the United States, soils represent an area in excess of 3.5 million square miles.

DEPTH

Soils of the world vary greatly in depth, especially if one uses a broad definition of the word "soil." Frequently, gravel deposits along some of the old river systems extend to depths of several hundred feet; thus, not only are soils widespread but they also occupy tremendous volume, a desirable characteristic for a purifier of large amounts of complex materials.

FILTERING CAPACITY

Sand and gravel have long been used for purifying water. Water passing through such materials is relatively free of solids, except those in the finest colloidal state.

EXCHANGE PROPERTY

Soils with significant quantities of fine silt, clay, and colloidal organic matter have the ability to exchange certain cations as well as anions. Thus, materials in solution react with the soil and impurities are removed in much the same manner as occurs in a modern water softener when "soft" sodium is exchanged for "hard" calcium in water.

BIOLOGICAL ACTIVITY (SOIL ECOLOGY)

Soils, especially in the surface horizons, contain billions of micro-organisms and macroorganisms which rapidly decompose organic matter. The organisms require food for energy, which they obtain from the break-down of the complex organic compounds into the simpler compounds of carbon dioxide, sulfur dioxide, water, and ammonia.

CHEMICAL PRECIPITATION

The precipitation of materials from solution represents a standard chemical laboratory technique. The same types of reactions occur in the soil under certain conditions of temperature, pressure, and concentration of materials. Thus, many relatively soluble materials, when added to the soil, become much less soluble and precipitate out of solution when they react with some of the naturally occurring minerals or organic fractions of the soil. When added to the soil, soluble phosphates are subject to precipitation. This partially explains the fact that very small amounts of phosphorus are normally found in drainage waters.

SUMMARY

Today, environmental pollution is a titanic and a growing problem. Some believe that it is so great and significant that an ecological catastrophe will occur unless the living standards of the more affluent societies such as our own are lowered.

Others, perhaps more optimistic, believe that many of the pollutants in reality are resources yet unused. Thus, it seems that whether wastes are considered as resources or resources produce waste, depends upon one's attitude.

Mismanaged soil itself may be a pollutant. Erosion from farm lands, new highways, and sites of industrial development results in the degradation of both water and air. Spawning beds of fish are destroyed. Dams become ineffective when old river channels are silted. Vision is obscured by dust, and accidents occur when finer soil particles are mixed with the atmosphere.

On the other hand, well-managed soil and soil materials can be used as a decontaminating mechanism. Sands and gravels are now used in the water filtration plants of many modern cities. Deep, moist, and well-aerated soil is a powerful waste disposal medium for organic materials because of the millions of decomposing organisms (soil ecology) present in each gram of soil. The more completely the organic wastes can be mixed into the soil, the faster will be the disintegration. The colloidal fraction of the soil acts as a purifying agent because of its exchange properties. Thus, many soluble materials, including gasses, can be trapped by certain soils so that the original quality of both air and water are not greatly polluted.

Soils are a complex mixture of many chemical compounds, both organic and inorganic. Several materials, when added to the soil, precipitate as less soluble chemical compounds, greatly reducing, if not eliminating, the hazard of pollution.

The mere fact that soils have wide distribution makes them readily available as a decontaminating agent. Public sentiment and health authorities in all areas of high population are insisting on safer waste disposal. Agricultural lands which usually are composed of deep soils may benefit greatly by serving as a medium for safely receiving many of the pollutants.

A word of caution is necessary. Because the use of soil as a decontaminating agent is a relatively new concept, certain precautions and regulations need yet to be worked out by massive research efforts. This is the responsibility of all of those interested in effective land use as well as all persons who want to live in a more productive and healthful environment.

QUESTIONS

1. Define the terms soil ecology and environmental pollution.
2. In what ways do soils contribute to the pollution problem?
3. In what ways can soil be used to solve some of the pollution problems?
4. What characteristics do agricultural soils have that make some of them especially suitable for organic waste disposal?
5. Which fertilizer elements are most likely to pollute waters?

REFERENCES

Agriculture and the Quality of Our Environment. Washington: American Association for the Advancement of Science, Publication No. 85, 1967.

Bartelli, L. J., A. A. Klingebiel, J. V. Baird, and M. R. Heddleson, *Soil Surveys and Land Use Planning.* Madison, Wisc.: Soil Science Society and American Society of Agronomy, 1966, 196 pp.

Jacobson, Jay S., and A. Clyde Hill (eds.), *Recognition of Air Pollution Injury to Vegetation: A Pictorial Atlas.* Pittsburgh, Penna.: Air Pollution Control Association, 1970, 50 pp.

National Academy of Sciences—National Research Council, *The Behavior of Radioactive Fallout in Soils and Plants.* Publication No. 1092, 1963, 32 pp.

Our Heritage of Land and Water Resources. Madison, Wisc.: American Society of Agronomy, Special Publication No. 7, 1966.

Palm, Einar, "What Air Pollution Does to Your Plants," *Crops and Soils Magazine.* Jan., 1971, pp. 14–17.

Soil: The Yearbook of Agriculture (1957). United States Department of Agriculture.

Stanford, G., C. B. England, and A. W. Taylor, *Fertilizer Use and Water Quality.* Agricultural Research Service, ARS 41–168, United States Department of Agriculture, Oct., 1970, 19 pp.

Wadleigh, C. H., *Wastes in Relation to Agriculture and Forestry,* Miscellaneous Publication No. 1065, United States Department of Agriculture, 1968.

U. S. Department of Interior, *Water Pollution Aspects of Urban Runoff,* WP–20–15, Project No. 120. American Public Works Association (Jan., 1969), 272 pp.

APPENDICES

A. Atomic Weight and Valence of Common Elements

B. Definition of Units
   Table 1. Length
   Table 2. Area
   Table 3. Volume
   Table 4. Capacity
   Table 5. Mass

C. Conversion Factors
   Table 1. Length
   Table 2. Area
   Table 3. Volume
   Table 4. Mass
   Table 5. Degrees F to Degrees C
   Table 6. Degrees C to Degrees F
   Table 7. Oxide to Elemental and Elemental to Oxide for
            P and K

D. Glossary

APPENDIX A

ATOMIC WEIGHT AND VALENCE OF COMMON ELEMENTS

Name	Symbol	Atomic Weight	Valence
Barium	Ba	137.36	2
Boron	B	10.82	3
Calcium	Ca	40.08	2
Carbon	C	12.00	2, 4
Chlorine	Cl	35.46	1, 3, 5, 7
Cobalt	Co	58.94	2, 3
Copper	Cu	63.57	1, 2
Fluorine	F	19.00	1
Hydrogen	H	1.01	1
Iodine	I	126.92	1, 3, 5, 7
Iron	Fe	55.84	2, 3
Magnesium	Mg	24.32	2
Manganese	Mn	54.93	2, 4, 6, 7
Molybdenum	Mo	96.00	3, 4, 6
Nitrogen	N	14.01	3, 5
Oxygen	O	16.00	2
Phosphorus	P	31.02	3, 5
Potassium	K	39.10	1
Silicon	Si	28.06	4
Sodium	Na	23.00	1
Sulfur	S	32.06	2, 4, 6
Zinc	Zn	65.38	2

APPENDIX B

DEFINITIONS OF UNITS [1]

TABLE 1 LENGTH

Fundamental Units

A meter (m) is a unit of length equal to 1 650 763.73 times the wavelength of the orange-red radiation of krypton 86 under specified conditions.
A yard (yd) is a unit of length equal to 0.914 4 meter.

Multiples and Submultiples

1 kilometer (km) = 1000 meters
1 hectometer (hm) = 100 meters
1 dekameter (dkm) = 10 meters
1 decimeter (dm) = 0.1 meter
1 centimeter (cm) = 0.01 meter
1 millimeter (mm) = 0.001 meter
1 micron (μ) = 0.000 001 meter = 0.001 millimeter
1 millimicron (mμ) = 0.000 000 001 meter = 0.001 micron

$$1 \text{ angstrom (A)} \text{[2]} \begin{cases} = 0.000\ 000\ 1 & \text{millimeter} \\ = 0.000\ 1 & \text{micron} \\ = 0.1 & \text{millimicron} \end{cases}$$

$$1 \text{ statute mile} \begin{cases} = 8 \text{ furlongs} = 320 \text{ rods} \\ = 1760 \text{ yards} = 5280 \text{ feet} \end{cases}$$

1 furlong = ⅛ mile = 40 rods = 220 yards = 660 feet
1 rod = 5½ yards = 16½ feet = 25 links
1 foot = ⅓ yard = 12 inches
1 hand = 4 inches
1 inch = $\frac{1}{36}$ yard = $\frac{1}{12}$ foot

---

[1] *Source: Units of Weight and Measure (United States Customary and Metric),* United States Department of Commerce, National Bureau of Standards, Miscellaneous Publication 233 (1960).

[2] The angstrom is defined basically as the unit such that 6 438.469 6 of these units equal the wavelength of the red radiation of cadmium under specified conditions.

1 line (button) $= \frac{1}{40}$ inch

1 point (printers) $= 0.013\ 837$ inch $= \frac{1}{72}$ inch (nearly)

1 mil $= \frac{1}{1000}$ inch

1 chain (Gunter's) $= 4$ rods $= 22$ yards $= 66$ feet $= 100$ links

1 link (Gunter's) $= \frac{1}{100}$ chain $= 7.92$ inches

1 international nautical mile $= 1\ 852$ meters $= 6\ 076.115\ 49$ feet (approximately) [3]

1 fathom $= 6$ feet $= 8$ spans

1 span $= \frac{1}{8}$ fathom $= 9$ inches

TABLE 2   AREA

Fundamental Units

A square meter (m^2) is a unit of area equal to the area of a square the sides of which are 1 meter.

A square yard (yd^2) is a unit of area equal to the area of a square the sides of which are 1 yard.

Multiples and Submultiples

1 square kilometer (km^2) $= 1\ 000\ 000$ square meters

1 hectare (ha), or square hectometer (hm^2) $= 10\ 000$ square meters

1 are (a), or square dekameter (dkm^2) $= 100$ square meters

1 centare (ca) $= 1$ square meter

1 square decimeter (dm^2) $= 0.01$ square meter

1 square centimeter (cm^2) $= 0.000\ 1$ square meter

1 square millimeter (mm^2) $= 0.000\ 001$ square meter

1 square mile (mi^2) $\begin{cases} = 640 \text{ acres} = 102\ 400 \text{ square rods} \\ = 3\ 097\ 600 \text{ square yards} = 27\ 878\ 400 \text{ square feet} \end{cases}$

1 acre (acre) $= 10$ square chains $\begin{cases} = 160 \text{ square rods} = 4840 \text{ square yards} \\ = 43\ 560 \text{ square feet} \end{cases}$

1 square chain (ch^2) $\begin{cases} = 16 \text{ square rods} = 484 \text{ square yards} = \\ \quad 4356 \text{ square feet} \\ = 10\ 000 \text{ square links} \end{cases}$

1 square link (li^2) $\begin{cases} = 0.000\ 1 \text{ square chain} = 0.048\ 4 \text{ square yard} \\ = 0.435\ 6 \text{ square foot} = 62.726\ 4 \text{ square inches} \end{cases}$

1 square rod (rd^2) $= 30.25$ square yards $= 272.27$ square feet $= 625$ square links

1 square foot (ft^2) $= \frac{1}{9}$ square yard $= 144$ square inches

1 square inch ($in.^2$) $= \frac{1}{1296}$ square yard $= \frac{1}{144}$ square foot

---

[3] Prior to July 1, 1954, the U.S. nautical mile of 1 853.248 meters $= 6\ 080.20$ feet was in use in this country.

TABLE 3   VOLUME

Fundamental Units

A cubic meter (m^3) is a unit of volume equal to a cube the edges of which are 1 meter.

A cubic yard (yd^3) is a unit of volume equal to a cube the edges of which are 1 yard.

Multiples and Submultiples

1 cubic kilometer (km^3) = 1 000 000 000 cubic meters
1 cubic hectometer (hm^3) = 1 000 000 cubic meters
1 cubic dekameter (dkm^3) = 1000 cubic meters
1 stere (s) = 1 cubic meter
1 cubic decimeter (dm^3) = 0.001 cubic meter
1 cubic centimeter (cm^3) = 0.000 001 cubic meter = 0.001 cubic decimeter
1 cubic millimeter (mm^3) = 0.000 000 001 cubic meter = 0.001 cubic centimeter
1 cubic foot (ft^3) = $\frac{1}{27}$ cubic yard
1 cubic inch (in.^{3}) = $\frac{1}{46656}$ cubic yard = $\frac{1}{1728}$ cubic foot
1 board foot (fbm) = 144 cubic inches = $\frac{1}{12}$ cubic foot
1 cord (cd) = 128 cubic feet

TABLE 4   CAPACITY

Fundamental Units

A liter (liter) is a unit of capacity equal to the volume occupied by 1 kilogram of pure water at its maximum density (at a temperature of 4° C, practically) and under the standard atmospheric pressure (of 760 mm). It is equivalent in volume to 1.000 028 cubic decimeters.

A gallon (gal) is a unit of capacity equal to the volume of 231 cubic inches. It is used for the measurement of liquid commodities only.

A bushel (bu) is a unit of capacity equal to the volume of 2 150.42 cubic inches. It is used in the measurement of dry commodities only.[4]

---

[4] This is the so-called stricken or struck bushel. A heaped bushel for apples of 2 747.715 cubic inches was established by the U. S. Court of Customs Appeals on Feb. 15, 1912, in United States *v.* Weber (no. 757). A heaped bushel, equivalent to 1¼ stricken bushels is also recognized.

Multiples and Submultiples

1 hectoliter (hl) = 100 liters
1 dekaliter (dkl) = 10 liters
1 deciliter (dl) = 0.1 liter
1 centiliter (cl) = 0.01 liter
1 milliliter (ml) = 0.001 liter = 1.000 028 cubic centimeters
1 liquid quart (liq qt) = $\frac{1}{4}$ gallon = 57.75 cubic inches
1 liquid pint (liq pt) = $\frac{1}{8}$ gallon = $\frac{1}{2}$ liquid quart = 28.875 cubic inches
1 gill (gi) = $\frac{1}{32}$ gallon = $\frac{1}{4}$ liquid pint = 7.218 75 cubic inches
1 fluid ounce (fl oz) = $\frac{1}{128}$ gallon = $\frac{1}{16}$ liquid pint
1 fluid dram (fl dr) = $\frac{1}{8}$ fluid ounce = $\frac{1}{128}$ liquid pint
1 minim (min) = $\frac{1}{60}$ fluid dram = $\frac{1}{480}$ fluid ounce
1 peck (pk) = $\frac{1}{4}$ bushel = 537.605 cubic inches
1 dry quart (dry qt) = $\frac{1}{32}$ bushel = $\frac{1}{8}$ peck = 67.200 625 cubic inches
1 dry pint (dry pt) = $\frac{1}{64}$ bushel = $\frac{1}{2}$ dry quart = 33.600 312 5 cubic inches
1 barrel, for fruits, vegetables, and other dry commodities, other than cranberries,[5] = 7 056 cubic inches = 105 dry quarts
1 barrel for cranberries = 5 826 cubic inches [6]

TABLE 5 MASS

Fundamental Units

A kilogram (kg) is a unit of mass equal to the mass of the International Prototype Kilogram.

A gram (g) is a unit of mass equal to 0.001 of the mass of the International Prototype Kilogram.

An avoirdupois pound (lb avdp) is a unit of mass equal to 0.453 592 37 kilogram.

A troy pound (lb t) is a unit of mass equal to $\frac{5760}{7000}$ of that of the avoirdupois pound.

Multiples and Submultiples

1 metric ton (t) = 1000 kilograms
1 hectogram (hg) = 100 grams
1 dekagram (dkg) = 10 grams
1 decigram (dg) = 0.1 gram
1 centigram (cg) = 0.01 gram

---

[5] As fixed by United States standard barrel act (United States Code, Title 15, Ch. 6, Sec. 234).

[6] As computed from dimensions fixed by United States standard barrel act, see footnote No. 5 above.

1 milligram (mg) = 0.001 gram

1 metric carat [7] (c) = 200 milligrams = 0.2 gram

1 avoirdupois ounce (oz avdp) = $\frac{1}{16}$ avoirdupois pound = 437.5 grains

1 avoirdupois dram (dr avdp) = $\frac{1}{256}$ avoirdupois pound = $\frac{1}{16}$ avoirdupois ounce

1 grain (grain) = $\frac{1}{7000}$ avoirdupois pound = $\frac{1}{5760}$ troy pound

1 apothecaries pound (lb ap) = 1 troy pound = $\frac{5760}{7000}$ avoirdupois pound

1 apothecaries or troy ounce (oz ap *or* oz t) = $\frac{1}{12}$ troy pound = $\frac{480}{7000}$ avoirdupois pound = 480 grains

1 apothecaries dram (dr ap) = $\frac{1}{96}$ apothecaries pound = $\frac{1}{8}$ apothecaries ounce = 60 grains

1 pennyweight (dwt) = $\frac{1}{20}$ troy ounce = 24 grains

1 apothecaries scruple (s ap) = $\frac{1}{3}$ apothecaries dram = 20 grains

1 short hundredweight (sh cwt) = 100 avoirdupois pounds

1 long hundredweight = 112 avoirdupois pounds

1 short ton = 2000 avoirdupois pounds

1 long ton = 2240 avoirdupois pounds

---

[7] See National Bureau of Standards Circular C43, The Metric Carat, for a discussion of the metric carat and tables of the relations of the metric carat to the old carat of 205.3 milligrams in use in this country previous to July 1, 1913.

APPENDIX C
CONVERSION FACTORS

TABLE 1 LENGTH *

Multiply	By	To Obtain
Miles	1.609	Kilometers
Miles	1,760	Yards
Miles	5,280	Feet
Miles	63,360	Inches
Meters	1.09	Yards
Meters	3.28	Feet
Meters	39.37	Inches
Meters	100	Centimeters
Meters	0.001	Kilometers
Yards	0.91	Meters
Yards	3.0	Feet
Yards	36	Inches
Feet	0.305	Meters
Feet	0.33	Yards
Feet	12.0	Inches
Inch	0.08	Feet
Inch	0.028	Yards
Inch	2.54	Centimeters
Inches	25.40	Millimeters
Feet	30.48	Centimeters

* Source: *Water Measurement Manual,* 2nd ed. (Washington: Bureau of Reclamation, United States Department of the Interior, 1967), pp. 223–25.

TABLE 2   AREA *

Multiply	By	To Obtain
Square mile	27,878,400	Square feet
Square mile	3,097,600	Square yards
Square mile	640	Acres
Square mile	259	Hectares
Square mile	2.59	Square kilometers
Acre	208.71	Feet square
Acre	0.405	Hectares
Acre	0.004	Square kilometers
Acre	4,840	Square yards
Acre	43,560	Square feet
Acre	4,047	Square meters
Square yards	0.836	Square meters
Square yards	9.00	Square feet
Square yards	1,296	Square inches
Square feet	0.093	Square meters
Square feet	0.111	Square yards
Square feet	144.0	Square inches
Square inches	6.4516	Square centimeters
Square inches	0.0069	Square feet

* *Source: Water Measurement Manual,* 2nd ed. (Washington: Bureau of Reclamation, United States Department of the Interior, 1967), pp. 223–25.

TABLE 3   VOLUME *

Multiply	By	To Obtain
Acre-feet	325,851	U.S. gallons
Acre-feet	43,560	Cubic feet
Acre-feet	1,613.6	Cubic yards
Acre-feet	1,233.5	Cubic meters
Cubic yard	27.0	Cubic feet
Cubic yard	46,656	Cubic inches
Cubic yard	0.765	Cubic meters
Cubic feet	1,728.0	Cubic inches
Cubic feet	7.48	U.S. gallons
Cubic feet	28.317	Liters
Cubic feet	0.037	Cubic yards
Cubic feet	0.028	Cubic meters
U.S. gallon	231.0	Cubic inches
U.S. gallon	3.785	Liters
U.S. gallon	0.134	Cubic feet
Cubic inches	16.387	Cubic centimeters
Cubic inches	0.004	U.S. gallons

* *Source: Water Measurement Manual,* 2nd ed. (Washington: Bureau of Reclamation, United States Department of the Interior, 1967), pp. 223–25.

TABLE 4   MASS *

Multiply	By	To Obtain
Pounds, avoirdupois	0.4536	Kilograms
Pounds, avoirdupois	16	Ounces
Pounds, avoirdupois	1.215	Pounds, troy
Kilograms	1000	Grams
Kilograms	2.2046	Pounds, avoirdupois
Ton (short, 2000 lb)	907.185	Kilograms
Ton (short, 2000 lb)	0.9072	Metric tons
Ton (long, 2240 lb)	0.9842	Metric tons
Ton (metric, 2204.6 lb)	0.977	Short tons
Ton (metric, 2204.6 lb)	1.016	Long tons

Source: Water Measurement Manual, 2nd ed. (Washington: Bureau of Reclamation, United States Department of the Interior, 1967), pp. 223–25.

TABLE 5   DEGREES FAHRENHEIT TO DEGREES CENTIGRADE *

Fahrenheit	Centigrade
1	− 17.2
2	− 16.7
3	− 16.1
4	− 15.6
5	− 15.0
6	− 14.4
7	− 13.9
8	− 13.3
9	− 12.8
10	− 12.2
20	− 6.7
30	− 1.1
40	+ 4.4
50	+ 10.0
60	+ 15.6
70	+ 21.1
80	+ 26.7
90	+ 32.2
100	+ 37.8
200	+ 93.3
300	+148.9
400	+204.4
500	+260.0

* To convert degrees Fahrenheit to degrees Centigrade, subtract 32 from the °F and multiply by 5/9. For example:
$$°C = 5/9 \ (°F - 32)$$
When F = 50, F − 32 = 18, 5/9 × 18 = 10°C.

TABLE 6   DEGREES CENTIGRADE TO DEGREES FAHRENHEIT *

Centigrade	Fahrenheit
0	32.0
1	33.8
2	35.6
3	37.4
4	39.2
5	41.0
6	42.8
7	44.6
8	46.4
9	48.2
10	50.0
20	68.0
30	86.0
40	104.0
50	122.0
60	140.0
70	158.0
80	176.0
90	194.0
100	212.0
200	392.0
300	572.0
400	752.0
500	932.0

* To convert degrees Centigrade to degrees Fahrenheit multiply the °C by 9/5 and add 32. For example,
$$°F = 9/5°C + 32$$
When °C = 50, 9/5 $\times$ 50 = 90, plus 32 = 122.0°F.

TABLE 7   OXIDE TO ELEMENTAL AND ELEMENTAL TO OXIDE CONVERSIONS FOR PHOSPHORUS AND POTASSIUM *

Per Cent or Pounds		Multiply by Factor		To Convert to per Cent or Pounds
P_2O_5	$\times$	0.44	$=$	P
K_2O	$\times$	0.83	$=$	K
P	$\times$	2.29	$=$	P_2O_5
K	$\times$	1.20	$=$	K_2O

* Based upon the following approximate atomic weights:
$$P = 31$$
$$K = 39$$
$$O = 16$$

APPENDIX D

GLOSSARY [1]

A

A Horizon. See soil horizon.

AASHO Classification (soil engineering). The official classification of soil materials and soil aggregate mixtures for highway construction used by the American Association of State Highway Officials.

ABC Soil. A soil with a distinctly developed profile, including A, B, and C horizons. See diagnostic horizons; soil horizons.

Abrasion. The wearing away by friction, the chief agents being currents of water or wind laden with sand and other rock debris and glaciers.

Absorption Loss (irrigation). The initial loss of water from a canal or reservoir by wetting of the soil at the time water is first turned into the structure.

Accelerated Erosion. See erosion.

Accretion. The gradual addition of new land to old by the deposition of sediment carried by the water of a stream.

Acidity, Active. The activity of hydrogen ions in the aqueous phase of a soil, measured and expressed as a pH value.

Acidity, Free. The titratable acidity in the aqueous phase of a soil, expressed in milliequivalents per unit mass of soil or in other suitable units.

Acidity, Potential. The amount of exchangeable hydrogen ions in a soil that can be rendered free or active in the soil solution by cation exchange, usually expressed in milliequivalents per unit mass of soil.

Acid Soil. A soil with a preponderance of hydrogen ions, and probably of aluminum in proportion to hydroxyl ions. Specifically, soil with a pH value less than 7.0. The term is usually applied to the surface layer or to the root zone unless specified otherwise.

Acre-Foot. The volume of water that will cover 1 acre to a depth of 1 foot.

---

[1] This glossary comprises selections from, *Resource Conservation Glossary*, published by the Soil Conservation Society of America, 1970. The complete 52-page glossary is for sale by the society at $5.00 at this address: 7515 North East Ankeny Road, Ankeny, Iowa, 50021. Used with permission.

Acre-Inch. The volume of water that will cover 1 acre to a depth of 1 inch.

AC Soil. A soil having a profile containing A and C horizons with no clearly developed B horizon.

Actual Use (range). The actual grazing use of a grazing unit, usually expressed as animal months or animal unit months, per year.

Adhesion. Molecular attraction which holds the surfaces of two substances in contact, such as water and rock particles.

Aeolian Soil Material (obsolete). See eolian soil material.

Aeration, Zone of. The zone between the land surface and the water table.

Aeration, Soil. The process by which air in the soil is replenished by air from the atmosphere. In a well-aerated soil the soil air is similar in composition to the atmosphere above the soil. Poorly aerated soils usually contain a much higher percentage of carbon dioxide and a correspondingly lower percentage of oxygen. The rate of aeration depends largely on the volume and continuity of pores in the soil.

Aerial Photograph. A photograph of the earth's surface taken from airborne equipment, sometimes called aerial photo or air photograph.

Aerobic. 1: Having molecular oxygen as a part of the environment. 2: Growing only in the presence of molecular oxygen, as aerobic organisms. 3: Occurring only in the presence of molecular oxygen (said of certain chemical or biochemical processes, such as aerobic decomposition).

Afforestation. The artificial establishment of forest crops by planting or sowing on land that has not previously, or not recently, grown tree crops.

Aftermath. The regrowth of forage crops after harvesting.

Aggregation, Soil. The cementing or binding together of several soil particles into a secondary unit, aggregate, or granule. Water-stable aggregates, which will not disintegrate easily, are of special importance to soil structure.

Agricultural Land. Land in farms regularly used for agricultural production. The term includes all land devoted to crop or livestock enterprises, for example, the farmstead lands, drainage and irrigation ditches, water supply, cropland, and grazing land of every kind in farms.

Agronomic Practices. The soil and crop activities employed in the production of farm crops, such as selecting seed, seedbed preparation, fertilizing, liming, manuring, seeding, cultivation, harvesting, curing, crop sequence, crop rotations, cover crops, stripcropping, and pasture development.

Air Porosity. The proportion of the bulk volume of soil that is filled with air at any given time or under a given condition such as a specified moisture condition. Commonly considered to be the larger pores, that is, those filled with air when the soil is at field capacity. Sometimes called noncapillary pore space when determined as the bulk volume of pores that are unable to hold water when subjected to a tension of 60 centimeters of water.

Alfisols. See soil classification.

Alkali. In chemistry, any substance having marked basic properties in contradistinction with acid, that is, being capable of furnishing to its solution or other substances the hydroxyl ion (OH negative). The important alkali metals are sodium and potassium. In a less scientific sense the term is applied to the soluble salts, especially the sulfates and chlorides of sodium, potassium, and magnesium and the carbonates of sodium and potassium, which are present in some soils of arid and semiarid regions in sufficient quantities to be detrimental to ordinary agriculture.

Alkaline. An adjective applied to an alkali; opposite of acidic.

Alkaline Soil. Generally, a soil that is alkaline throughout most or all of the parts of it occupied by plant roots, although the term is commonly applied to only the surface layer or horizon of a soil. Precisely, any soil horizon having a pH value greater than 7.0.

Alkali Soil. 1: A soil with a high degree of alkalinity (pH of 8.5 or higher) or with a high exchangeable sodium content (15 per cent or more of the exchange capacity) or both. 2: A soil that contains sufficient alkali (sodium) to interfere with the growth of most crop plants. See saline-alkali soil; sodic soil.

Alkalinity. The quality or state of being alkaline. The concentration of OH negative ions.

Alluvial. Pertaining to material which has been transported and deposited by running water.

Alluvial Fan. A sloping, fan-shaped mass of sediment deposited by a stream where it emerges from an upland onto a plain.

Alluvial Land. Areas of unconsolidated alluvium, generally stratified and varying widely in texture, recently deposited by streams, and subject to frequent flooding. A miscellaneous land type.

Alluvial Soils. A former great soil group of azonal soils, developed from transported and relatively recently deposited material (alluvium) characterized by a weak modification (or none) of the original material by soil-forming processes.

Alluvium. A general term for all detrital material deposited or in transit by streams, including gravel, sand, silt, clay, and all variations and mixtures of these. Unless otherwise noted, alluvium is unconsolidated.

Alpine Meadow Soils. A former great soil group of the intrazonal order, comprised of dark soils of grassy meadows at altitudes above the timberline.

Ammonification. The biochemical process whereby ammoniacal nitrogen is released from nitrogen-containing organic compounds.

Ammonium Fixation. The adsorption or absorption of ammonium ions by the mineral or organic fractions of the soil in a manner that they are relatively insoluble in water and relatively unexchangeable by the usual methods of cation exchange.

Anaerobic. 1: The absence of molecular oxygen. 2: Growing in the absence of molecular oxygen (such as anaerobic bacteria). 3: Occurring in the absence of molecular oxygen (as a biochemical process).

Ando Soils. A former great soil group of zonal soils that are dark colored, high in organic matter, and developed in volcanic ash deposits.

Angle of Repose. Angle between the horizontal and the maximum slope that a soil assumes through natural processes.

Anion. Negatively charged ion; ion which during electrolysis is attracted to the anode.

Annual Flood. The highest peak discharge in a water year.

Annual Plant. A plant that completes its life cycle and dies in 1 year or less.

Ap. The surface layer of a soil disturbed by cultivation or pasturing.

Apparent Specific Gravity. Ratio of the mass of oven-dry soil to the mass of an equal volume of water.

Apron. A floor or lining to protect a surface from erosion, for example, the pavement below chutes, spillways, or at the toes of dams.

Aquifer. A geologic formation or structure that transmits water in sufficient quantity to supply the needs for a water development. The term water-bearing is sometimes used synonymously with aquifer when a stratum furnishes water for a specific use. Aquifers are usually saturated sands and gravels, and fractured, cavernous, and vesicular rock.

Arable Land. Areas of land so located that production of cultivated crops is economical and practical.

Argillic Horizon. See diagnostic horizons.

Argillan. See clay film.

Arid. A term applied to regions or climates that lack sufficient moisture for crop production without irrigation. The limits of precipitation vary considerably according to temperature conditions, with an upper annual limit for cool regions of 10 inches or less and for tropical regions as much as 15 to 20 inches. Contrast with semiarid.

Aridisols. See soil classification.

Artesian Water. Water confined under enough pressure to cause it to rise above the level where it is encountered in drilling. Flowing artesian wells are produced when the pressure is sufficient to force the water above the land surface.

Aspect (forestry). The direction that a slope faces.

Atom. The smallest portion of an element that can take part in a chemical reaction.

Automated System (irrigation). An irrigation system using timers or self propulsion to reduce labor requirements in the application of irrigation water.

Available Nutrient. That portion of any element or compound in the soil that can be absorbed readily and assimilated by growing plants. Not to be confused with "exchangeable."

Available Water. The portion of water in a soil that can be absorbed by plant roots, usually considered to be that water held in the soil against a tension between $\frac{1}{3}$ and approximately 15 atmospheres (bars). See field capacity.

Available Water-Holding Capacity (soils). The capacity to store water available for use by plants, usually expressed in linear depths of water per unit depth of soil. Commonly defined as the difference between the percentage of soil water at field capacity and the percentage at wilting point. This difference multiplied by the bulk density and divided by 100 gives a value in surface inches of water per inch depth of soil. See field capacity; wilting point.

B

B Horizon. See soil horizon.

Bacteria. Microscopic, single-celled or noncellular plants, usually saprophytic or parasitic.

Badland. A land type consisting of steep or very steep barren land, usually broken by an intricate maze of narrow ravines, sharp crests, and pinnacles resulting from serious erosion of soft geologic materials. Most common in arid and semiarid regions. A miscellaneous land type.

Band Seeding. Seeding of grasses and legumes in a row 1 to 2 inches directly above a band of fertilizer.

Base Exchange Capacity (obsolete). See cation exchange capacity.

Base Level. The theoretical limit toward which erosion constantly tends to reduce the land. Sea level is the general base level, but in the reduction of the land there may be many temporary base levels which, for the time being, the streams cannot reduce. These temporary base levels may be controlled by the level of a lake or river into which the stream flows or by a particularly resistant stratum of rock that the stream has difficulty in eroding.

Base Period. A period of time from which comparisons of other time periods are made, normally used with reference to price, population, and production.

Base Saturation Percentage. The ratio of bases (calcium, magnesium, potassium, and sodium) extracted from the soil by an extraction agent to the capacity of the soil to hold extractable bases, expressed as a percentage.

Basic Crops. Crops such as corn, wheat, and cotton, that are most important in the agricultural economy due to acreage, value, or climate.

Basin: Hydrology—The area drained by a river. Irrigation—A level plot or field, surrounded by dikes, which may be irrigated by flooding.

Basin Irrigation. A method of irrigation in which a level or nearly level area surrounded by an earth ridge or dike is flooded with water.

Basin Lister. See lister.

Bedding. 1: Method of surface drainage consisting of narrow-width plowlands in which the dead furrows run parallel to the prevailing land slope and are used as field drains. Also known as crowning or ridging. 2: The process of laying a drain or other conduit in its trench and tamping earth around the conduit to form its bed. The manner of bedding may be specified to conform to the earth load and conduit strength. 3: The arrangement of sediment and sedimentary rock in layers, strata, or beds more than 1 centimeter thick. See lamination. 4: Method of preparing seedbeds for row crop culture, also used for irrigation or drainage.

Bedload. The sediment that moves by sliding, rolling, or bounding on or very near the streambed; sediment moved mainly by tractive or gravitational forces or both but at velocities less than the surrounding flow.

Bedrock. The more or less solid rock in place either on or beneath the surface of the earth. It may be soft or hard and have a smooth or irregular surface.

Bench Terrace. See terrace.

Bentonite. A highly plastic clay consisting of the minerals montmorillonite and beidellite that swells extensively when wet.

Biennial Plant. A plant that requires 2 years to complete its life cycle.

Biological Oxygen Demand (B.O.D.). The amount of oxygen required for bacteria to decompose organic matter, as a measure of the amount of pollution.

Biomass. The amount of living matter in a given area.

Biota. The flora and fauna of a region.

Biotic Influence. The influence of animals and plants on associated plant or animal life as contrasted with climatic influences and edaphic (soil) influences.

Biotype. A group of individuals occurring in nature, all with essentially the same genetic constitution. A species usually consists of many biotypes.

Blocky Soil Structure. See soil structure types.

Blowout. 1: An excavation in areas of loose soil, usually sand, produced by wind. 2: A break–through or rupture of a soil surface attributable to hydraulic pressure, usually associated with sand boils.

Bog Soil. A former great soil group of the intrazonal order and hydromorphic suborder, including Fibric, Hemic, and Sapric soil material (muck and peat), developed under swamp or marsh types of vegetation. See organic soil materials and Table 2, page 554.

Border Dikes. Earth ridges built to guide or hold irrigation water within prescribed limits in a field; a small levee.

Border Ditch. Ditch used as a border of an irrigated strip or plot, water being spread from one or both sides of the ditch along its entire length.

Border Irrigation. A surface method of irrigation by flooding between border dikes.

Border Strip (irrigation). The area of land bounded by two border ridges or dikes that guide the irrigation stream from the point or points of application to the end of the strips.

Brackish. Slightly salty. Term applied to water with a saline content that is intermediate between that of streams and sea water; neither fresh nor salty but in between.

Breccia. A rock consisting of consolidated granular rock fragments larger than sand grains.

Broad-Base Terrace. See terrace.

Broadcast Seeding. Scattering seed on the surface of the soil. Contrast with drill seeding which places the seed in rows in the soil.

Broad-Crested Weir. An overflow structure for measuring water, often rectangular in cross section, in which the water adheres to the surface of the crest rather than springing clear. Contrast with sharp-crested weir.

Brown Forest Soils. A former great soil group of the intrazonal order and calcimorphic suborder formed on calcium-rich parent materials under deciduous forest and possessing a high base status but lacking a pronounced illuvial horizon. Developed under a deciduous forest in temperate humid regions from parent materials relatively rich in bases. See Table 2, page 554.

Brown Podzolic Soils. A former zonal great soil group similar to Podzols but lacking distinct A2 horizon characteristic of the Podzol group, developed under deciduous or mixed deciduous and coniferous forest in temperate or cool-temperate humid regions. See Table 2, page 554.

Brown Soils. A former great soil group of the temperate to cool arid regions, composed of soils with a brown surface and a light-colored transitional subsurface horizon over a calcium carbonate accumulation. See Table 2, page 554.

Brunizem (Prairie) Soils. The former zonal group of soils having a very dark brown or grayish brown surface horizon, grading through brown soil to the lighter colored parent material at 2 to 5 feet, developed under tall grasses in a temperate, relatively humid climate. These include only those dark-colored soils of the treeless plains in which carbonates have not been concentrated in any part of the profile by soil-forming processes. See Table 2, page 554.

Buffer Strips. Strips of grass or other erosion-resisting vegetation between or below cultivated strips or fields.

Bulk Density, Soil. The mass of dry soil per unit bulk volume. The bulk volume is determined before drying to constant weight at 105°C. A unit of measure expressed as grams per cubic centimeter or pounds per cubic foot. Contrast with specific gravity.

Bulk Volume. The volume, including the solids and the pores, of an arbitrary soil mass.

C

C Horizon. See soil horizons.

Calcareous Soil. Soil containing sufficient calcium carbonate (often with magnesium carbonate) to effervesce visibly when treated with cold 0.1 normal hydrochloric acid.

Calcic Horizon. See diagnostic horizons.

Caliche. 1: A layer near the surface, more or less cemented by secondary carbonates of calcium or magnesium precipitated from the soil solution. It may occur as a soft, thin soil horizon, as a hard, thick bed just beneath the solum, or as a surface layer exposed by erosion. Not a geologic deposit. 2: Alluvium cemented with sodium nitrate, chloride, and/or other soluble salts in the nitrate deposits of Chile and Peru. See hardpan.

Cambic Horizon. See diagnostic horizons.

Canal (irrigation). Constructed open channel for transporting water from the source of supply to the point of distribution.

Canopy. The cover of leaves and branches formed by the tops or crowns of plants.

Capability, Land. See land capability.

Capillary Fringe. A zone just above the water table (zero gauge pressure) that remains almost saturated. The extent and the degree of definition of the capillary fringe depends upon the size distribution of pores.

Capillary Porosity. The small pores or the bulk volume of small pores that hold water in soils against a tension usually greater than 60 centimeters of water. These pores are commonly filled with water when the soil is at field capacity.

Capillary Water. The water held in the "capillary" or small pores of a soil, usually with tension greater than 60 centimeters of water. Much of this water is considered to be readily available to plants.

Carbonaceous. Pertaining to or containing carbon derived from plant and/or animal residues.

Carbon Cycle. The sequence of transformation undergone by carbon utilized by organisms wherein it is used by one organism, later liberated upon the death and decomposition of the organism, and returned to its original state to be reused by another organism.

Carbon-Nitrogen Ratio. The ratio of the weight of organic carbon to the weight of total nitrogen in the soil or in organic material, obtained by dividing the percentage of organic carbon (C) by the percentage of total nitrogen (N).

Catch Crop. 1: A crop produced incidental to the main crop of the farm and usually occupying the land for a short period. 2: A crop grown to replace a main crop which has failed.

Catena. A sequence of soils of about the same geologic age, derived from similar parent material and occurring under similar climatic conditions but having different characteristics due to variations in relief and in natural drainage.

Cation. Positively charged ion; ion which, during electrolysis, is attracted to the cathode. Common soil cations are calcium, magnesium, sodium, potassium, ammonium, and hydrogen.

Cation Exchange. The interchange between a cation in solution and another cation on the surface of any surface-active material, such as clay colloid or organic colloid.

Cation-Exchange Capacity (sometimes called total-exchange capacity, base-exchange capacity, or cation-adsorption capacity). The sum total of exchangeable cations that a soil can adsorb expressed in milliequivalents per 100 grams of soil or clay.

Census of Agriculture. A census taken by the Bureau of Census every 5 years. It includes number of farms, land in farms, crop acreage and production, livestock numbers and production, farm spending, farm facilities and equipment, farm tenure, value of farm products sold, and farm size. Data are given by states and counties.

Chalk. Composed mainly of the calcareous shells of various marine microorganisms, but the matrix consists of fine particles of calcium carbonate, some of which may have been chemically precipitated.

Channel. A natural stream that conveys water; a ditch excavated for the flow of water.

Channel Improvement. The improvement of the flow characteristics of a channel by clearing, excavation, realignment, lining, or other means in order to increase its capacity. Sometimes used to connote channel stabilization.

Channel Stabilization. Erosion prevention and stabilization of velocity distribution in a channel, using jetties, drops, revetments, vegetation, and other measures.

Channel Storage. Water temporarily stored in channels while enroute to an outlet.

Channery. An adjective formerly incorporated into the soil textural class designations of horizons when the soil mass contains between 15 and 90 per cent by volume of rock fragments. In Scotland and Ireland the term may refer to gravel.

Check Dam. Small dam constructed in a gully or other small watercourse to decrease the streamflow velocity, minimize channel scour, and promote deposition of sediment.

Check Irrigation. A method of irrigation in which an area is practically or entirely surrounded by earth ridges.

Chernozem Soils. A former zonal great soil group consisting of soils with a thick, nearly black, organic-matter-rich A horizon, high in exchangeable calcium, underlain by a lighter colored transitional horizon that is above a zone of calcium carbonate accumulation. These soils occur in a cool sub-humid climate under a vegetation of tall and midgrass prairie. See Table 2, page 554.

Cherty. An adjective incorporated into the soil textural class designation of horizons when the soil mass contains between 15 and 90 per cent by volume of chert fragments. See chert fragments and coarse chert fragments as defined under coarse fragments and Table 1, page 511.

Chestnut Soils. A former zonal great soil group consisting of soils with a moderately thick, dark brown A horizon over a lighter colored horizon that is above a zone of calcium carbonate accumulation. See Table 2, page 554.

Chiseling. Breaking or loosening the soil, without inversion, with a chisel cultivator or chisel plow.

Chroma. The relative purity, strength, or saturation of a color; directly related to the dominance of the determining wavelength of the light and inversely related to grayness; one of the three variables of color. See Munsell color system; hue; value.

Classification. The assignment of objects or units to groups within a system or categories distinguished by their properties or characteristics.

Clastic. Composed of broken fragments of rocks and minerals.

Clay (soils). 1: A mineral soil separate consisting of particles less than 0.002 millimeter is equivalent diameter. 2: A soil textural class. 3: (engineering) A fine-grained soil that has a high plasticity index in relation to the liquid limits.

Clayey. See particle size classes for family groupings.

Clayey-Skeletal. See particle size classes for family groupings.

Clay Film. A thin coating of well-oriented clay particles on the surface of a soil aggregate, particle, or pore. Synonyms: clay coat, clay skin, argillan, tonhautchen.

Clay Mineral. 1: Naturally occuring inorganic crystalline material composed of fragments of hydrous aluminum silicate minerals found in soils and other earthy deposits, the particles being of clay size, that is, less than 0.002 millimeter in diameter. 2: Material as described under but not limited by particle size.

Claypan. A dense, compact layer in the subsoil having a much higher clay content than the overlying material from which it is separated by a sharply defined boundary; formed by downward movement of clay or by synthesis of clay in place during soil formation. Claypans are usually hard when dry and plastic and sticky when wet. They usually impede the movement of water and air. See hardpan.

Clean Tillage. Cultivation of a field so as to cover all plant residues and to prevent the growth of all vegetation except the particular crop desired.

Cleavage. The tendency of minerals to split along crystallographic planes. As applied to rocks, it is the property of splitting into thin parallel sheets which may be highly inclined to the bedding planes, as in slate.

Climate. The sum total of all atmospheric or meteorological influences, principally temperature, moisture, wind, pressure, and evaporation, which combine to characterize a region and give it individuality by influencing the nature of its land forms, soils, vegetation, and land use. Average weather.

Climate, Continental. The type of climate characteristic of land areas separated from the moderating influence of oceans by distance, direction, or mountain barriers, marked by relatively large daily and seasonal change in temperature.

Climate, Oceanic. The type of climate characteristic of land areas near oceans which contribute to the humidity and at the same time have a moderating influence on temperature and range of temperature variation. Synonym: marine climate.

Climatic Year. A continuous 12-month period arbitrarily selected for the analysis and presentation of climatological or streamflow data, generally beginning March 1 or April 1.

Climax Vegetation. Relatively stable vegetation in equilibrium with its environment and with good reproduction of the dominant plants.

Clod. A compact, coherent mass of soil ranging in size from 5 to 10 millimeters (0.2 to 0.4 inch) to as much as 200 to 250 millimeters (8 to 10 inches); produced artificially by the activity of man by plowing and digging when the soils are either too wet or too dry for normal tillage operations.

Closed Drain. Subsurface drain, tile, or perforated pipe that receives surface water only through surface inlets.

Coarse Cherty. Similar to cherty but fragments are coarse, chert in size. See coarse fragments and Table 1, page 511.

Coarse Fragments. Rock or mineral particles greater than 2.0 millimeters in diameter. See Table 1 for the names used for coarse fragments in soils.

Coarse-Loamy. See particle size classes for family groupings.

Coarse Sand. See soil separates; soil texture.

Coarse Sandy Loam. See soil texture.

Coarse-Silty. See particle size classes for family groupings.

Coarse Texture. The texture exhibited by sands and loamy sands. A soil containing large quantities of these textural classes (United States usage). See sand; sandy; moderately coarse texture.

Cobblestone. See coarse fragments and Table 1.

Cobbly. An adjective incorporated into the soil textural class descriptions of horizons when the soil mass contains between 15 and 90 per cent by volume of cobblestones. See cobblestone as defined under coarse fragments in Table 1.

Cohesion. Holding together; force holding a solid or liquid together, owing to attraction between like molecules; decreases with rise in temperature.

COLE. Coefficient of linear extensibility; the ratio of the difference between the moist and dry lengths of a clod to its dry length; (Lm-Ld)/Ld, wherein Lm is the moist length (at $\frac{1}{3}$ atmosphere) and Ld is the air-dry length. The measurement correlates with the volume change of a soil upon wetting and drying.

Colloid. A substance that, when suspended in water, diffuses not at all or very slowly through a semipermeable membrane, and usually has little effect on freezing point, boiling point, or osmotic pressure of the suspension; a substance in a state of fine subdivision with particles from 0.00001 to 0.0000001 centimeter (1 micron to 1 millimicron).

Colluvial (geology). Consisting of alluvium, talus, and cliff debris; material of avalanches.

Colluvial Soil Material. Soil material that has moved downhill and has ac-

TABLE 1 NAMES APPLIED TO COARSE FRAGMENTS IN SOILS

Fragments		Descriptive Terms Applied to Fragments Having		
Shape	Material	Diameters Less Than 3 in.	Diameters from 3 to 10 in.	Diameters More Than 10 in.
Rounded or sub- rounded	All kinds of rock	Pebble, gravel	Cobblestone	Stone *
Irregular and angular	Chert	Chert fragment	Coarse chert fragment	Stone *
	Other than chert	Angular pebble (Lengths up to 6 in.)	Angular cobble- stone (Lengths from 6 to 15 in.)	Stone * (Lengths over 15 in.)
Thin and flat	Limestone, sandstone, or schist	Fragment	Flagstone	Stone
	Slate	Slate fragment	Flagstone	Stone
	Shale	Shale fragment	Flagstone	Stone

* The word Bouldery is sometimes used when stones are larger than 24 in.

cumulated on lower slopes and/or at the bottom of the hill. Colluvial material is moved downhill by the force of gravity and to some extent by frost action and local wash. Synonym: colluvium.

Color. See Munsell color system.

Columnar Soil Structure. See soil structure types.

Compaction. To unite firmly; the act or process of becoming compact, usually applied in geology to the changing of loose sediments into hard, firm rock. With respect to construction work with soils, engineering compaction is any process by which the soil grains are rearranged to decrease void space and bring them into closer contact with one another, thereby increasing the weight of bulk material per cubic foot.

Competition. The general struggle for existence within an area in which the living organisms compete for a limited supply of the necessities of life.

Compost. Organic residues or a mixture of organic residues and soil that have been piled and allowed to undergo biological decomposition.

Compression. A system of forces or stresses that tends to decrease the volume or compact a substance, or the change of volume produced by such a system of forces.

Concretion (soils). A local concentration of a chemical compound, such as calcium carbonate or iron oxide, in the form of an aggregate or nodule, of varying size, shape, hardness, and/or color.

Conglomerate. The consolidated equivalent of gravel.

Conifer. A tree belonging to the order *Coniferae,* usually evergreen, with cones and needle-shaped or scalelike leaves and producing wood known commercially as "soft wood."

Conservation. The protection, improvement, and use of natural resources according to principles that will assure their highest economic or social benefits.

Conservation District. A public organization created under state enabling laws as a special-purpose district to develop and carry out a program of soil, water, and related resource conservation, use, and development within its boundaries; usually a subdivision of state government with a local governing body and always with limited authorities. Often called a soil conservation district or a soil and water conservation district.

Consistence (soil). 1: The resistance of a material to deformation or rupture. 2: The degree of cohesion or adhesion of the soil mass. Terms used for describing consistence of soil materials at various soil moisture contents and degrees of cementation are:

Wet. Nonsticky, slightly sticky, sticky, very sticky, nonplastic, slightly plastic, plastic, and very plastic.

Moist. Loose, very friable, friable, firm, very firm, and extremely firm.

Dry. Loose, soft, slightly hard, hard, very hard, and extremely hard.

Cementation. Weakly cemented, strongly cemented, and indurated.

Consolidate. Any or all of the processes whereby loose, soft, or liquid earth materials become firm and hard.

Consumptive Use. The quantity of water used and transpired by vegetation plus that evaporated. See evapotranspiration.

Contamination. The act of polluting or making impure, used here to indicate chemical, sediment, or bacteriological impurities.

Continuous Grazing. Allowing domestic livestock to graze a specific area throughout the grazing season. Not necessarily synonymous with year-long grazing.

Contour. 1: An imaginary line on the surface of the earth connecting points of the same elevation. 2: A line drawn on a map connecting points of the same elevation.

Contour Ditch. Irrigation ditch laid out approximately on the contour.

Contour Farming. Conducting field operations, such as plowing, planting, cultivating and harvesting on the contour.

Contour Flooding. Method of irrigating by flooding from contour ditches.

Contour Furrows. Furrows plowed approximately on the contour on pasture or rangeland to prevent soil loss and increase water infiltration. Also, furrows laid out approximately on the contour for irrigation purposes.

Contour Interval. The vertical distance between contour lines.

Contour Stripcropping. Layout of crops in comparatively narrow strips in which the farming operations are performed approximately on the contour. Usually strips of grass, close-growing crops, or fallow are alternated with those in cultivated crops.

Contrasting Textures (As used in the Soil Classification System of the National Cooperative Soil Survey in the United States). If two widely different particle size classes occur within a vertical distance of 5 inches in the control section from which the soil family is derived, both particle size classes are listed in the name. For example, if the upper part of the control section is loamy sand and the lower part is clay, the particle size class is sandy over clayey.

Control. In research, something under study, either untreated or given a standard treatment, which is used as a basis for comparison with other treatments.

Control Section (As used in the Soil Classification System of the National Cooperative Soil Survey in the United States). Arbitrary depths of soil material within which certain diagnostic horizons, features, and other characteristics are used as differentiae in the classification of soils. The thickness is specific for each characteristic being considered but may be different for different characteristics.

Control Structure. A regulating structure to maintain irrigation water at a desired elevation, usually installed in gravity flow systems.

Conveyance Loss. Loss of water from delivery systems during conveyance, including operational losses and losses due to seepage, evaporation, and transpiration by plants growing in or near the channel.

Cool-Season Plant. A plant that makes its major growth during the cool portion of the year, primarily in the spring but in some localities in the winter.

Corrosion. The solution of rocks, minerals, and other materials by chemical action.

Corrugation Irrigation. A partial surface flooding method of irrigation, normally used with drilled crops, whereby water is applied in small graded channels or furrows so spaced that an adequate lateral spread is obtained by the time the desired amount of water has entered the soil.

Cover Crop. A close-growing crop grown primarily for the purpose of protecting and improving soil between periods of regular crop production or between trees and vines in orchards and vineyards.

Cover, Ground. Any vegetation producing a protecting mat on or just above the soil surface. In forestry, low-growing shrubs, vines, and herbaceous plants under the trees.

Cover, Vegetative. All plants of all sizes and species found on an area, irrespective of whether they have forage or other value. Synonym: plant cover.

Creep (soil). Slow mass movement of soil and soil material down relatively steep slopes, primarily under the influence of gravity but facilitated by saturation with water and by alternate freezing and thawing.

Cropland. Land used primarily for the production of adapted cultivated, close-growing, fruit, or nut crops for harvest, alone or in association with sod crops.

Crop Residue. The portion of a plant or crop left in the field after harvest.

Crop Residue Management. Use of that portion of the plant or crop left in the field after harvest for protection or improvement of the soil.

Crop Rotation. The growing of different crops in recurring succession on the same land.

Crumb Structure. See soil structure type.

Crust. A dry surface layer of soil that is much more compact, hard, and brittle than the material immediately beneath it.

Cumulative Infiltration. Summation of the depth of water absorbed by a soil in a specified elapsed time in reference to the time of initial water application. Synonym: intake.

D

Dam. A barrier to confine or raise water for storage or diversion, to create a hydraulic head, to prevent gully erosion, or for retention of soil, rock, or other debris.

Dead Furrow. Double furrow left between two areas or lands due to plowing in opposite directions.

Debris. A term applied to the loose material arising from the disintegration of rocks and vegetative material; transportable by streams, ice, or floods.

Debris Dam. A barrier built across a stream channel to retain rock, sand, gravel, silt, or other material.

Debris Guard. Screen or grate at the intake of a channel, drainage, or pump structure for the purpose of stopping debris.

Deciduous Plant. A plant that sheds all its leaves every year at a certain season.

Decomposer. An organism, usually a bacterium or a fungus, that breaks down the bodies or parts of dead plants and animals into simpler compounds.

Deep Percolation. Water that percolates below the root zone.

Deferred Grazing. Discontinuance of grazing livestock on an area for a specified period of time during the growing season to promote plant reproduction, establishment of new plants, or restoration of vigor of old plants.

Deferred-Rotation Grazing. A systematic rotation of deferred grazing.

Deficiency. The amount by which a series of quantities falls short of adequacy; opposite of excess.

Deflocculate. 1: To separate the individual components of compound particles by chemical and/or physical means. 2: To cause the particles of the disperse phase of a colloidal system to become suspended in the dispersion medium.

Deformation of Rocks. Any change in the original form or volume of rock masses produced by tectonic (earth) forces; folding, faulting, and solid flow are common modes of deformation.

Degradation. To wear down by erosion, especially through stream action.

Degraded Chernozem. A former zonal great soil group consisting of soils with a very dark brown or black A1 horizon underlain by a dark gray, weakly expressed A2 horizon and a brown B–like horizon, formed in the forest-prairie transition of cool climates. See Table 2, page 554.

Degree of Grazing. The degree of utilization of selected plant species in a designated area at the time of measurement.

Delta. An alluvial deposit formed where a stream or river drops its sediment load on entering a body of more quiet water, formed largely beneath the water surface and in an area often resembling the shape of the Greek letter Delta, with the point of entry of the stream at one corner.

Denitrification. The biochemical reduction of nitrate or nitrite to gaseous nitrogen, either as molecular nitrogen or as an oxide of nitrogen.

Density. Number of organisms per unit mass at a given time.

Deposit. Material left in a new position by a natural transporting agent, such as water, wind, ice, or gravity, or by the activity of man.

Depth, Effective Soil. The depth of soil material that plant roots can penetrate readily to obtain water and plant nutrients. It is the depth to a layer that differs sufficiently from the overlying material in physical and/or chemical properties to prevent or seriously retard the growth of roots.

Desalinization. Removal of salts from saline soil, usually by leaching.

Desert. An area of land that has an arid, hot-to-cool climate with vegetation that is sparse and usually shrubby.

Desert Soils. A former zonal great soil group consisting of soils with a very thin, light-colored surface horizon that may be vesicular and is ordinarily underlain by calcareous material; formed in arid regions under sparse shrub vegetation.

Desiccation. A drying out of material.

Desilting Area. An area of grass, shrubs, or other vegetation used for inducing deposition of silt and other debris from flowing water, located above a stock tank, pond, field, or other area needing protection from sediment accumulation. See filter strip.

Detachment. The removal of transportable fragments of soil material from a soil mass by an eroding agent, usually falling raindrops, running water, or wind. Through detachment, soil particles or aggregates are made ready for transport. See soil erosion.

Detrital. Clastic. Rocks and minerals occurring in sedimentary rocks that were derived from pre-existing igneous, sedimentary, or metamorphic rocks.

Detritus. Matter worn from rocks by mechanical means, usually alluvial deposits.

Deviation, Standard (statistics). A measure of the average variation of a series of observations or items of a population about their mean. In normally distributed sets of moderate size the interval of the mean plus or minus the standard deviation includes about two-thirds of the items.

Diagnostic Horizons (As used in the Soil Classification System of the National Cooperative Soil Survey in the United States). Combinations of specific soil characteristics that are indicative of certain classes of soils. Those which occur at the soil surface are called epipedons, those below the surface, endopedons.

Argillic Horizon. A subsurface horizon into which clay has moved. It has at least 20 per cent more clay than the horizons above. The presence of clay films on ped surfaces and in soil pores is evidence of clay movement.

Calcic Horizon. A surface horizon more than 6 inches thick that has more than 15 per cent calcium carbonate equivalent and at least 5 per cent more carbonates than the C horizon.

Cambic Horizon. A subsurface horizon that has textures finer than loamy fine sand and in which materials have been altered or removed but not accumulated. Evidences of alteration include the elimination of fine stratifications; changes caused by wetness, such as gray colors and mottling; redistribution of carbonates; and yellower or redder colors than in the underlying horizons.

Duripan. A subsurface horizon that is cemented by silica.

Fragipan. See fragipan in alphabetical listing.

Histic Epipedon. A surface horizon that is saturated with water at some season unless artificially drained, generally between 8 and 12 inches thick and containing from 20 to 30 per cent organic matter if not plowed or from 14 to 28 per cent organic matter if plowed. In each case the limiting organic matter content depends on the amount of the mineral portion that is clay. The lower percentage is used if the horizon has no clay and the higher percentage if the horizon has 50 per cent or more clay.

Mollic Epipedon. A surface horizon that is dark colored, contains more than 1 per cent organic matter, and is generally more than 7 inches thick. It has more than 50 per cent base saturation and is not both hard and massive when dry. Dark colors have Munsell values darker than 3.5 when moist and 5.5 when dry, and Munsell chromas of less than 3.5 when moist and have in common soil color names, such as black, very dark brown, very dark gray, or very dark grayish brown.

Natric Horizon. A subsurface horizon that is a special kind of argillic horizon, containing much exchangeable sodium.

Ochric Epipedon. A surface horizon that is too light in color (higher value or chroma than mollic epipedon), too low in organic matter, or too thin to be either a mollic or an umbric epipedon.

Oxic Horizon. A subsurface horizon that is a mixture principally of kaolin, hydrated iron and aluminum oxides, quartz, and other highly insoluble primary minerals, and containing very little water–dispersible clay.

Petrocalcic Horizon. An indurated subsurface horizon cemented by carbonates.

Spodic Horizon. A subsurface horizon in which amorphous materials consisting of organic matter plus compounds of aluminum and usually iron have accumulated.

Umbric Epipedon. A surface horizon similar to a mollic epipedon but having less than 50 per cent base saturation.

Disintegration. The reduction of rock to smaller pieces mainly by mechanical means.

Disperse. 1: To break up compound particles, such as aggregates, into the individual component particles. 2: To distribute or suspend fine particles, such as clay, in or throughout a dispersion medium, such as water.

Dispersion. The act of dispersing; to separate, spread, scatter. That which is dispersed could be open textured and porous. On the other hand, colloidal particles may be dispersed and held in suspension in a fluid state as a gel and move in a mass as a mud flow.

Dispersion medium. The portion of a colloidal system in which the disperse phase is distributed.

Dispersion Ratio. The ratio of silt plus clay remaining in suspension after limited shaking and settling, using specific procedures, to the total silt plus clay as determined by mechanical analysis. The greater this ratio, the more easily the soil can be dispersed.

Dispersion, Soil. The breaking down of soil aggregates into individual particles, resulting in single-grain structure. Ease of dispersion is an important factor influencing the erodibility of soils. Generally speaking, the more easily dispersed the soil, the more erodible it is.

Disposal Field. Area used for spreading liquid effluent for separation of wastes from water, degradation of impurities, and improvement of drainage waters. Synonym: infiltration field.

Dissolved Solids. The total dissolved mineral constituents of water.

Distributary. Smaller conduit taking water from a canal for delivery to fields; any system of secondary conduits; river channel flowing away from the main stream and not rejoining it, as contrasted to a tributary.

Distribution System (irrigation). 1: System of ditches and their appurtenances which convey irrigation water from the main canal to the farm units. 2: Any system that distributes water within a farm.

Diversion. Channel constructed across the slope for the purpose of intercepting surface runoff; changing the accustomed course of all or part of a stream. See terrace.

Diversion Dam. A barrier built to divert part or all of the water from a stream into a different course.

Diversion Terrace. Diversions, which differ from terraces in that they consist of individually designed channels across a hillside, may be used to protect bottomland from hillside runoff or may be needed above a terrace system for protection against runoff from an unterraced area. They may also divert water out of active gullies, protect farm building from runoff, reduce the number of waterways, and are sometimes used in connection with strip-cropping to shorten the length of slope so that the strips can effectively control erosion. See terrace.

Drain. 1: A buried pipe or other conduit (closed drain). 2: A ditch (open drain) for carrying off surplus surface water or groundwater. 3: To provide channels, such as open ditches or closed drains, so that excess water can be removed by surface flow or by internal flow. 4: To lose water (from the soil) by percolation.

Drainage. 1: The removal of excess surface water or ground water from land by means of surface or subsurface drains. 2: Soil characteristics that affect natural drainage.

Drainage District. A cooperative, self-governing public corporation created under state law to finance, construct, operate, and maintain a drainage system involving a group of land owners.

Drainage, Soil. As a natural condition of the soil, soil drainage refers to the frequency and duration of periods when the soil is free of saturation; for example, in well-drained soils the water is removed readily but not rapidly; in poorly drained soils the root zone is waterlogged for long periods unless artificially drained, and the roots of ordinary crop plants cannot get enough oxygen; in excessively drained soils the water is removed so completely that most crop plants suffer from lack of water. Strictly speaking, excessively drained soils are a result of excessive runoff due to steep slopes or low available water-holding capacity due to small amounts of silt and clay in the soil material.

Drawdown. Lowering of the water surface (in open channel flow); water table (in ground water flow) resulting from a withdrawal of water.

Drift Fence. A fence without closure used to influence animal movement.

Drift, Glacial. Rock debris transported by glaciers and deposited either directly from the ice or from the meltwater.

Drill Seeding. Planting seed with a drill in relatively narrow rows, generally less than a foot apart. Contrast with broadcast seeding.

Dryland Farming. The practice of crop production in low rainfall areas without irrigation.

Dry Weight (soils). The equilibrium weight of the solid soil particles after the water has been vaporized by heating to 105°C.

Duckfoot. An implement with horizontally spreading, V-shaped tillage blades or sweeps which are normally adjusted to provide shallow cultivation without turning over the surface soil or burying surface crop residues.

Duff. The organic layer on top of mineral soil, consisting of freshly fallen vegetative matter in the process of decomposition.

Dugout Pond. An excavated pond as contrasted with a pond formed by constructing a dam.

Duripan. See diagnostic horizons.

Dust Mulch. A loose, finely granular or powdery condition on the surface of the soil, usually produced by shallow cultivation when the soil is dry.

E

Earth Dam. Dam constructed of compacted soil materials.

Ecological Niche. The role of an organism in an ecosystem.

Ecology. The study of the interrelationships of organisms to one another and to the environment.

Ecosystem. Energy-driven complex of a community of organisms and its controlling environment.

Ecotone. A transition line or strip of vegetation between two plant communities, having characteristics of both kinds of neighboring vegetation.

Ecotype. A locally adapted population of a species which has a distinctive limit of tolerance to environmental factors. See biotype.

Effective Precipitation. That portion of total precipitation that becomes available for plant growth. It does not include precipitation lost to deep percolation below the root zone or to surface runoff.

Effluent. 1: The discharge or outflow of water from ground or subsurface storage. 2: The fluids discharged from domestic, industrial, and/or municipal waste collection systems or treatment facilities.

Elevated Ditch (irrigation). Earth-fill, constructed to specifications similar to those for earth-fill dams, to provide normal grade as a substitute for flumes or siphons. Synonym: raised ditch.

Electrolysis. Chemical decomposition of certain substances by an electric current passing through a substance.

Eluvial Horizon. A soil horizon formed by the process of eluviation. See eluviation; illuvial horizon.

Eluviation. The removal of soil material in suspension (or in solution) from a layer or layers of a soil. (Usually, the loss of material in solution is described by the term "leaching.") See illuviation; leaching.

Emergency Spillway. A spillway used to carry runoff exceeding a given design flood.

Entisols. See soil classifications.

Environment. The sum total of all the external conditions that may act upon an organism or community to influence its development or existence.

Eolian. A term applied to deposits arranged or transported by the wind.

Eolian Soil Material. Soil material accumulated through wind action. The most extensive areas in the United States are silty deposits (loess), but large areas of sandy deposits, called sand dunes, also occur.

Ephemeral Stream. A stream or portion of a stream that flows only in direct response to precipitation. It receives little or no water from springs and no long continued supply from snow or other sources. Its channel is at all times above the water table.

Epipedon. See diagnostic horizons.

Erodible (geology and soils). Susceptible to erosion.

Erosion. 1: The wearing away of the land surface by running water, wind, ice, or other geological agents, including such processes as gravitational creep. 2: Detachment and movement of soil or rock fragments by water, wind, ice, or gravity. The following terms are used to describe different types of water erosion:

Accelerated Erosion. Erosion much more rapid than normal, natural, or geologic, primarily as a result of the influence of the activities of man or other animals, or natural catastrophies such as fires.

Geological Erosion. The normal or natural erosion caused by geological processes acting over long geologic periods and resulting in the wearing away of mountains, the building up of floodplains and coastal plains. Synonym: natural erosion.

Gully Erosion. The erosion process whereby water accumulates in narrow channels and, over short periods, removes the soil from this narrow area to considerable depths, ranging from 1 to 2 feet to as much as 75 to 100 feet.

Natural (Normal) Erosion. Wearing away of the earth's surface by water, ice, or other natural agents under natural environmental conditions of climate and vegetation, undisturbed by man. Synonym: geological erosion.

Rill Erosion. An erosion process in which numerous small channels several inches deep are formed; occurs mainly on recently cultivated soils. See rill.

Sheet Erosion. The removal of a fairly uniform layer of soil from the land surface by runoff water.

Splash Erosion. The spattering of small soil particles caused by the impact of raindrops on wet soils. The loosened and spattered particles may or may not be subsequently removed by surface runoff.

Erosion Classes (soil survey). A grouping of erosion conditions based on the degree of erosion or on characteristic patterns. Applied to accelerated erosion, not to normal, natural, or geological erosion. Four erosion classes are recognized for water erosion and three for wind erosion. Specific definitions for each vary somewhat from one climatic zone or major soil group to another.

Erosive. Refers to wind or water having sufficient velocity to cause erosion. Not to be confused with erodible as a quality of soil.

Escarpment. A steep face or a ridge of high land; the escarpment of a mountain range is generally on the side nearest the sea.

Esker. A narrow ridge of gravelly or sandy drift deposited by a stream in association with glacier ice.

Essential Element (plant nutrition). A chemical element required for the normal growth of plants.

Eutrophication. A means of aging of lakes whereby aquatic plants are abundant and waters are deficient in oxygen. The process is usually accelerated by enrichment of waters with surface runoff containing nitrogen and phosphorus.

Evaporation. The process by which a liquid is changed to a vapor or gas.

Evapotranspiration. Water transpired by vegetation plus that evaporated from the soil. Synonym: consumptive use.

Excessive Precipitation. Standard U.S. Weather Bureau term for "rainfall in which the rate of fall is greater than certain adopted limits, chosen with regard to the normal precipitation (excluding snow) of a given place or area." Not the same as excess rainfall.

Excess Rainfall. Direct runoff at the place where it strikes the soil.

F

Fallow. Allowing cropland to lie idle, either tilled or untilled, during the whole or greater portion of the growing season.

Family (soil). See soil classification.

Fan. An accumulation of debris brought down by a stream on a steep gradient and coming to rest on a gently sloping plain in the shape of a fan.

Farm. Places of less than 10 acres where gross sales of agricultural products equal or exceed $250 annually and places of 10 acres or more where gross sales of agricultural products equal or exceed $50 per year (U.S. Bureau of Census).

Farm Management. The organization and administration of farm resources, including land, labor, crops, livestock, and equipment.

Farm Manager. A salaried person who operates land for others and is paid a salary and/or commission for his services.

Farm Operator. A person who operates a farm either by performing the labor himself or directly supervising it.

Farm Pond. A water impoundment made by constructing a dam or embankment or by excavating a pit or "dug out." See tank, earth.

Farm Tenancy. The leasing or renting of farm land together with improvements and sometimes equipment by nonowners for the purpose of occupying and operating.

Fault. A fracture along which there has been displacement of one block of earth or rock with respect to the other.

Fauna. The animal life of a region.

Feed. Harvested forage, such as hay or fodder or grain, grain products, and other foodstuffs processed for feeding livestock.

Ferritic. See soil mineralogy classes for family groupings.

Fertility, Soil. The quality of a soil that enables it to provide nutrients in adequate amounts and in proper balance for the growth of specified plants when other growth factors, such as light, moisture, temperature, and the physical condition of the soil, are favorable.

Fertilizer. Any organic or inorganic material of natural or synthetic origin that is added to a soil to supply elements essential to plant growth.

Fertilizer Analysis. The exact percentage composition, expressed in terms of nitrogen, phosphoric acid, and potash.

Fertilizer Grade. The guaranteed minimum analysis in whole numbers, in per cent, of the major plant nutrient elements contained in a fertilizer material or in a mixed fertilizer. For example, a fertilizer with a grade of 20–10–5 contains 20 per cent nitrogen (N), 10 per cent available phosphoric acid (P_2O_5), and 5 per cent water-soluble potash (K_2O). Minor elements may also be included. Recent trends are to express the percentages in terms of the elemental fertilizer [nitrogen (N), phosphorus (P), and potassium (K)].

Fertilizer Unit. One per cent (20 pounds) of a short ton of fertilizer.

Fibers (As used in the Soil Classification System of the National Cooperative Soil Survey in the United States). Fragments or pieces of plant tissue larger than 0.15 millimeter (0.006 inches) but exclusive of fragments of wood that are larger than 20 millimeters (0.8 inches) in cross section and so undecomposed that they cannot be crushed and shredded with the hands.

Fibric Materials. See organic soil materials.

Fibrous Root System. A plant root system having a large number of small, finely divided, widely spreading roots but no large individual roots. Typified by grass root system. Contrast with taproot system.

Field Capacity (field moisture capacity). The amount of soil water remaining in a soil after the free water has been allowed to drain away for a day or two if the root zone has been previously saturated. It is the greatest amount of water that the soil will hold under conditions of free drainage, usually expressed as a percentage of the oven-dry weight of soil.

Field Crops. General grain, hay, root, and fiber crops.

Field Strip Cropping. A system of strip cropping in which crops are grown in parallel strips laid out across the general slope but which do not follow the exact contour. Strips of grass or close-growing crops are alternated with strips of cultivated crops.

Field Test. An experiment conducted under field conditions. Ordinarily, less subject to control than a formal experiment and maybe less precise. Synonym: field trial.

Filter Strip. Strip of permanent vegetation above farm ponds, diversion terraces, and other structures to retard flow of runoff water, causing deposition of transported material, thereby reducing sediment flow.

Fine. See particle size classes for family groupings.

Fine-Loamy. See particle size classes for family groupings.

Fine Sand. 1: A soil separate. See soil separates. 2: A soil textural class. See soil texture.

Fine Sandy Loam. See soil texture.

Fine-Silty. See particle size classes for family groupings.

Fine Texture. Consisting of or containing large quantities of the fine fractions, particularly silt and clay. Includes sandy clay, silty clay, and clay textural classes. See soil texture.

Firm. A term describing the consistence of a moist soil that offers distinctly noticeable resistance to crushing but can be crushed with moderate pressure between the thumb and forefinger. See consistence.

Fixation. The process or processes in a soil by which certain chemical elements essential for plant growth are converted from a soluble or exchangeable form to a much less soluble or to a nonexchangeable form; for example, phosphate fixation.

Fixed Phosphorus. Soluble phosphorus that has become attached to the solid phase of the soil in forms highly unavailable to crops; unavailable phosphorus; phosphorus in other than readily or moderately available forms.

Flaggy. An adjective incorporated into the soil textural class designations of horizons when the soil mass contains between 15 and 90 per cent by volume of flagstones. See flagstone as defined under coarse fragment, Table 1.

Flagstone. See coarse fragment, Table 1, page 511.

Flocculate. To aggregate or clump together individual soil particles, especially fine clay, into small clumps or granules. Opposite of deflocculate or disperse.

Flocculation. The process by which suspended colloidal or very fine particles are assembled into larger masses or floccules which eventually settle out of suspension.

Flood. An overflow or inundation that comes from a river or other body of water and causes or threatens damage.

Flood Control. Methods or facilities for reducing flood flows.

Flood Control Project. A structural system installed for protection of land and improvements from floods by the construction of dikes, river embankments, channels, or dams.

Flood irrigation. The application of irrigation water whereby the entire surface of the soil is covered by a sheet of water; called "controlled flooding" when water is impounded or the flow directed by border dikes, ridges, or ditches.

Flood Peak. The highest value of the stage or discharge attained by a flood, thus, peak stage or peak discharge.

Floodplain. Nearly level land situated on either side of a channel which is subject to overflow flooding.

Flood Stage. The stage at which overflow of the natural banks of a stream begins to cause damage.

Flora. The sum total of the kinds of plants in an area at one time.

Fluvial. Of or pertaining to rivers; growing or living in streams or ponds; produced by river action, as a fluvial plain.

Fluvioglacial. Pertaining to streams flowing from glaciers or to the deposits made by such streams.

Fodder. The dried, cured plants of tall, coarse grain crops, such as corn and sorghum, including the grain, stems, and leaves. Grain parts not snapped off or threshed. Contrast with stover and hay.

Fold (geology). A bend or flexure in a layer or layers of rock.

Food Chain. A series of plant or animal species in a community, each of which is related to the next as a source of food.

Forage. All browse and herbaceous food that is available to livestock or game animals, used for grazing or harvested for feeding.

Forb. An herbaceous plant, such as clover, which is not a grass, sedge, or rush.

Forest. A plant association predominantly of trees and other woody vegetation.

Forest Influences. The effects of forests on soil, water supply, climate, and environment.

Formation (geology). Any assembly of rocks, that have some characteristic in common, whether of origin, age, or composition.

Fracture. A manner of breaking or appearance of a mineral when broken that is distinctive for certain minerals, as conchoidal fracture.

Fragipan. A natural subsurface horizon with high bulk density relative to the solum above, seemingly cemented when dry but showing a moderate to weak brittleness when moist. The layer is low in organic matter, mottled, slowly or very slowly permeable to water, and usually shows occasional or frequent bleached cracks, forming polygons. It may be found in profiles of either cultivated or virgin soils but not in calcareous material.

Fragmental. See particle size classes for family groupings.

Friable. Easy to break, crumble, or crush.

Frigid. See soil temperature classes for family groupings.

Fringe Water. Water occurring in the capillary fringe.

Frost Heave. The raising of the surface soil due to the accumulation of ice in the underlying soil.

Furrow Dams. Small earth dams used to impound water in furrows.

Furrow Irrigation. A partial surface flooding method of irrigation normally used with clean-tilled crops where water is applied in furrows or rows of sufficient capacity to contain the designed irrigation stream.

G

Gage or **Gauge.** Device for registering precipitation, water level, discharge velocity, pressure, or temperature.

Gaging Station. A selected section of a stream channel equipped with a gage, recorder, or other facilities for determining stream discharge.

Gate (irrigation). Structure or device for controlling the rate of flow into or from a canal, ditch, or pipe.

Gated Pipe. Portable pipe with small gates installed along one side for distributing water to corrugations or furrows.

Gel. A jellylike material formed by the coagulation of a colloidal suspension or sol.

Geochemistry. All parts of geology that involve chemical changes. It is the study of (1) the relative and absolute abundance of the elements and the atomic species (isotopes) in the earth. (2) The distribution and migration of the individual elements in the various parts of the earth (the atmosphere, hydrosphere, and lithosphere) and in minerals and rocks, with the object of discovering principles governing their distribution and migration.

Geological Erosion. See erosion.

Geomorphology. That branch of both physiography and geology that deals with the form of the earth, the general configuration of its surface, and the changes that take place in the evolution of land forms.

Glacial Drift. See drift, glacial.

Glacial Till. See till.

Glaciofluvial Deposits. Material moved by glaciers and subsequently sorted and deposited by streams flowing from the melting ice. The deposits are stratified and may occur in the form of outwash plains, deltas, kames, eskers, and kame terraces. See drift, glacial; till.

Gradation (geology). The bringing of a surface or a stream bed to grade, by running water. As used in connection with sedimentation and fragmental products for engineering evaluation, the term gradation refers to the frequency distribution of the various sized grains that constitute a sediment, soil, or other material.

Grade. 1: The slope of a road, channel, or natural ground. 2: The finished surface of a canal bed, roadbed, top of embankment, or bottom of excavation; any surface prepared for the support of construction such as paving or laying a conduit. 3: To finish the surface of a canal bed, roadbed, top of embankment, or bottom of excavation.

Graded Stream. A stream in which, over a period of years, the slope is delicately adjusted to provide, with available discharge and with prevailing channel characteristics, just the velocity required for transportation of the load (sediment) supplied from the drainage basin. The graded profile is the slope of transportation.

Graded Terrace. See terrace.

Grade Stabilization Structure. A structure for the purpose of stabilizing the grade of a gully or other watercourse, thereby preventing further head-cutting or lowering of the channel grade.

Gradient. Change of elevation, velocity, pressure, or other characteristics per unit length. Synonym: slope.

Granular Structure. See soil structure.

Grass. A member of the botanical family *Gramineae,* characterized by blade-like leaves arranged on the culm or stem in two ranks.

Grassed Waterway. A natural or constructed waterway, usually broad and shallow, covered with erosion-resistant grasses, used to conduct surface water from cropland.

Grassland. Land on which the existing plant cover is dominated by grasses.

Grasslike Plants. A plant that resembles a true grass, for example, sedges and rushes, but is taxonomically different.

Gravel. A mass of pebbles. See coarse fragments.

Gravel Envelope. Selected aggregate placed around the screened pipe section of well casing or a subsurface drain to facilitate the entry of water into the well or drain.

Gravel Filter. Graded sand and gravel aggregate placed around a drain or well screen to prevent the movement of fine materials from the aquifer into the drain or well.

Gravelly. An adjective incorporated into the soil textural class designations of horizons when the soil mass contains between 15 and 90 per cent pebbles by volume. See Table 1, page 511

Gravitational Water. Water that moves into, through, or out of the soil under the influence of gravity.

Gravity Irrigation. Irrigation in which the water is not pumped but flows and is distributed by gravity, includes sprinkler systems when gravity furnishes the desired head.

Gray-Brown Podzolic Soil. A former zonal great soil group consisting of soils with a thin, moderately dark A1 horizon and a grayish brown A2 horizon underlain by a B horizon containing a high percentage of cations and an appreciable quantity of illuviated silicate clay, formed on relatively young land surfaces, mostly glacial deposits, from material relatively rich in calcium, under deciduous forests in humid temperate regions. See Table 2, page 554.

Gray Wooded Soils. A former zonal great soil group consisting of soils with a thin A1 horizon over a light-colored, bleached A2 horizon underlain by a B horizon containing a high percentage of bases and an appreciable quantity of illuviated silicate clay. These soils occur in subhumid to semiarid, cool climatic regions under coniferous, deciduous, or mixed forest cover.

Grazing. The eating of any kind of standing vegetation by domestic livestock or wild animals.

Grazing Capacity. The maximum stocking rate of livestock possible without inducing damage to vegetation or soil.

Grazing Land. Land used regularly for grazing. The term is not confined to land suitable only for grazing, but cropland and pasture used in connection with a system of farm crop rotation are usually not included.

Grazing Permit. A document authorizing the use of public or other lands for grazing purposes under specified conditions, issued to the livestock operator by the agency administering the lands.

Grazing Season. The portion of the year that livestock graze or are permitted to graze on a given range or pasture. Sometimes called grazing period.

Grazing System. The manipulation of grazing animals to accomplish a desired result.

Grazing Unit. Any division of the range or pasture used to facilitate administration or the handling of livestock.

Great Group. See Great Group, page 557.

Green Manure Crop. Any crop grown for the purpose of being turned under while green or soon after maturity for soil improvement.

Groundwater. Subsurface water in the zone of saturation.

Ground-Water Podzol Soil. A former great soil group of the intrazonal order and hydromorphic suborder, consisting of soils with an organic mat on the surface over a very thin layer of acid humus material underlain by a whitish gray leached layer which may be as much as 2 or 3 feet in thickness, and is underlain by a brown or very dark brown cemented hardpan layer; formed under various types of forest vegetation in cool to tropical, humid climates under conditions of poor drainage. See Table 2, page 554.

Grove. A group of trees without understory; woods of small extent, for example, sugar grove.

Growing Stock (forestry). The sum, by number or volume, of all the trees in a forest or a specified part of it.

Grumusol. A former great soil group of the intrazonal order of dark clay soils, developed under widely varying climates but usually with alternating wet and dry seasons, composed of clays with a high shrink-swell potential.

Gully. A channel or miniature valley cut by concentrated runoff but through which water commonly flows only during and immediately after heavy rains or during the melting of snow. A gully may be dendritic or branching or it may be linear, rather long, narrow, and of uniform width. The distinction between gully and rill is one of depth. A gully is sufficiently deep that it would not be obliterated by normal tillage operations, whereas a rill is of lesser depth and would be smoothed by ordinary farm tillage. Synonym: arroyo. See erosion; rill.

Gully Erosion. See erosion.

Gully Control Plantings. The planting of forage, legume, or woody plant seeds, seedlings, cuttings, or transplants in gullies to establish or reestablish a vegetative cover adequate to control runoff and erosion and incidentally produce useful products.

H

Habitat. The environment in which the life needs of a plant or animal are supplied.

Half-Bog Soils. A former great soil group of the intrazonal order and hydromorphic suborder, consisting of soil with dark brown or black peaty (Fibric) material over grayish and rust-mottled mineral soil, formed under conditions of poor drainage under forest, sedge, or grass vegetation in cool to tropical humid climates. See Table 2, page 554.

Half-Shrub. A perennial plant with a woody base whose annually produced stems die back each year.

Halophyte. A plant adapted to existence in a saline environment, such as greasewood (*Sarcobatus*), saltgrass (*Distichlis*), and the saltbushes (*Atriplex* spp.).

Hardpan. A hardened soil layer in the lower A or B horizon caused by cementation of soil particles with organic matter or with materials such as silica, sesquioxides, or calcium carbonate. The hardness does not change appreciably with changes in moisture content, and pieces of the hard layer do not slake in water. See caliche; duripan; ortstein.

Hayland. Land used primarily for the production of hay from long-term stands of adapted forage plants.

Headwater. 1: The source of a stream. 2: The water upstream from a structure or point on a stream.

Heaving. The partial lifting of plants out of the ground, frequently breaking their roots, as a result of freezing and thawing of the surface soil during the winter.

Heavy Soil. A commonly used term to describe various fine-textured soils.

Hemic Materials. See organic soil materials.

Herb. Any flowering plant except those developing persistent woody bases and stems above ground.

Herbage. The sum total of all herbaceous plants.

Herbicide. A chemical substance used for killing plants, especially weeds.

Herd. A group of animals, especially cattle or big game, collectively considered as a unit in grazing practices.

Heterogeneous. Differing in kind; having unlike qualities; possessed of different characteristics; opposed to homogeneous.

Highway Erosion Control. The prevention and control of erosion in ditches, at cross drains, and on fills and road banks within a highway right-of-way. Includes vegetative practices and structural practices.

Histic Epipedon. See diagnostic horizons.

Histosols. See soil classification, organic soil materials, bog soils.

Hogback. A sharp-crested ridge formed by a hard rock ledge.

Homogeneous. Of the same kind or nature; consisting of similar parts or of elements of a like nature; opposed to heterogeneous.

Horizon. See soil horizon.

Hue. One of the three variables of color, caused by light of certain wavelengths and changes with the wavelength. See Munsell color system; chroma; value.

Humic Gley Soils. A former intrazonal great soil group of poorly drained, hydromorphic soils with dark-colored, organic mineral horizons of moderate thickness, underlain by gray mineral horizons developed under wet conditions. See Table 2, page 554.

Humid. A term applied to regions or climates where moisture, when distributed normally throughout the year, should not be a limiting factor in the production of most crops. The lower limit of precipitation under cool climates may be as little as 20 inches annually. In hot climates, it may be as much as 60 inches. Natural vegetation is generally forest. Contrast with subhumid.

Humidity, Absolute. The actual quantity or mass of water vapor present in a given volume of air, generally expressed in grains per cubic foot or in grams per cubic meter.

Humidity, Relative. The ratio of the actual amount of water vapor present in the portion of the atmosphere under consideration to the quantity that would be there if it were saturated, expressed in percentage.

Humus. 1: That more or less stable fraction of the soil organic matter remaining after the major portion of added plant and animal residues have decomposed, usually amorphous and dark colored. 2: Includes the F and H layers in undisturbed forest soils. See soil organic matter; soil horizons 01 and 02.

Humus Layer. The top portion of the soil which owes its characteristic features to the decomposed organic matter contained in it.

Hydration. The chemical combination of water with another substance.

Hydrologic Cycle. The circuit of water movement from the atmosphere to the earth and return to the atmosphere through various stages or processes, as precipitation, interception, runoff, infiltration, percolation, storage, evaporation, and transpiration.

Hydrophyte. A plant that grows in water or in wet or saturated soils, as distinguished from the opposite, xerophyte, and the intermediate, mesophyte.

Hydrous Mica. A hydrous aluminum silicate clay mineral with 2:1 lattice structure and containing a considerable amount of potassium which serves as an additional bonding between the crystal units, resulting in particles larger than normal in montmorillonite. It has a smaller cation exchange capacity than montmorillonite. Sometimes referred to as illite or mica. See clay mineral; montmorillonite.

Hydroxide. A compound of an element with the radicle or ion OH negative, as sodium hydroxide, NaOH.

Hygroscopic Coefficient. The weight percentage of water held by, or remaining in, the soil (1) after the soil has been air-dried or (2) after the soil has reached equilibrium with an unspecified environment of high relative humidity, usually near saturation, or with a specified relative humidity at a specified temperature.

Hygroscopic Water. Water so tightly held by the attraction of soil particles that it cannot be removed except as a vapor by raising the temperature above the boiling point of water. It is unavailable to plants.

Hyperthermic. See soil temperature for family groupings.

I

Igneous Rock. Formed by solidification from a molten or partially molten state. Synonym: Primary rock.

Illite. See hydrous mica.

Illitic. See soil mineralogy classes for family groupings.

Illuvial Horizon. A soil layer or horizon in which material carried from an overlying layer has been precipitated from solution or deposited from suspension. The layer of accumulation. In contrast to eluvial horizon.

Illuviation. The process of deposition of soil material removed from one horizon to another in the soil, usually from an upper to a lower horizon in the soil profile. See eluviation for contrasting term.

Impervious Soil. A soil through which water, air, or roots cannot penetrate.

Impoundment. Generally, an artificial collection or storage of water, as a reservoir, pit, dugout, or sump. See reservoir.

Inceptisols. See soil classification.

Indicator. An organism, species, or community that shows the presence of certain environmental conditions.

Indigenous. Refers to an organism that is native, not introduced, in an area.

Indurated (soil). Soil material cemented into a hard mass that will not soften on wetting. See hardpan; consistence.

Infiltration. The flow of a liquid *into* a substance through pores or other openings, connoting flow into a soil in contradistinction to the word percolation which connotes flow *through* a porous substance.

Infiltration Rate. A soil characteristic determining or describing the maximum rate at which water can enter the soil under specified conditions, including the presence of an excess of water. See infiltration velocity.

Infiltration Velocity. The actual rate at which water is entering the soil at any given time. It may be less than the maximum (the infiltration rate) because of a limited supply of water (rainfall or irrigation). It has the same units as the infiltration rate. See infiltration rate.

Infiltrometer. A device for measuring the rate of entry of fluid into a porous body, for example, water into soil.

Influent Water. Water which flows into sink holes, open cavities, and porous materials and disappears into the ground.

Inlet (hydraulics). 1: A surface connection to a closed drain. 2: A structure at the diversion end of a conduit. 3: The upstream end of any structure through which water may flow.

Inoculation. The process of introducing pure or mixed cultures or microorganisms into natural or artificial culture media.

Intake. 1: The headworks of a conduit; the place of diversion, 2: Entry of water into soil. See infiltration.

Intake Rate. The rate of entry of water into soil. See infiltration rate.

In Situ. In place. Rocks, soil, and fossils that are situated in the place where they were originally formed or deposited.

Interception (hydraulics). The process by which precipitation is caught and held by foliage, twigs, and branches of trees, shrubs, and other vegetation. Often used for "interception loss" or the amount of water evaporated from the intercepted precipitation.

Intermittent Grazing. Alternate grazing and resting a pasture or range for variable periods of time. See rotation grazing.

Intermittent Stream. A stream or portion of a stream that flows only in direct response to precipitation. It receives little or no water from springs and no long-continued supply from melting snow or other sources. It is dry for a large part of the year, ordinarily more than 3 months.

Internal Soil Drainage. The downward movement of water through the soil profile. The rate of movement is determined by the texture, structure, and other characteristics of the soil profile and underlying layers and by the height of the water table, either permanent or perched. Relative terms for expressing internal drainage are none, very slow, slow, medium, rapid, and very rapid.

Interplanting. In woodland, planting of young trees among existing trees or brushy growth. In orchards, planting of farm crops among the trees, especially while the trees are too small to occupy the land completely. In cropland, planting of several crops together on the same land, for example, the planting of beans with corn.

Interstices. The pore space or voids in soil and rock.

Intrusive. Denoting igneous rocks in a molten state which have invaded other rock formations and cooled below the surface of the earth.

Invasion. The migration of organisms from one area to another area and their establishment in the latter.

Ion. An atom or group of atoms with an electrical charge.

Irrigable Lands. Lands having soil, topographic, drainage, and climatic conditions favorable for irrigation, and located in a position where a water supply is or can be made available.

Irrigation. Application of water to lands for agricultural purposes.

Irrigation Application Efficiency. Percentage of irrigation water applied to an area that is stored in the soil for crop use.

Irrigation District. A cooperative, self-governing public corporation set up as a subdivision of the state, with definite geographic boundaries, organized to obtain and distribute water for irrigation of lands within the district. It is created under authority of the state legislature with the consent of a designated fraction of the landowners or citizens and has taxing power.

Irrigation Frequency. Time interval between irrigations.

Irrigation Water Management. The use and management of irrigation water where the quantity of water used for each irrigation is determined by the water-holding capacity of the soil and the needs of the crop, and where the water is applied at a rate and in such a manner that the crop can use it efficiently, and significant erosion does not occur.

Irrigation Water Requirement. Quantity of water, exclusive of effective precipitation, that is required for crop production.

Isofrigid. See soil temperature classes for family groupings.

Isohyperthermic. See soil temperature classes for family groupings.

Isomesic. See soil temperature classes for family groupings.

Isothermic. See soil temperature classes for family groupings.

J

Joint. A fracture or parting that abruptly interrupts the physical continuity of a rock mass.

K

Kame. A conical hill or short irregular ridge of gravel or sand deposited in contact with glacier ice.

Kaolin. A rock consisting of clay minerals of the kaolinite group.

Kaolinite. 1: Hydrous aluminum silicate clay mineral of the 1:1 crystal lattice group, that is, consisting of one silicon tetrahedral layer and one aluminum oxide-hydroxide octahedral layer. 2: The 1:1 group or family of aluminosilicates.

Kaolinitic. See soil mineralogy classes for family groupings.

L

Land. The total natural and cultural environment within which production takes place; a broader term than soil. In addition to soil, its attributes include other physical conditions, such as mineral deposits, climate, and water supply; location in relation to centers of commerce, populations, and other land; the size of the individual tracts or holdings; and existing plant cover, works of improvement, and the like. Some use the term loosely in other senses: as defined above but without the economic or cultural criteria; especially in the expression "natural land"; as a synonym for "soil"; for the solid surface of the earth; and also for earthy surface formations, especially in the geomorphological expression "land form."

Land Capability. The suitability of land for use without permanent damage. Land capability, as ordinarily used in the United States, is an expression of the effect of physical land conditions, including climate, on the total suitability for use without damage for crops that require regular tillage, for grazing, woodland, and wildlife. Land capability involves consideration of (1) the risks of land damage from erosion and other causes and (2) the difficulties in land use owing to physical land characteristics, including climate.

Land Capability Class. One of eight classes of land in the land capability classification of the Soil Conservation Service. These eight land capability classes, distinguished according to the risk of land damage or the difficulty of land use, are:

Land suitable for cultivation and other uses.

    I. Soils in class I have few limitations that restrict their use.

    II. Soils in class II have some limitations that reduce the choice of plants or require moderate conservation practices.

    III. Soils in class III have severe limitations that reduce the choice of plants or require special conservation practices, or both.

    IV. Soils in class IV have very severe limitations that restrict the choice of plants, require very careful management, or both.

Land generally not suitable for cultivation (without major treatment).

    V. Soils in class V have little or no erosion hazard but have other limitations, impractical to remove, that limit their use largely to pasture, range, woodland, or wildlife food and cover.

    VI. Soils in class VI have severe limitations that make them generally unsuited for cultivation and limit their use largely to pasture or range, woodland, or wildlife food and cover.

    VII. Soils in class VII have very severe limitations that make them unsuited to cultivation and that restrict their use largely to grazing, woodland, or wildlife.

    VIII. Soils and landforms in class VIII have limitations that preclude their use for commercial plant production and restrict their use to recreation, wildlife, water supply, or esthetic purposes.

Land Capability Classification. A grouping of kinds of soil into special units, subclasses, and classes according to their capability for intensive use and the treatments required for sustained use, prepared by the Soil Conservation Service, U.S.D.A.

Land Capability Map. A map showing land capability units, subclasses, and classes or a soil survey map colored to show land capability classes.

Land Capability Subclass. Groups of capability units within classes of the land capability classification that have the same kinds of dominant limitations for agricultural use as a result of soil and climate. Some soils are subject to erosion if they are not protected, while others are naturally wet and must be drained if crops are to be grown. Some soils are shallow or droughty or have other soil deficiencies. Still other soils occur in areas where climate limits their use. The four kinds of limitations recognized at the subclass level are: risks of erosion, designated by the symbol (e); wetness, drainage, or overflow (w); other root zone limitations (s); and climatic limitations (c). The subclass provides the map user information about both the degree and kind of limitation. Capability class I has no subclasses.

Land Capability Unit. Capability units provide more specific and detailed information for application to specific fields on a farm or ranch than the subclass of the land capability classification. A capability unit is a group of soils that are nearly alike in suitability for plant growth and responses to the same kinds of soil management.

Land Classification. The arrangement of land units into various categories based on the properties of the land or its suitability for some particular purpose.

Land Form. A discernible natural landscape, such as a floodplain, stream terrace, plateau, or valley.

Land Leveling. Process of shaping the land surface for better movement of water and machinery over the land. Also called land forming, land shaping, or land grading.

Land, Marginal. Land that returns barely enough to meet expenses in a specific use.

Land Reclamation. Making land capable of more intensive use by changing its general character, as by drainage of excessively wet land; irrigation of arid or semiarid land; or recovery of submerged land from seas, lakes, and rivers.

Land Resting. Temporary discontinuance of cultivation of a piece of land.

Landscape. All the natural features, such as fields, hills, forests, and water, that distinguish one part of the earth's surface from another part, usually that portion of land or territory which the eye can comprehend in a single view, including all of its natural characteristics.

Landslide. The failure of a slope in which the sliding movement of the soil mass takes place along an interior surface.

Land, Submarginal. Land that does not return enough to pay costs of operation in a specific use.

Land, Supermarginal. Land that returns a profit after all expenses are paid.

Land Use Plan. A community plan outlining proposed future land uses and their distribution. Zoning is the most frequently used method for carrying out the land use plan.

Land Use Planning. The process by which decisions are made on future land uses over extended time periods that are deemed to best serve the general welfare. Decision-making authorities on land uses are usually vested in state and local governmental units, but citizen participation in the planning process is essential for proper understanding and implementation, usually through zoning ordinances.

Laterite Soils. Similar to Latosols. See Table 2, page 554.

Latosols. A former great soil group of zonal soils including soils formed under forested, tropical, humid conditions and characterized by low silica-sesqui-oxide ratios of the clay fractions, low base-exchange capacity, low activity of the clay, low content of most primary minerals, low content of soluble constituents, and a high degree of aggregate stability; usually having a red color. See Table 2, page 554.

Leached Layer. A soil layer from which the soluble materials ($CaCO_3$ and $MgCO_3$ and material more soluble) have been dissolved and washed away by percolating water.

Leached Saline Soils. 1: Soils from which the soluble salts have been removed by leaching. 2: Soils that have been saline and still possess the major physical characteristics of saline soils but from which the soluble salts have been leached, generally for reclamation.

Leached Soil. A soil from which most of the soluble materials ($CaCO_3$ and $MgCO_3$ and more soluble materials) have been removed from the entire profile or have been removed from one part of the profile and have accumulated in another part.

Leaching. The removal of materials in solution from the soil. See eluviation.

Legume. A member of the legume or pulse family, *Leguminosae*. One of the most important and widely distributed plant families. The fruit is a "legume" or pod that opens along two sutures when ripe. Leaves are alternate, have stipules, and are usually compound. Includes many valuable food and forage species, such as peas, beans, peanuts, clovers, alfalfas, sweet clovers, lespedezas, vetches, and kudzu. Practically all legumes are nitrogen-fixing plants.

Legume Inoculation. The addition of nitrogen-fixing bacteria to legume seed or to the soil in which the seed is to be planted.

Length of Run (irrigation). Distance water must run in furrows or between borders over the surface of a field from one head ditch to another, or to the end of the field.

Levee. An embankment along a river to prevent inundation. In irrigation, a low continuous embankment for dividing areas to be flooded.

Level Terrace. See terrace.

Light Soil. A coarse-textured soil with a low drawbar pull and, hence, easy to cultivate.

Lime. Lime, from the strictly chemical standpoint, refers to only one compound, calcium oxide (CaO); however, the term lime is commonly used in agriculture to include a great variety of materials which are usually composed of the oxide, hydroxide, or carbonate of calcium or of calcium and magnesium. The most commonly used forms of agricultural lime are ground limestone, marl, and oyster shells (carbonates), hydrated lime (hydroxides), and burnt lime (oxides).

Lime, Agricultural. A soil amendment consisting principally of calcium carbonate but including magnesium carbonate and perhaps other materials, used to furnish calcium and magnesium as essential elements for the growth of plants and to neutralize soil acidity.

Lime (calcium) Requirement. The amount of agricultural limestone, or the equivalent of other specified liming material, required per acre to a soil depth of 6 inches (or on 2 million pounds of soil) to raise the pH of the soil to a desired value under field conditions.

Lime Concretion. An aggregate of precipitated calcium carbonate, or other material cemented by precipitated calcium carbonate.

Limestone. A sedimentary rock composed of calcium carbonate, $CaCO_3$. There are many impure varieties.

Liming. The application of lime to land, primarily to reduce soil acidity and supply calcium for plant growth. Dolomitic limestone supplies both calcium and magnesium. May also improve soil structure, organic matter content, and nitrogen content of the soil by encouraging the growth of legumes and soil microorganisms. Liming an acid soil to a pH value of 6.5 to 7.0 is desirable for maintaining a high degree of availability of most of the nutrient elements required by plants.

Limnic Materials (As used in the Soil Classification System of the National Cooperative Soil Survey in the United States). Includes both organic and inorganic materials either (1) deposited in water by precipitation or action of aquatic organisms, such as algae or diatoms, or (2) derived from underwater and floating aquatic plants subsequently modified by aquatic animals. Examples are marl and diatomaceous earth.

Lister. A double plow, the shares of which throw the soil in opposite directions, leaving the field with a series of alternate ridges and furrows. Row crops may be seeded in the bottoms of the furrows or on top of the ridge as they are opened up. When no seed is planted, the operation is sometimes referred to as blank listing.

Lithic Contact (As used in the Soil Classification System of the National Cooperative Soil Survey in the United States). A boundary between soil and continuous, coherent underlying material which has a hardness of 3 or more (Mohs scale). When moist, the underlying material cannot be dug with a spade, and chunks will not disperse in water with 15 hours of shaking. Example: basalt rock.

Lithification. The process of converting a sedimentary deposit into an indurated rock.

Lithology. The study of rocks based on the megascopic examination of samples.

Lithosols. A former great soil group of azonal soils characterized by an incomplete solum or no clearly expressed soil morphology and consisting of freshly and imperfectly weathered rock or rock fragments. See Table 2, page 554.

Litter (forestry). A surface layer of loose organic debris in forests, consisting of freshly fallen or slightly decomposed organic materials.

Livestock Pond. An impoundment, the principal purpose of which is to supply water to livestock. Includes reservoirs, pits, and tanks.

Loam. A soil textural class. See soil texture.

Loamy. Intermediate in texture and properties between fine-textured and coarse-textured soils. Includes all textural classes with the word "loam" as a part of the class name, such as clay loam. See loam; soil texture. See particle size classes for family groupings for its use in the Soil Classification System of the National Cooperative Soil Survey in the United States.

Loamy Coarse Sand. See soil texture.

Loamy Fine Sand. See soil texture.

Loamy Sand. See soil texture.

Loamy Skeletal. See particle size classes for family groupings.

Loamy Very Fine Sand. See soil texture.

Loess. Material transported and deposited by wind and consisting of predominantly silt-sized particles.

Loose. A soil consistence term. See consistence.

Loose Rock Dam. A dam built of rock without the use of mortar; a rubble dam. See rock-fill dam.

Low Humic Gley Soils. A former intrazonal group of somewhat poorly to poorly drained soils with very thin surface horizons moderately high in organic matter over gray and brown mineral horizons, which are developed under wet conditions. See Table 2, page 554.

Lysimeter. Device to measure the quantity or rate of water movement through a block of soil, usually undisturbed or *in situ;* or to collect such percolated water for analysis.

M

Macronutrient. A chemical element necessary in large amounts (usually greater than 1 part per million in the plant) for the growth of plants, usually applied artificially in fertilizer or liming materials. "Macro" refers to quantity and not the essentiality of the element. See micronutrient.

Made Land. Areas filled with earth or earth and trash mixed, usually made by or under the control of man. A miscellaneous land type.

Malthusian Theory of Population. Thomas Malthus asserted that man could increase his subsistence only arithmetically, whereas population tended to increase geometrically. Thus, population always tended toward the limit set by subsistence and was contained within that limit by the operation of positive and preventive checks, such as famine, pestilence, and premature mortality.

Manometer. Instrument that measures fluid pressure by fluid displacement. Can be a differential or U-tube manometer.

Manure. The excreta of animals, with or without the admixture of bedding or litter, in varying stages of decomposition.

Mapping Unit. See soil mapping unit.

Marble. A metamorphosed form of limestone or dolomite in which the grains are recrystallized.

Margin. The point at which the value of the added output just equals the value of the unit of input that produced it.

Marl. An earthy, unconsolidated deposit formed in freshwater lakes, consisting chiefly of calcium carbonate mixed with clay or other impurities in varying proportions.

Marsh. Periodically wet or continually flooded area with the surface not deeply submerged. Covered dominantly with sedges, cattails, rushes, or other hydrophytic plants. Subclasses include freshwater and saltwater marshes. See swamp; miscellaneous land type.

Marsh, Tidal. A low, flat area traversed by interlacing channels and tidal sloughs and periodically inundated by high tides. Vegetation usually consists of salt-tolerant plants.

Masonry Dam. A dam built of rock and mortar.

Matrix (geology). Natural material in which larger particles are embedded.

Meadow. An area of natural or planted vegetation dominated by grasses and grasslike plants used primarily for hay production.

Mean (statistics). The average of a group of items obtained by adding together all items and dividing by the total number of items used.

Mechanical Analysis. See particle size analysis and particle size distribution.

Mechanical Practices. Soil and water conservation practices that primarily change the surface of the land or that store, convey, regulate, or dispose of runoff water without excessive erosion.

Median. The value of the middle item when items are arrayed according to size.

Medium Texture. Intermediate between fine- and coarse-textured soils, containing moderate amounts of sand, silt, and clay. Includes the following textural classes: very fine sandy loam, loam, silt loam, and silt.

Megascopic. Large enough to be distinguished by the naked eye without the aid of a microscope.

Mellow Soil. A very soft, very friable, porous soil without any tendency toward hardness or harshness.

Mesic. See soil temperature classes for family groupings.

Mesophyte. A plant that grows under intermediate moisture conditions.

Metamorphic Rock. That which has formed in the solid state in response to pronounced changes in temperature, pressure, and chemical environment. The process takes place, in general, deep in the crust of the earth below the zone of weathering and cementation.

Mho. The reciprocal of ohm which is a unit of electrical inductance.

Micaceous. See soil mineralogy classes for family groupings.

Microclimate. The climatic nature of the air space near the soil surface.

Micronutrient. A chemical element necessary in only extremely small amounts (less than 1 part per million in the plant) for the growth of plants. "Micro" refers to the amount used rather than to its essentiality. Examples are boron, chlorine, copper, iron, manganese, and zinc. See macronutrient.

Microorganisms. Forms of life that are either too small to be seen with the unaided eye or are barely discernible.

Microrelief. Minor differences in surface configuration of the land surface.

Mine Dumps. Areas covered with overburden and other waste materials from ore and coal mines, quarries, and smelters, usually with little or no vegetative cover. A miscellaneous land type.

Mineral Soil. A soil consisting predominantly of, and having its properties determined predominantly by, mineral matter, usually containing less than 20 per cent organic matter but sometimes containing an organic surface layer up to 30 centimeters thick. See organic soil.

Minimum Tillage. The least amount of tillage required to create the proper soil condition for seed germination, plant establishment, and prevention of competitive plant growth.

Minor Element. See micronutrient.

Miscellaneous Land Type. A mapping unit for areas of land that have little or no natural soil or that are too nearly inaccessible for orderly examination or that occur where, for other reasons, it is not feasible to classify the soil. Examples are alluvial land, badlands, made land, marsh, mine dump, mine wash, river wash, rock land, rough broken land, rubble land, swamp, and urban land. See individual definitions.

Mixed. See soil mineralogy classes for family groupings.

Mode (statistics). The most frequent or most common value, provided that a sufficiently large number of items are available to give a smooth distribution.

Moderately Coarse Texture. Intermediate between coarse and medium texture and consisting predominantly of coarse particles. In soil textural classification it includes all the sandy loams except the very fine sandy loam. See coarse texture.

Moderately Fine Texture. Intermediate between fine and medium texture and consisting predominantly of intermediate-size soil particles or relatively small amounts of fine or coarse particles. In soil textural classification it includes clay loam, sandy clay loam, and silty clay loam. See fine texture.

Mohs' Scale of Hardness. Relative hardness of minerals ranging from a rating of 1 for the softest (talc) to 10 for the hardest (diamond). Calcite has a hardness of 3 and can be scratched with a copper coin.

Moisture Volume Percentage. The ratio of the volume of water in a soil to the total bulk volume of the soil.

Moisture Weight Percentage. The water content expressed as a percentage of the oven-dry weight of soil.

Mole Drain. Unlined drain formed by pulling a bullet-shaped cylinder through the soil at a depth of 1 to 3 feet.

Mollic Epipedon. See diagnostic horizons.

Mollisols. See soil classification.

Monolithic. Of or pertaining to a structure formed from a single mass of stone.

Montmorillonite. A hydrous, aluminosilicate clay mineral with 2:1 expanding crystal lattice, that is, with two silicon tetrahedral layers enclosing an aluminum octahedral layer. Considerable expansion may be caused along the C axis by water moving between silica layers of contiguous units.

Montmorillonitic. See soil mineralogy classes for family groupings.

Moraine. An accumulation of drift, with an initial topographic expression of its own, built within a glaciated region chiefly by the direct action of glacial ice. Examples are ground, lateral, recessional, and terminal moraines.

Mottled (soils). Soil horizons irregularly marked with spots of color. A common cause of mottling is impeded drainage, although there are other causes, such as soil development from an unevenly weathered rock. The differential weathering of various kinds of minerals may cause mottling.

Muck. Highly decomposed organic material in which the original plant parts are not recognizable. Contains more mineral matter and is usually darker than peat. See muck soil; peat; peat soil; organic soil (Sapric) materials.

Muck Soil. 1: An organic soil in which the organic matter is well decomposed (U.S.A. usage). 2: A soil containing 20 to 50 per cent organic matter. See organic soil (Sapric) materials.

Mulch. A natural or artificial layer of plant residue or other materials, such as leaves, sand, plastic, or paper, on the soil surface.

Munsell Color System. A color designation system that specifies the relative degrees of the three simple variables of color: hue, value, and chroma. For example: 10YR 6/4 is a color (of soil) with hue 10YR, value 6, and chroma 4. See hue; value; chroma.

N

Native Species. A species that is a part of an area's original fauna or flora.

Natric Horizon. See diagnostic horizons.

Natural Erosion. See erosion.

Natural Grassland. An area in which the natural potential plant community is dominated by grasses and grasslike plants. Associated species include forbs and woody plants.

Natural Revegetation. Natural reestablishment of plants; propagation of new plants over an area by natural processes.

Net Duty of Water. The amount of water delivered to the land to produce a crop, measured at the point of delivery to the field.

Neutral Soil. A soil in which the surface layer, at least to normal plow depth, is neither acid nor alkaline in reaction. For most practical purposes it is a soil with a pH range from 6.6 through 7.3. See acid soil; alkaline soil; pH; reaction, soil.

Nitrate Reduction. The biochemical reduction of nitrates to the nitrite form.

Nitrification. The biological oxidation of ammonium salts to nitrites and the further oxidation of nitrites to nitrates.

Nitrogen Assimilation. The incorporation of nitrogen compounds into cell substances by living organisms.

Nitrogen Cycle. The sequence of biochemical changes undergone by nitrogen, wherein it is used by a living organism, liberated upon the death and decomposition of the organism, and converted to its original state of oxidation.

Nitrogen Fixation. The conversion of elemental nitrogen (N_2) to organic combinations or to forms readily usable in biological processes.

Nitrogen-Fixing Plant. A plant that can assimilate and fix the free nitrogen of the atmosphere with the aid of bacteria living in the root nodules. Legumes with the associated *Rhizobium* bacteria in the root nodules are the most important nitrogen-fixing plants.

Nodule. A structure developed on the roots of most legumes and some other plants in response to the stimulus of root-nodule bacteria. Legumes bearing these nodules are nitrogen-fixing plants, utilizing atmospheric nitrogen instead of depending solely on nitrogen compounds in the soil.

Noncalcic Brown Soils. A former zonal group of soils with slightly acid, light pinkish or light reddish brown A horizons over light reddish brown or dull red B horizons developed under mixed grass and forest vegetation in a subhumid, wet-dry climate. See Table 2, page 554.

Nonrenewable. Natural resources that, once used, are gone forever.

Nonsaline-Alkali Soil. See sodic soil.

Normal. A mean or average value established from a series of observations for purposes of comparison, for example, normal precipitation, normal temperature, normal flow.

Normal Depth. Depth of flow in an open conduit during uniform flow for the given conditions.

Normal Erosion. See erosion.

Notch. The opening in a dam or spillway for the passage of water.

No-Tillage. A method of planting crops that involves no seedbed preparation other than opening the soil for the purpose of placing the seed at the intended depth. This usually involves opening a small slit or punching a hole into the soil. There is usually no cultivation during crop production. Chemical weed control is normally used. Also referred to as slot planting or zero tillage.

Nurse Crop. See companion crop.

Nurse Tree. A tree that protects or fosters the growth of a younger tree.

Nursery. A place where plants, such as trees, shrubs, vines, and grasses, are propagated for transplanting or for use as stocks for grafting; a planting of young trees or other plants, the young plants being called nursery stock or planting stock.

Nutritive Ratio. The ratio or proportion between digestible protein and digestible nonnitrogenous nutrients (carbohydrates and fats) in a livestock feed.

O

Observation Well. Hole bored to a desired depth below the ground surface, used for observing the water table level.

Ochric Epipedon. See diagnostic horizons.

Open Drain. Natural watercourse or constructed open channel that conveys drainage water.

Open Range. An extensive grazing area on which the movement of livestock is unrestricted.

Order. See soil classification.

Organic Matter. See soil organic matter.

Organic Soil. A soil that contains a high percentage (greater than 20 or 30 per cent) of organic matter in the solum.

Organic Soil Materials (As used in the Soil Classification System of the National Cooperative Soil Survey in the United States). 1: Saturated with water for prolonged periods unless artificially drained and having more than 30 per cent organic matter if the mineral fraction is more than 50 per cent clay, or more than 20 per cent organic matter if the mineral fraction has no clay. 2: Never saturated with water for more than a few days and having more than 34 per cent organic matter. See soil classification, Histosols, peat, muck, and bog soils.

Kinds of organic materials:

Fibric Materials. The least decomposed of all the organic soil materials, containing very high amounts of fiber that are well preserved and readily identifiable as to botanical origin. These materials have a bulk density of less than 6.25 pounds per cubic foot and a fiber content (unrubbed) that exceeds 66% of the organic volume, more than 40% after rubbing. When saturated, the maximum water content of the material ranges from 850 to 3000 per cent on an oven-dry basis. See peat.

Hemic Materials. Intermediate in degree of decomposition between the less decomposed fibric and the more decomposed sapric materials. These materials have a bulk density of 6.25 to 12.5 pounds per cubic foot, and fiber content (unrubbed) is between 33% and 66% of the organic volume, more than 10% after rubbing. When saturated, the maximum water content of the material ranges from 450 to 850 per cent on an oven-dry basis. See peat, muck.

Sapric Materials. The most highly decomposed of the organic materials, having the highest bulk density, least amount of plant fiber, and lowest water content at saturation. These materials have a bulk density of more than 12.5 pounds per cubic foot and a fiber content (unrubbed) of less than 33% of the organic volume. When saturated, the maximum water content of the material averages less than 450 per cent on an oven-dry basis. See muck.

Ortstein. The organic—and sesquioxide—cemented subsoil layer in podzols or groundwater podzols. It does not soften appreciably when immersed in water. Synonym: hardpan. See soil classification (Spodosol).

Outlet. Point of water disposal from a stream, river, lake, tidewater, or artificial drain.

Outlet Channel. A waterway constructed or altered primarily to carry water from man-made structures, such as terraces, tile lines, and diversions.

Overgrazed Range. A range deteriorated from its productive potential due to continued overuse.

Overgrazing. Grazing so heavy that it impairs future forage production and causes deterioration through damage to plants or soil or both.

Overstocking. Placing a number of animals on a given area that will result in overuse at the end of the planned grazing period.

Overuse. Excessive use of the current year's growth which will result in range deterioration or overgrazing, if continued.

Oxic Horizon. See diagnostic horizons.

Oxidation. Combination with oxygen; addition of oxygen or other atom or group; removal of hydrogen or other atom or group.

Oxisols. See soil classification.

P

Palatability. Plant characteristics or conditions that stimulate a selective response by animals.

Pan. Horizon or layer in soils that is strongly compacted, indurated, or very high in clay content. See caliche; claypan; duripan; fragipan; hardpan; orstein; petrocalcic horizon.

Pan, Pressure or Induced. A subsurface horizon or soil layer having a high bulk density and a lower total porosity than the soil directly above or below it as a result of pressure applied by normal tillage operations or by other artificial means. Frequently referred to as plow pan, plowsole, tillage pan, or traffic pan.

Paralithic Contact (As used in the Soil Classification System of the National Cooperative Soil Survey in the United States). A boundary between soil and continuous coherent underlying material that has a hardness of less than 3 (Mohs scale). When moist, the underlying material can be dug with a spade and chunks will disperse in water with 15 hours' shaking. Example, shale. See lithic contact.

Parent Material (soils). The unconsolidated, chemically weathered mineral or organic matter from which the solum of soils has developed by pedogenic processes. The C horizon may or may not consist of materials similar to those from which the A and B horizons developed.

Particle Size. The effective diameter of a particle measured by sedimentation, sieving, or micrometric methods.

Particle-Size Analysis. Determination of the amounts of different particle sizes in a soil sample, usually by sedimentation, sieving, micrometry, or a combinations of these methods.

Particle-Size Classes for Family Groupings (As used in the Soil Classification System of the National Cooperative Soil Survey in the United States). Various particle-size classes are applied to arbitrary control sections that vary according to the depth of the soil, presence or absence of argillic horizons, depth to paralithic or lithic contacts, fragipans, duripans, and petrocalcic horizons. No single set of particle-size classes is appropriate as a family grouping for all kinds of soils. The classification tabulated below provides a choice of either seven or eleven particle-size classes. This choice permits relatively fine distinctions in soils if texture is important, and broader groupings if texture is not susceptible to precise measurement or if the use of narrowly defined classes produces meaningless groupings.

1. Fragmental. Stones, cobbles, gravel, and very coarse sand, with fines too few to fill interstices larger than 1 millimeter.
2. Sandy-skeletal. More than 35 per cent, by volume, coarser than 2 millimeters, with enough fines to fill interstices larger than 1 millimeter; fraction less than 2 millimeters is as defined for particle-size class 5.
3. Loamy-skeletal. More than 35 per cent, by volume, coarser than 2 millimeters, with enough fines to fill interstices larger than 1 millimeter; fraction less than 2 millimeters is as defined for particle-size class 6.
4. Clayey-skeletal. More than 35 per cent, by volume, coarser than 2 millimeters, with enough fines to fill interstices larger than 1 millimeter; fraction less than 2 millimeters is as defined for particle-size class 7.
5. Sandy. Sands, except very fine sand; and loamy sands, except loamy very fine sand.
6. Loamy
   6a. Coarse-loamy. With less than 18 per cent clay and more than 15 per cent coarser than very fine sand (including coarse fragments up to 7.5 centimeters).
   6b. Fine-loamy. With more than 18 per cent clay but less than 35 per cent clay and more than 15 per cent coarser than very fine sand (including coarse fragments up to 7.5 centimeters).
   6c. Coarse-silty. With less than 18 per cent clay and less than 15 per cent coarser than very fine sand (including coarse fragments up to 7.5 centimeters).
   6d. Fine-silty. With more than 18 per cent clay and less than 35 per cent clay and less than 15 per cent coarser than very fine sand (including coarse fragments up to 7.5 centimeters).
7. Clayey
   7a. Fine. With more than 35 per cent clay but less than 60 per cent clay.
   7b. Very-fine. With more than 60 per cent clay.

Particle-Size Distribution. The amount of the various soil separates in a soil sample, usually expressed as weight percentages. See soil texture; particle-size classes for family groupings.

Pasture. An area devoted to the production of forage (introduced or native) and harvested by grazing.

Pasture Improvement. Any practice of grazing, mowing, fertilizing, liming, seeding, scattering droppings, contour furrowing, or other methods of management designed to improve vegetation for grazing purposes.

Pastureland. Land use primarily for the production of adapted domesticated forage plants to be grazed by livestock.

Pasture Management. The application of practices to keep pasture plants growing actively over as long a period as possible so that they will provide palatable feed of high nutritive value; to encourage the growth of desirable grasses and legumes while crowding out weeds, brush, and inferior grasses.

Peat. Unconsolidated soil material consisting largely of undecomposed or only slightly decomposed organic matter accumulated under conditions of excessive moisture. See organic soil (Fibric) materials.

Peat Soil. 1: An organic soil in which the organic matter is not yet decomposed or is slightly decomposed (U.S.A. usage). 2: An organic soil containing more than 50 per cent organic matter. See peat; muck; muck soil; Histolsols.

Pebble. See coarse fragments.

Ped. A unit of soil structure; an aggregate such as crumb, prism, block, or granule, formed by natural processes. Contrast with clod, which is formed artificially by man.

Pedon (As used in the Soil Classification System of the National Cooperative Soil Survey in the United States). The smallest volume that can be called "a soil." It has three dimensions. It extends downward to the depth of plant roots or to the lower limit of the genetic soil horizons. Its lateral cross section is roughly hexagonal and ranges from 1 to 10 square meters in size depending on the variability in the horizons.

Percolation, Soil Water. The downward movement of water through soil, especially the downward flow of water in saturated or nearly saturated soil at hydraulic gradients of the order of 1.0 or less.

Perennial Plant. A plant that normally lives for 3 or more years.

Permafrost. 1: Permanently frozen material underlying the solum. 2: A perennially frozen soil horizon.

Permanent Pasture. Grazing land occupied by perennial pasture plants or by self-seeding annuals, usually both, which remains unplowed for many years. Contrast with rotation pasture.

Permanent Wilting Percentage. See wilting point.

Permeability. Capacity for transmitting a fluid. It is measured by the rate at which a fluid of standard viscosity can move through material in a given interval of time under a given hydraulic gradient.

Permeability, Soil. The quality of a soil horizon that enables water or air to move through it. The permeability of a soil may be limited by the presence of a less permeable horizon even though the others are permeable.

Permissible Velocity (hydraulics). The highest velocity at which water may be carried safely in a channel or other conduit. The highest velocity that can exist through a substantial length of a conduit and not cause scour of the channel. Synonyms: safe or noneroding velocity.

Pesticide. A chemical agent used to control pests.

Petrocalcic Horizon. See diagnostic horizons.

pH, Soil. A numerical measure of the acidity or hydrogen ion activity of a soil. The neutral point is pH 7.0. All pH values below 7.0 are acid and all above 7.0 are alkaline.

Phase, Soil. A subdivision of a soil taxon, usually a soil series or other unit of classification based on characteristics that affect the use and management of the soil but which do not vary sufficiently to differentiate it as a separate soil series. A variation in a property or characteristic, such as degree of slope, degree of erosion, content of stones, and texture of the surface. Phases of soil series are the major components of the soil mapping units shown on detailed soil maps in the U.S.

Phenology. The study of the time of appearance of characteristic, periodic events in the life cycles of organisms in nature and how these events are influenced by environmental factors.

Phosphate or Potash Fixation (soils). The process or processes by which these two elements are converted from a soluble or exchangeable form to a much less soluble or nonexchangeable form in a soil.

Phosphorus Fixation. See fixed phosphorus.

Phreatophyte. A plant deriving its water from the water table; commonly used to describe nonbeneficial, water-loving vegetation that transpires excessive amounts of water.

Piezometer. A tube for measuring the pressure of a fluid, such as water.

Piezometric Surface. The imaginary surface to which water in a well will rise above an aquifer.

Piping. Removal of soil material through subsurface flow channels or "pipes" developed by seepage water through soil cracks.

Pit. See dugout pond.

Pitting. 1: Making shallow pits of suitable capacity and distribution to retain water from rainfall or snowmelt on rangeland or pasture. 2: Small cavities in a surface created by corrosion or cavitation.

Planosols. A former intrazonal great soil group of soils having one or more horizons abruptly separated from and sharply contrasting to an adjacent horizon because of cementation, compaction, or high clay content. They are formed under forest or grass vegetation in mesothermal to tropical, perhumid to semiarid climates, usually with a fluctuating water table.

Plant Food. The organic compounds elaborated within the plant to nourish its cells. The term is a frequent synonym for plant nutrients, particularly in the fertilizer trade.

Plant Indicator. See indicator.

Plant Nutrients. The elements or groups of elements taken in by a plant which are essential to its growth and used in elaboration of its food and tissues. Includes nutrients obtained from fertilizer ingredients. See essential element; macronutrients; micronutrients.

Plant Residue. See crop residue; humus; litter; organic matter.

Plant Succession. The process of vegetation development whereby an area becomes successively occupied by different plant communities of higher ecological order.

Plastic Soil. A soil capable of being molded or deformed continuously and permanently by relatively moderate pressure. See consistence.

Platy. See soil structure types.

Playa. A shallow, flat basin on a plain where water gathers after a rain and is evaporated.

Plow Layer. The soil ordinarily moved in tillage; equivalent to surface soil.

Plow Pan. See pan, pressure or induced.

Plow-Plant. Plowing and planting a crop in one operation, with no additional seedbed preparation.

Plowsole. See pan, pressure or induced.

Podzol. A former great soil group of the zonal order consisting of soils formed in cool-temperate to temperate, humid climates under coniferous or mixed coniferous and deciduous forest and characterized particularly by a highly leached, whitish gray A2 horizon. Iron oxide and alumina, and organic matter, have been removed from the A2 horizon and deposited in the B horizon.

Point Row. A row which forms an angle with another row instead of paralleling it to the end of the field. A row that "comes to a point," ending part way across the field instead of at the edge of the field.

Pollution, Water. Any change in the character of water adversely affecting its usefulness.

Polypedon (As used in the Soil Classification System of the National Cooperative Soil Survey in the United States). Two or more contiguous pedons, all of which are within the defined limits of a single soil series. In early stages of development this was called a soil individual. See soil mapping unit.

Pore Space. Total space not occupied by soil particles in a bulk volume of soil, commonly expressed as a percentage. The per cent pore space is equal to

$$100 - \frac{\text{Bulk density}}{\text{Particle density}} \times 100$$

Porosity. The degree to which the total volume of a soil, sediment, or rock is permeated with pores or cavities, generally expressed as a percentage of the whole volume unoccupied by solid particles. See air porosity; capillary porosity.

Post-Emergence (crop production). Application of chemicals, fertilizers, or other materials and operations associated with crop production after the crop has emerged through the soil surface.

Potassium Fixation. The process of converting exchangeable or water-soluble potassuim to moderately soluble potassium, that is, to a form not easily exchanged from the adsorption complex with a cation of a neutral salt solution.

Prairie. See natural grassland.

Prairie Soils. See Brunizem soils.

Pre-Emergence. (crop production). Application of chemicals, fertilizers, or other materials and operations associated with crop production before the crop has emerged through the soil surface.

Prismatic Soil Structure. See soil structure types.

Productivity, Soil. See soil productivity.

Proper Grazing Use. Grazing ranges and pastures in a manner that will maintain adequate cover for soil protection and maintain or improve the quality and quantity of desirable vegetation.

Proper Stocking. Stocking the grazing unit to obtain proper grazing use.

Protection Forest. An area wholly or partly covered with woody growth, managed primarily for its beneficial effects on soil and water conservation rather than for wood or forage production.

Puddled Soil. A dense soil dominated by massive or blocky structure, almost impervious to air and water. This condition results from handling a clay soil when it is in a wet, plastic condition so that when it dries it becomes hard and cloddy.

Puddling. The act of destroying soil structure by intensive tillage when saturated. Puddling reduces porosity and permeability. This process is sometimes used to reduce leakage of reservoirs and canals. Also used in wetland rice fields, outside the U.S., in preparation for planting rice plants.

Q

Quicksand. Sand which is unstable because of upward pressure of water.

R

R Horizon. See soil horizon.

Rainfall. A fall of rain; the amount of water that falls as rain expressed in inches in depth.

Rainfall Intensity. The rate at which rain is falling at any given instant, usually expressed in inches per hour.

Range. 1: All land producing native forage for animal consumption and land that is revegetated naturally or artificially to provide a forage cover that is managed like native vegetation. Generally considered as land that is not cultivated. 2: (wildlife). The geographic area occupied by an organism.

Range Management. The art and science of planning and directing range use to obtain sustained maximum animal production consistent with perpetuation of the natural forage resources.

Reaction, soil. The degree of acidity or alkalinity of a soil, usually expressed as a pH value. Descriptive terms commonly associated with certain ranges in pH are: extremely acid, less than 4.5; very strongly acid, 4.5–5.0; strongly acid, 5.1–5.5; medium acid, 5.6–6.0; slightly acid, 6.1–6.5; neutral, 6.6–7.3; mildly alkaline, 7.4–7.8; moderately alkaline, 7.9–8.4; strongly alkaline, 8.5–9.0; and very strongly alkaline, more than 9.0.

Red Desert Soils. A former zonal group of soils having light reddish brown, friable soil over a reddish-brown or dull red, heavy horizon grading into an accumulation of carbonate of lime; found in warm-temperate and tropical deserts and characterized by scant desert shrub vegetation. See page 554.

Reddish Brown Soils. A former zonal group of soils with a light brown surface horizon of a slightly reddish cast which grades into dull reddish brown or red material heavier than the surface soil, then into a horizon of whitish or pinkish lime accumulation; developed under shrub and short-grass vegetation of warm-temperate to tropical regions of semiarid climate. See page 554.

Reddish Brown Lateritic Soils (of U.S.). A former zonal group of soils with dark reddish brown granular surface soils, red friable clay B horizons, and red or reticulately mottled lateritic parent material, developed under humid tropical climate with wet-dry seasons and tropical forest vegetation.

Red-Yellow Podzolic Soils. A former group of well-developed, well-drained acid soils having thin A1 horizons over light-colored, bleached A2 horizons over a red, yellowish-red, or yellow and more clayey B horizon containing illuviated silicate clay. See Table 2.

Reforestation. Restocking an area with forest trees.

Regolith. The layer or mantle of loose, noncohesive or cohesive rock material, of whatever origin, that nearly everywhere forms the surface of the land and rests on bedrock. It comprises rock waste of all sorts; volcanic ash; glacial drift; alluvium; wind-blown deposits; accumulations of vegetation, such as peat; and soil.

Regosol. Any soil of the former azonal order without definite genetic horizons and developing from or on deep, unconsolidated, soft mineral deposits, such as sands, loess, or glacial drift.

Regression. A statistical method for studying and expressing the change in one variable associated with and dependent on changes in another related variable or set of variables.

Rendzina. A former great soil group of the intrazonal order and calcimorphic suborder consisting of soils with brown or black friable surface horizons underlain by light gray to pale yellow calcareous material, developed from soft, highly calcareous parent material under grass or mixed grasses and forests in humid to semiarid climates. See Table 2, page 554.

Renewable Natural Resources. A resource that can be restored and improved to produce the things man needs.

Reservoir. Impounded body of water or controlled lake in which water is collected or stored.

Residual Material. Unconsolidated and partly weathered mineral materials accumulated by disintegration of consolidated rock in place.

Residual Soil. A soil formed in material weathered from bedrock without transportation from the original location. See residual material.

Retention. The amount of precipitation on a drainage area that does not escape as runoff. It is the difference between total precipitation and total runoff.

Return Flow. That portion of the water diverted from a stream which finds its way back to the stream channel either as surface or underground flow.

Rhizobia. The bacteria capable of living in symbiotic relationship with leguminous plants in nodules on the roots, the association usually being capable of fixing nitrogen (from the generic name *Rhizobium*).

Rhizome. A horizontal underground stem, usually sending out roots and aboveground shoots at the nodes.

Rill. A small, intermittent water course with steep sides, usually only a few inches deep and, hence, no obstacle to tillage operations.

Rill Erosion. See erosion.

River Basin. The United States has been divided into 20 major water resource regions (river basins).

Riverwash. Barren alluvial land, usually coarse-textured, exposed along streams at low water and subject to shifting during normal high water. A miscellaneous land type.

Roadside Erosion Control. See highway erosion control.

Rock-Fill Dam. A dam composed of loose rock usually dumped in place, often with the upstream part constructed of handplaced or derrick-placed rock and faced with rolled earth or with an impervious surface of concrete, timber, or steel.

Rock Land. Areas containing many rock outcrops and shallow soils. Rock outcrops usually occupy from 25 to 90 per cent of the area. A miscellaneous land type.

Root Nodule. A swelling formed on the roots of leguminous plants, caused by the symbiotic nitrogen-fixing bacteria.

Root Zone. The part of the soil that is penetrated or can be penetrated by plant roots.

Rotation (forest). The period of years required to establish and grow woodland tree crops to a specified condition of maturity for harvest. See crop rotation.

Rotation-Deferred Grazing. See deferred-rotation grazing.

Rotation Grazing. Grazing two or more pastures or parts of a range in regular order, with definite recovery periods between grazing periods. Where only two fields are involved, sometimes called alternate grazing. Contrast with continuous grazing.

Rotation Irrigation. A system by which each irrigator receives his allotted quantity of water, not at a continuous rate, but at stated intervals. For example, a number of irrigators receiving water from the same lateral may agree among themselves to rotate the water, each taking the entire flow in turn for a limited period.

Rotation Pasture. A cultivated area used as a pasture for one or more years as a part of crop rotation. Contrast with permanent pasture.

Row Crop. A crop planted in rows, normally to allow cultivation between rows during the growing season.

Roughage. Feed with high fiber content and low total digestible nutrients, such as hay and stover.

Rough Broken Land. Land with very steep topography and numerous intermittent drainage channels, usually covered with vegetation. See miscellaneous land type and badlands.

Rubble Land. Land areas with 90 per cent or more of the surface covered with stones and boulders. A miscellaneous land type.

Runoff (hydraulics). That portion of the precipitation on a drainage area that is discharged from the area in stream channels. Types include surface runoff, ground water runoff, or seepage.

Runoff Plots. Areas of land, usually small, arranged so the portion of rainfall or other precipitation flowing off and perhaps carrying soluble materials and soil may be measured. Usually the flow from runoff plots includes only surface flow.

S

Saline-Alkali Soil. 1: A soil containing sufficient exchangeable sodium to interfere with the growth of most crop plants and containing appreciable quantities of soluble salts. The exchangeable-sodium percentage is greater than 15, the conductivity of the saturation extract greater than 4 millimhos per centimeter (25°C), and the pH is usually 8.5 or less in the saturated soil. 2: A saline-alkali soil has a combination of harmful quantities of salts and either a high alkalinity or high content of exchangeable sodium, or both, so distributed in the profile that the growth of most crop plants is reduced. Often called saline-sodic soil.

Saline Soil. A nonalkali soil containing sufficient soluble salts to impair its productivity but not containing excessive exchangeable sodium. This name was formerly applied to any soil containing sufficient soluble salts to interfere with plant growth commonly greater than 3000 parts per million.

Saltation. Particle movement in water or wind where particles skip or bounce along the stream bed or soil surface.

Salting. 1: Providing salt as a mineral supplement for animals. 2: Placing salt on the range in such a manner as to improve distribution of livestock.

Sample Plot. An area of land, usually small, used for measuring or observing performance under existing or applied treatments. It may be temporary or permanent.

Sample, Random. A sample drawn without bias from a population in which every item has an equal chance of being drawn.

Sample, Representative. A sample drawn in such a way that it gives a true value for the population from which it is drawn.

Sand. 1: A soil particle between 0.05 and 2.0 millimeters in diameter. 2: Any one of five soil separates: very coarse sand, coarse sand, medium sand, fine sand, and very fine sand. See soil separates. 3: A soil textural class. See soil texture.

Sand Lens. Lenticular band of sand in distinctly sedimentary banded material.

Sand Trap (irrigation, drainage). A device, often a simple enlargement in a ditch or conduit, for arresting the heavier particles of sand and silt carried by the water. Means for removing such material may be included.

Sandy. See coarse-textured and particle-size classes for family groupings.

Sandy Clay. A soil textural class. See soil texture.

Sandy Clay Loam. A soil textural class. See soil texture.

Sandy Loam. A soil textural class. See soil texture.

Sandy-Skeletal. See particle-size classes for family groupings.

Sapric Materials. See organic soil materials.

Saturate. 1: To fill all the voids between soil particles with liquid. 2: To form the most concentrated solution possible under a given set of physical conditions in the presence of an excess of the substance.

Saturation Point. 1: That point at which a soil or aquifer will no longer absorb any amount of water without losing an equal amount. 2: (wildlife). The maximum density under which a species will normally live.

Savanna (savannah). A grassland with scattered trees, either as individuals or clumps. Often a transitional type between true grassland and forest.

Scarify. To abrade, scratch, or modify the surface, for example, to scratch the impervious seed coat of hard seed or to break the surface of the soil with a narrow-bladed implement.

Scour. To abrade and wear away. 1: The wearing away of terrace or diversion channels or stream beds; 2: To brighten a tillage implement.

Seasonal Grazing. Grazing restricted to a specific season.

Sediment. Solid material, both mineral and organic, that is in suspension, is being transported, or has been moved from its site of origin by air, water, gravity, or ice and has come to rest on the earth's surface either above or below sea level.

Sediment Discharge. The quantity of sediment, measured in dry weight or by volume, transported through a stream cross section in a given time. Sediment discharge consists of both suspended load and bedload.

Sedimentary Rocks. Formed by lithification of sediments.

Seedbed. The soil prepared by natural or artificial means to promote the germination of seed and the growth of seedlings.

Seed Inoculation. The process of adding microorganisms to seed, used frequently to designate the treatment of leguminous seed with symbiotic nitrogen-fixing bacteria *(Rhizobia)*.

Seedling. A young plant grown from seed.

Seepage. 1: Water escaping through or emerging from the ground along an extensive line or surface as contrasted with a spring where the water emerges from a localized spot. 2: The process by which water percolates through the soil. 3: (percolation) The slow movement of gravitational water through the soil.

Self-Mulching Soil. A soil with a high shrink-swell potential in the surface layer so that it cracks into a granular mulch when it dries. See Vertisol.

Semiarid. A term applied to regions or climates where moisture is normally greater than under arid conditions but still definitely limits the growth of most crops. Dryland farming methods or irrigation generally are required for crop production. The upper limit of average annual precipitation in the cool semiarid regions is as low as 15 inches, whereas in tropical regions it is as high as 45 or 50 inches. Contrast with arid.

Separate, Soil. See soil separate.

Series. See soil classification.

Settling Basin. An enlargement in the channel of a stream to permit the settling of debris carried in suspension.

Shaly. An adjective incorporated into the soil textural class designations of horizons when the soil mass contains between 15 and 90 per cent by volume of shale fragments. See Table 1, page 511.

Sheet Erosion. See erosion.

Sheet Flow. Water, usually storm runoff, flowing in a thin layer over the ground surface. Synonym: overland flow.

Shelterbelt. A wind barrier of living trees and shrubs, established and maintained for protection of farm fields. Synonym: windbreak.

Shrink-Swell Potential. Susceptibility to volume change due to loss or gain in moisture content.

Shrub. A woody perennial plant differing from a perennial herb by its more woody stems and from a tree by its low stature and habit of branching from the base. There is no definite line between herbs and shrubs or between shrubs and trees; all possible intergradations occur.

Sierozem. A former zonal great soil group consisting of soils with pale grayish A horizons grading into calcareous material at a depth of 1 foot or less, formed in temperate to cool, arid climates under a vegetation of desert plants, short grass, and scattered brush. See Table 2, page 554.

Significant (statistics). A term applied to differences and correlations to indicate that they are probably not due to chance alone. Significant ordinarily indicates a probability of not less than 95 per cent, whereas highly significant indicates a probability of not less than 99 per cent.

Siliceous. See soil mineralogy classes for family groupings.

Silt. 1: A soil separate consisting of particles between 0.05 and 0.002 millimeter in equivalent diameter. See soil separates. 2: A soil textural class. See soil texture.

Silting. See sediment.

Silt Loam. A soil textural class containing a large amount of silt and small quantities of sand and clay. See soil texture.

Silty Clay. A soil textural class containing a relatively large amount of silt and clay and a small amount of sand. See soil texture.

Silty Clay Loam. A soil textural class containing a relatively large amount of silt, a lesser quantity of clay, and a still smaller quantity of sand. See soil texture.

Single Grain. Lack of soil structure in incoherent materials. See soil structure grades.

Sink. Depression in the land surface.

Siphon (hydraulics). A closed conduit, a part of which rises above the hydraulic grade line, utilizing atmospheric pressure to cause the flow of water.

Site (ecology). 1: An area considered for its ecological factors with reference to capacity to produce vegetation; the combination of biotic, climatic, and soil conditions of an area. 2: An area sufficiently uniform in soil, climate, and natural biotic conditions to produce a particular climax vegetation.

Site Index (forestry). A numerical expression commonly accepted as an indicator of the productivity of a site. It is an expression of the height-age relationship of the tallest trees (dominants and codominants) in normal stands at some designated age, such as 50 years.

Slick Spots. Small areas in a field that are slick when wet due to a high content of clay with exchangeable sodium.

Slip. The downslope movement of a soil mass under wet or saturated conditions; a microlandslide that produces a microrelief in soils.

Slope. Degree of deviation of a surface from the horizontal, usually expressed in per cent or degrees.

Slope Characteristics. Slopes may be characterized as concave (decrease in steepness in lower portion), uniform, or convex (increase in steepness at base). Erosion is strongly affected by shape, ranked in order of increasing erodibility from concave to uniform to convex.

Slough. Wet or marshy area.

Sod Grasses. Stoloniferous or rhizomatous grasses that form a sod or turf.

Sodic Soil. 1: A soil that contains sufficient sodium to interfere with the growth of most crop plants. 2: A soil in which the exchangeable-sodium percentage is 15 or more. Sodic soils because of dispersion of the organic matter have been called "black alkali" soils; sometimes also called nonsaline-alkali soils.

Soil. 1: The unconsolidated mineral and organic material on the immediate surface of the earth that serves as a natural medium for the growth of land plants. 2: The unconsolidated mineral matter on the surface of the earth that has been subjected to and influenced by genetic and environmental factors of parent material, climate (including moisture and temperature effects), macro- and microorganisms, and topography, all acting over a period of time and producing a product—soil—that differs from the material from which it is derived in many physical, chemical, biological, and morphological properties and characteristics. 3: A kind of soil is the collection of soils that are alike in specified combinations of characteristics. Kinds of soil are given names in the system of soil classification. The terms "the soil" and "soil" are collective terms used for all soils, equivalent to the word "vegetation" for all plants.

Soil Alkalinity. The degree or intensity of alkalinity of a soil, expressed by a value greater than 7.0 on the pH scale.

Soil Association. 1: A group of defined and named taxonomic soil units occurring together in an individual and characteristic pattern over a geographic region, comparable to plant associations in many ways. Sometimes called "natural land type." 2: A mapping unit used on reconnaissance or generalized soil maps in which two or more defined taxonomic units occurring together in a characteristic pattern are combined because the scale of the map or the purpose for which it is being made does not require delineation of the individual soils.

Soil Auger. A tool for boring into the soil and withdrawing a small sample for field or laboratory observation. Soil augers may be classified as (1) those with worm-type bits, uninclosed, or (2) those with worm-type bits at the bottom of a hollow cylinder.

Soil Classification. The systematic arrangement of soils into classes in one or more categories or levels of classification for a specific objective. Broad groupings are made on the basis of general characteristics and subdivisions

TABLE 2   A COMPARISON OF THE NEW UNITED STATES SOIL CLASSIFICATION
SYSTEM ADOPTED IN 1965 WITH THE APPROXIMATE EQUIVALENTS UNDER
THE SOIL CLASSIFICATION SYSTEM IN USE BEFORE 1965

Soil Order (Adopted in 1965)	Approximate Equivalents (In Use Before 1965)
1. Alfisols	Gray Brown Podzolic, Gray Wooded soils, Noncalcic Brown soils, Degraded Chernozem, and associated Planosols and some Half-Bog soils
2. Aridisols	Desert, Reddish Desert, Sierozem, Solonchak, some Brown and Reddish Brown soils, and associated Solonetz
3. Entisols	Azonal soils and some Low-Humic Gley soils
4. Histosols	Bog soils
5. Inceptisols	Ando, Sol Brun Acide, some Brown Forest, Low-Humic Gley, and Humic Gley soils
6. Mollisols	Chestnut, Chernozem, Brunizem (Prairie), Rendzina, some Brown, Brown Forest, and associated Solonetz and Humic Gley soils
7. Oxisols	Laterite soils, Latosols
8. Spodosols	Podzols, Brown Podzolic soils, and Ground Water Podzols
9. Ultisols	Red Yellow Podzolic soils, Reddish Brown Lateritic soils of the U.S., and associated Planosols and Half-Bog soils
10. Vertisols	Grumusols

Source: Soil Survey Staff, Soil Conservation Service, U.S.D.A. *Soil Classification: A Comprehensive System—Seventh Approximation,* United States Government Printing Office, Washington, D.C. (1960), p. 13.

on the basis of more detailed differences in specific properties. The categories of the system used in the United States since 1965 are briefly discussed below. The relationship between the orders of the present system and approximate equivalents of the previous system used in the United States are shown in Table 2.

Order. The category at the highest level of generalization in the soil classification system. The properties selected to distinguish the orders are reflections of the degree of horizon development and the kinds of horizons present.

The 10 orders are:

Alfisols. Soils with gray to brown surface horizons, medium to high supply of bases, and B horizons of illuvial clay accumulation. These soils form mostly under forest or savannah vegetation in climates with slight to pronounced seasonal moisture deficit.

Aridisols. Soils with pedogenic horizons, low in organic matter, that are never moist as long as 3 consecutive months. They have an ochric epipedon that is normally soft when dry or that has distinct structure. In addition, they have one or more of the following diagnostic horizons: argillic, natric, cambic, calcic, petrocalcic, gypsic or salic, or a duripan.

Entisols. Soils that have no diagnostic pedogenic horizons. They may be found in virtually any climate on very recent geomorphic surfaces, either on steep slopes that are undergoing active erosion or on fans and floodplains where the recently eroded materials are deposited. They may also be on older geomorphic surfaces if the soils have been recently disturbed to such depths that the horizons have been destroyed or if the parent materials are resistant to alteration, as is quartz.

Histosols. Soils formed from organic soil materials.

Inceptisols. Soils that are usually moist with pedogenic horizons of alteration of parent materials but not of illuviation. Generally, the direction of soil development is not yet evident from the marks left by the various soil-forming processes or the marks are too weak to classify in another order.

Mollisols. Soils with nearly black, organic-rich surface horizons and high supply of bases. These are soils that have decomposition and accumulation of relatively large amounts of organic matter in the presence of calcium. They have mollic epipedons and base saturation greater than 50 per cent (NH_4OAc) in any cambic or argillic horizon. They lack the characteristics of Vertisols and must not have oxic or spodic horizons.

Oxisols. Soils with residual accumulations of inactive clays, free oxides, kaolin, and quartz. They are mostly in tropical climates.

Spodosols. Soils with illuvial accumulations of amorphous materials in subsurface horizons. The amorphous material is organic matter and compounds of aluminum and usually iron. These soils are formed in acid mainly coarse-textured materials in humid and mostly cool or temperate climates.

Ultisols. Soils that are low in supply of bases and have subsurface horizons of illuvial clay accumulation. They are usually moist, but during the warm season of the year, some are dry part of the time. The balance between liberation of bases by weathering and removal by leaching is normally such that a permanent agriculture is impossible without fertilizers or shifting cultivation.

Vertisols. Clayey soils with high shrink-swell potential that have wide, deep cracks when dry. Most of these soils have distinct wet and dry periods throughout the year.

Suborder. This category narrows the ranges in soil moisture and temperature regimes, kinds of horizons, and composition, according to which of these is most important. Moisture and/or temperature or soil properties associated with them are used to define suborders of Alfisols, Mollisols, Oxisols, Ultisols, and Vertisols. Kinds of horizons are used for Aridisols, composition for Histosols and Spodosols, and combinations for Entisols and Inceptisols. Some of the more important suborders in the United States are listed below; they are listed alphabetically under the respective orders.

Alfisols
    Aqualfs. Alfisols seasonally saturated with water
    Boralfs. Alfisols that are cool or cold

Udalfs. Alfisols in moist, warm-temperate climates

Ustalfs. Alfisols in warm climates that are intermittently dry for long periods during the year.

Xeralfs. Alfisols in warm climates that are continuously dry for long periods in the summer but moist in the winter

Aridisols

Argids. Aridisols with horizons of clay accumulation

Orthids. Aridisols without horizons of clay accumulation

Entisols

Aquents. Entisols permanently or seasonally saturated with water

Orthents. Entisols with loamy or clayey textures

Psamments. Entisols with sandy textures

Histosols

Fibrists. Histosols largely undecomposed fibrous organic materials

Hemists. Histosols intermediate between Fibrists and Saprists in decomposition of organic materials

Saprists. Histosols largely decomposed organic materials

Inceptisols

Andepts. Inceptisols with large amounts of amorphous or vitric pyroclastic materials

Aquepts. Inceptisols seasonally saturated with water

Ochrepts. Inceptisols that have thin or light-colored surface horizons with little organic matter and altered subsurface horizons

Umbrepts. Inceptisols with thick, dark-colored surface horizons rich in organic matter and altered subsurface horizons

Mollisols

Aquolls. Mollisols seasonally saturated with water

Borolls. Mollisols that are cool or cold

Udolls. Mollisols in moist, warm-temperate climates

Ustolls. Mollisols that are intermittently dry for long periods during the warm season of the year

Xerolls. Mollisols that are continuously dry for long periods during the warm season of the year

Spodosols

Aquods. Spodosols seasonally saturated with water

Orthods. Spodosols with subsurface accumulations of iron, aluminum, and organic matter

Ultisols

Aquults. Ultisols seasonally saturated with water

Humults. Ultisols with high or very high organic matter content

Udults. Ultisols with low organic matter content in moist, warm climates

Xerults. Ultisols with low to moderate organic matter content, continuously dry for long periods in the summer but moist in the winter

Vertisols

Uderts. Vertisols that crack open for only short periods, less than a total of 3 months in a year

Usterts. Vertisols in which cracks open and close more than once during the year but do not remain open continuously throughout the year

Great Group. The classes in this category contain soils that have the same kinds of horizons in the same sequence and have similar moisture and temperature regimes. Exceptions to the horizon sequences are made for horizons near the surface that may get mixed or lost by erosion if plowed.

Subgroup. The great groups are subdivided into subgroups that show the central properties of the great group, intergrade subgroups that show properties of more than one great group, and other subgroups for soils with atypical properties that are not characteristic of any great group.

Family. Families are defined largely on the basis of physical and mineralogic properties of importance to plant growth.

Series. The soil series is a group of soils having horizons similar in differentiating characteristics and arrangement in the soil profile, except for texture of the surface, slope, and erosion.

Soil Complex. A mapping unit used in detailed soil surveys where two or more defined taxonomic units are so intimately intermixed geographically that it is undesirable or impractical, because of the scale being used, to separate them. A more intimate mixing of smaller areas of individual taxonomic units than that described under soil association.

Soil Conditioner. Any material added to a soil for the purpose of improving its physical condition.

Soil-Conserving Crops. Crops that prevent or retard erosion and maintain or replenish rather than deplete soil organic matter.

Soil Correlation. The process of defining, mapping, naming, and classifying the kinds of soils in a specific soil survey area, the purpose being to insure that soils are adequately defined, accurately mapped, and uniformly named in all soil surveys made in the United States.

Soil-Depleting Crops. Crops that under the usual management tend to deplete nutrients and organic matter in the soil and permit deterioration of soil structure.

Soil Erosion. The detachment and movement of soil from the land surface by wind or water. See gully erosion; rill erosion; sheet erosion; splash erosion; wind erosion.

Soil Fertility. See fertility, soil.

Soil-Formation Factors. The variables, usually interrelated natural agencies, active in and responsible for the formation of soil. The factors are usually grouped as follows: parent material, climate, organisms, topography, and time. Many people believe that activities of man in his use and manipulation of soil becomes such an important influence on soil formation that he should be added as a sixth variable. Others consider man as an organism.

Soil Fumigation. Treatment of the soil with volatile or gaseous substances that penetrate the soil mass and kill one or more forms of soil organisms.

Soil Genesis. The mode of origin of the soil with special reference to the processes or soil-forming factors responsible for the development of the solum or true soil from the unconsolidated parent material.

Soil Granule. A cluster of soil particles behaving as a unit in soil structure.

Soil Horizon. A layer of soil or soil material approximately parallel to the land surface and differing from adjacent genetically related layers in physical, chemical, and biological properties or characteristics, such as color, structure, texture, consistence, amount of organic matter, and degree of acidity or alkalinity. Table 3 lists the designation and description of the major soil horizons. Few if any soils have all of these horizons well developed, but every soil has some of them.

TABLE 3 DESIGNATION AND DESCRIPTION OF MAJOR SOIL HORIZONS

Horizon Designation	Description
0	Organic horizons of mineral soils. Horizons (i) formed or forming in the upper part of mineral soils above the mineral part; (ii) dominated by fresh or partly decomposed organic material; and (iii) containing more than 30 per cent organic matter if the mineral fraction is more than 50 per cent clay, or more than 20 per cent organic matter if the mineral fraction has no clay. Intermediate clay content requires proportional organic matter content.
01	Organic horizons in which essentially the original form of most vegetative matter is visible to the naked eye. The 01 corresponds to the L (litter) and some F (fermentation) layers in forest soils designations and to the horizon formerly called Aoo.
02	Organic horizons in which the original form of most plant or animal matter cannot be recognized with the naked eye. The 02 corresponds to the H (humus) and some F (fermentation) layers in forest soils designations and to the horizon formerly called Ao.
A	Mineral horizons consisting of (i) horizons of organic matter accumulation formed or forming at or adjacent to the surface; (ii) horizons that have lost clay, iron, or aluminum with resultant concentration of quartz or other resistant minerals of sand or silt size; or (iii) horizons dominated by (i) or (ii), but transitional to an underlying B or C.
A1	Mineral horizons, formed or forming at or adjacent to the surface, in which the feature emphasized is an accumulation of humified organic matter intimately associated with the mineral fraction.
A2	Mineral horizons in which the feature emphasized is loss of clay, iron, or aluminum, with resultant concentration of quartz or other resistant minerals in sand and silt sizes.
A3	A transitional horizon between A and B, dominated by properties characteristic of an overlying A1 or A2 but having some subordinate properties of an underlying B.
AB	A horizon transitional between A and B, having an upper part dominated by properties of A and a lower part dominated by properties of B; the two parts cannot be conveniently separated into A3 and B1.
A & B	Horizons that would qualify for A2 except for included parts constituting less than 50 per cent of the volume that would qualify as B.
AC	A horizon transitional between A and C, having subordinate properties of both A and C but not dominated by properties characteristic of either A or C.
B & A	Any horizon qualifying as B in more than 50 per cent of its volume, including parts that qualify as A2.

TABLE 3 (Continued)

Horizon Designation	Description
B	Horizons in which the dominant feature or features is one or more of the following: (i) an illuvial concentration of silicate clay, iron, aluminum, or humus, alone or in combination; (ii) a residual concentration of sesquioxides or silicate clays, alone or mixed, that has formed by means other than solution and removal of carbonates or more soluble salts; (iii) coatings of sesquioxides adequate to give conspicuously darker, stronger, or redder colors than overlying and underlying horizons in the same sequum but without apparent illuviation of iron and not genetically related to B horizons that meet requirements of (i) or (ii) in the same sequum; or (iv) an alteration of material from its original condition in sequums lacking conditions defined in (i), (ii), and (iii) that obliterates original rock structure, that forms silicate clays, liberates oxides, or both, and that forms granular, blocky, or prismatic structure if textures are such that volume changes accompany changes in moisture.
B1	A transitional horizon between B and A1 or between B and A2 in which the horizon is dominated by properties of an underlying B2 but has some subordinate properties of an overlying A1 or A2.
B2	That part of the B horizon where the properties on which the B is based are without clearly expressed subordinate characteristics, indicating that the horizon is transitional to an adjacent overlying A or an adjacent underlying C or R.
B3	A transitional horizon between B and C or R in which the properties diagnostic of an overlying B2 are clearly expressed but are associated with clearly expressed properties characteristic of C or R.
C	A mineral horizon or layer, excluding bedrock, that is either like or unlike the material from which the solum is presumed to have formed, relatively little affected by pedogenic processes, and lacking properties diagnostic of A or B but including materials modified by (i) weathering outside the zone of major biological activity; (ii) reversible cementation, development of brittleness, development of high bulk density, and other properties characteristic of fragipans; (iii) gleying; (iv) accumulation of calcium or magnesium carbonate or more soluble salts; (v) cementation by accumulations, such as calcium or magnesium carbonate or more soluble salts; or (vi) cementation by alkali-soluble siliceous material or by iron and silica.
R	Underlying consolidated bedrock, such as granite, sandstone, or limestone. If presumed to be like the parent rock from which the adjacent overlying layer or horizon was formed, the symbol R is used alone. If presumed to be unlike the overlying material, the R is preceded by a Roman numeral denoting lithologic discontinuity, such as II R.

Soil Improvement. The processes for, or the results of, making the soil more productive for growing plants by drainage, irrigation, and the addition of fertilizers and soil amendments such as lime and manure.

Soil Individual. See polypedon.

Soil Loss Tolerance. The maximum average annual soil loss in tons per acre per year that should be permitted on a given soil.

Soil Management. The sum total of all tillage operations, cropping practices, fertilizer, lime, and other treatments conducted on, or applied to, a soil for the production of plants.

Soil Map. A map showing the distribution of soil types or other soil mapping units in relation to the prominent physical and cultural features of the earth's surface. The following kinds of soil maps are recognized in the U.S.: detailed, detailed reconnaissance, reconnaissance, generalized, and schematic.

Soil Mapping Unit. A kind of soil, a combination of kinds of soil, or miscellaneous land type or types, that can be shown at the scale of mapping for the defined purposes and objectives of the survey. (Combination of kinds of soil includes soil association, complexes, undifferentiated soils, or any class or combination of classes at the family level or higher categories of the soil classification system.) Soil mapping units are the basis for the delineations of a soil survey map. A soil survey identification legend lists all mapping units for the survey of an area (any size area from a small plot to a county, a nation, or the world). Mapping units normally contain inclusions of soils outside the limits of the taxonomic name, or names, used as the name for the mapping unit. Mapping units are generally designed to reflect significant differences in use and management. Mapping units are phases of series.

Soil Mineralogy Classes for Family Groupings (As used in the Soil Classification System of the National Cooperative Soil Survey in the United States). The family category includes mineralogy classes for specific control sections that are similar to those used for particle-size classes for family groupings. For example, the term micaceous denotes that more than 40 per cent by weight of the 0.02- to 20-millimeter fraction of the soil material within the control section is mica. Examples of some mineralogy classes are listed below:

Ferritic. For soils of any texture, the whole soil less than 2 millimeters in diameter in the control section contains more than 40 per cent by weight of iron oxide as (Fe_2O_3) extractable by citrate-dithionite.

Illitic. In clayey soils, more than half by weight of the clay-size fraction is composed of illite (hydrous mica) commonly with greater than 3 per cent K_2O.

Kaolinitic. In clayey soils, more than half by weight of the clay-size fraction is composed of kaolinite, dickite; and nacrite with smaller amounts of other 1:1 or nonexpanding 2:1 layer minerals or gibbsite.

Micaceous. Soil with more than 40 per cent by weight of mica.

Mixed. Soils that have a combination of minerals in which no single class of mineralogy is dominant.

Montmorillonitic. In clayey soils, more than half by weight of the clay-size fraction is composed of montmorillonite and nontronite, or a mixture with more montmorillonite than any other single clay mineral.

Siliceous. In the 0.02 to 2-millimeter fraction within the control section of sandy, silty, and loamy soils more than 90 per cent by weight of silica minerals or other minerals with a hardness of 7 or more in the Mohs scale.

Soil Moisture Tension. The force per unit area that must be exerted to remove water from soil.

Soil Monolith. A vertical section of a soil profile removed and mounted for display or study.

Soil Morphology. The constitution of the soil, including the texture, structure, consistence, color, and other physical, chemical, and biological properties of the various soil horizons that make up the soil profile.

Soil Organic Matter. The organic fraction of the soil that includes plant and animal residues at various stages of decomposition, cells and tissues of soil organisms, and substances synthesized by the soil population. Commonly determined as the amount of organic material contained in a soil sample passed through a 2-millimeter sieve. See humus.

Soil Pores. See pore space.

Soil Porosity. See porosity.

Soil Probe. A tool having a hollow cylinder with a cutting edge at the lower end, used for probing into the soil and withdrawing a small sample for field or laboratory study.

Soil Productivity. The capacity of a soil in its normal environment for producing a specified plant or sequence of plants under a specified system of management.

Soil Profile. A vertical section of the soil from the surface through all its horizons, including C horizons. See soil horizon.

Soil Separates. Mineral particles, less than 2.0 millimeters in equivalent diameter, ranging between specified size limits. The names and size limits of separates recognized by the National Cooperative Soil Survey in the United States are: very coarse sand, 2.0 to 1.0 millimeters (called fine gravel prior to 1947, now fine gravel includes particles between 2.0 millimeters and about 12.5 millimeters in diameter); coarse sand, 1.0 to 0.5 millimeter; medium sand, 0.5 to 0.25 millimeter; fine sand, 0.25 to 0.10 millimeter; very fine sand, 0.10 to 0.05 millimeter; silt, 0.05 to 0.002 millimeter; and clay, less than 0.002 millimeter. (Before 1937, clay included particles less than 0.005 millimeter in diameter and silt, those particles from 0.05 to 0.005 millimeter.) The separates recognized by the International Society of Soil Science are: coarse sand, 2.0 to 0.2 millimeters; fine sand, 0.2 to 0.02 millimeter; silt, 0.02 to 0.002 millimeter; and clay, less than 0.002 millimeter in diameter.

Soil Series. See soil classification.

Soil Structure. The combination or arrangement of primary soil particles into secondary particles, units, or peds (Table 4). The secondary units are characterized and classified on the basis of size, shape, and degree of distinctness into classes, types, and grades, respectively. See soil structure classes; soil structure grade; soil structure types. See Table 4, page 562.

Soil Structure Classes. A grouping of soil structural units or peds on the basis of size (Table 4, page 562). See soil structure; soil structure types.

Soil Structure Grades. A grouping or classification of soil structure on the basis of inter- and intra-aggregate adhesion, cohesion, or stability within the profile. Four grades used are structureless, weak, moderate, and strong, depending on observable degree of aggregation.

TABLE 4 CLASSES AND TYPES OF SOIL STRUCTURE *

Class	Platelike with one dimension (the vertical) limited and greatly less than the other two; arranged around a horizontal plane; faces mostly horizontal	Prismlike with two dimensions (the horizontal) limited and considerably less than the vertical; arranged around a vertical line; vertical faces well defined; vertices angular		Blocklike, polyhedronlike, or spheroidal, with three dimensions of the same order of magnitude, arranged around a point			
				Blocklike; blocks or polyhedrons having plane or curved surfaces that are casts of the molds formed by the faces of the surrounding peds		Spheroids of polyhedrons having plane or curved surfaces which have slight or no accommodation to faces of surrounding peds	
		Without rounded caps	With rounded caps	Faces flattened; most vertices sharply angular	Mixed rounded and flattened faces with many rounded vertices	Relatively non-porous peds	Porous peds
	Platy	Prismatic	Columnar	(Angular) Blocky †	Subangular Blocky ‡	Granular	Crumb
Very fine or very thin	Very thin platy; < 1 mm	Very fine prismatic; < 10 mm	Very fine columnar; < 10 mm	Very fine angular blocky; < 5 mm	Very fine subangular blocky; < 5 mm	Very fine granular; < 1 mm	Very fine crumb; < 1 mm
Fine or thin	Thin platy; 1–2 mm	Fine prismatic; 10–20 mm	Fine columnar; 10–20 mm	Fine angular blocky; 5–10 mm	Fine subangular blocky; 5–10 mm	Fine granular; 1–2 mm	Fine crumb; 1–2 mm
Medium	Medium platy; 2–5 mm	Medium prismatic; 20–50 mm	Medium columnar; 20–50 mm	Medium angular blocky; 10–20 mm	Medium subangular blocky; 10–20 mm	Medium granular; 2–5 mm	Medium crumb; 2–5 mm
Coarse or thick	Thick platy; 5–10 mm	Coarse prismatic; 50–100 mm	Coarse columnar; 50–100 mm	Coarse angular blocky; 20–50 mm	Coarse subangular blocky; 20–50 mm	Coarse granular; 5–10 mm	—
Very coarse or very thick	Very thick platy; > 10 mm	Very coarse prismatic; > 100 mm	Very coarse columnar; > 100 mm	Very coarse angular blocky; > 50 mm	Very coarse subangular blocky; > 50 mm	Very coarse granular; > 10 mm	—

Type (Shape and Arrangement of Peds)

* Source: Soil Survey Staff, Soil Conservation Service, U.S.D.A., Soil Survey Manual, Agricultural Handbook 18, Washington: U.S. Government Printing Office, 1962, p. 228.

† (a) Sometimes called nut. (b) The word "angular" in the name ordinarily can be omitted.

‡ Sometimes called nuciform, nut, or subangular nut. Since the size connotation of these terms is a source of great confusion to many, they are not as one

Soil Structure Types. A classification of soil structure based on the shape of the aggregates or peds and their arrangement in the profile (Table 4). Generally the shape of soil structure types is referred to as either platy, prismatic, columnar, blocky, granular, or crumb. See soil structure; soil structure classes; soil structure grades.

Soil Survey. A general term for the systematic examination of soils in the field and in laboratories; their description and classification; the mapping of kinds of soil; the interpretation of soils according to their adaptability for various crops, grasses, and trees; their behavior under use or treatment for plant production or for engineering purposes; and their productivity under different management systems.

Soil Taxonomic Unit. A unit of all soils that fall within the defined limits of a class at any categoric level in a system of soil classification. Commonly used as a member of the lowest class in the present classification scheme and in that use is equivalent to a series.

Soil Temperature Classes for Family Groupings (as used in the Soil Classification System of the National Cooperative Soil Survey in the United States). Classes are based on mean annual soil temperature and difference between mean summer and mean winter temperature. Soil temperature is determined at a depth of 50 centimeters (20 inches) or at a lithic or paralithic contact, whichever is shallower. Unless used in a higher category, soil temperature classes are used at the family level as follows: (1) Soils with 5°C (9°F) or more difference between mean summer (June, July, and August) and mean winter (December, January, and February) temperatures, and with mean annual soil temperatures as follows: less than 8°C (47°F), **frigid**; 8 to 15°C (47 to 59°F), **mesic**, 15 to 22°C (59 to 72°F), **thermic**; and more than 22°C (72°F), **hyperthermic**. (2) Soils with less than 5°C (9°F) difference between mean summer and winter soil temperatures, and with mean annual soil temperatures as follows: less than 8°C (47°F), **isofrigid**; 8 to 15°C (47 to 59°F), **isomesic**; 15 to 22°C (59 to 72°F), **isothermic**; and 22°C (72°F) or higher, **isohyperthermic**.

Soil Texture. The limits of the various classes and subclasses are as follows: (For a guide for textural classification in soil families see Figure G.1)

Sand. Soil material that contains 85 per cent or more of sand. The percentage of silt plus 1.5 times the percentage of clay shall not exceed 15.

    Coarse Sand. 25 per cent or more very coarse and coarse sand and less than 50 per cent any other one grade of sand.

    Sand. 25 per cent or more very coarse, coarse, and medium sand and less than 50 per cent fine or very fine sand.

    Fine Sand. 50 per cent or more fine sand, or less than 25 per cent very coarse, coarse, and medium sand, and less than 50 per cent very fine sand.

    Very Fine Sand. 50 per cent or more very fine sand.

Loamy Sand. Soil material that contains, at the upper limit, 85 to 90 per cent sand, and the percentage of silt plus 1.5 times the percentage of clay

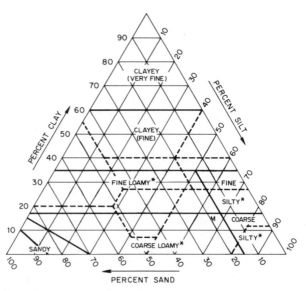

* VERY FINE SAND (0.05 – 0.1) IS TREATED AS SILT FOR FAMILY GROUPINGS;
COARSE FRAGMENTS ARE CONSIDERED THE EQUIVALENT OF COARSE SAND
IN THE BOUNDARY BETWEEN THE SILTY AND LOAMY CLASSES.

COMPARISON OF PARTICLE – SIZE SCALES

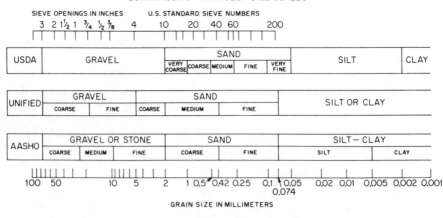

FIG. G.1. Guide for textural classification in soil families. *Source:* Supplement to *Soil Classification: A Comprehensive System, Seventh Approximation,* Soil Survey Staff, Soil Conservation Service, U.S. Dept. of Agr. (Mar., 1967), p. 40.

is not less than 15. At the lower limit, it contains not less than 70 to 85 per cent sand, and the percentage of silt plus twice the percentage of clay does not exceed 30.

Loamy Coarse Sand. 25 per cent or more very coarse and coarse sand and less than 50 per cent any other one grade of sand.

Loamy Sand. 25 per cent or more very coarse, coarse, and medium sand and less than 50 per cent fine or very fine sand.

Loamy Fine Sand. 50 per cent or more fine sand, or less than 25 per cent very coarse, coarse, and medium sand and 50 per cent very fine sand.

Loamy Very Fine Sand. 50 per cent or more very fine sand.

Sandy Loam. Soil material that contains either 20 per cent or less clay, and the percentage of silt plus twice the percentage of clay exceeds 30, and 52 per cent or more sand; or less than 7 per cent clay, less than 50 per cent silt, and between 43 and 52 per cent sand.

Coarse Sandy Loam. 25 per cent or more very coarse and coarse sand and less than 50 per cent any other one grade of sand.

Sandy Loam. 30 per cent or more very coarse, coarse, and medium sand but less than 25 per cent very coarse sand and less than 30 per cent very fine or fine sand.

Fine Sandy Loam. 30 per cent or more fine sand and less than 30 per cent very fine sand, or between 15 and 30 per cent very coarse, coarse, and medium sand.

Very Fine Sandy Loam. 30 per cent or more very fine sand, or more than 40 per cent fine and very fine sand, at least half of which is very fine sand and less than 15 per cent very coarse, coarse, and medium sand.

Loam. Soil material that contains 7 to 27 per cent clay, 28 to 50 per cent silt, and less than 52 per cent sand.

Silt Loam. Soil material that contains 50 per cent or more silt and 12 to 27 per cent clay, or 50 to 80 per cent silt and less than 12 per cent clay.

Silt. Soil material that contains 80 per cent or more silt and less than 12 per cent clay.

Sandy Clay Loam. Soil material that contains 20 to 35 per cent clay, less than 28 per cent silt, and 45 per cent or more sand.

Clay Loam. Soil material that contains 27 to 40 per cent clay and 20 to 45 per cent sand.

Silty Clay Loam. Soil material that contains 27 to 40 per cent clay and less than 20 per cent sand.

Sandy Clay. Soil material that contains 35 per cent or more clay and 45 per cent or more sand.

Silty Clay. Soil material that contains 40 per cent or more clay and 40 per cent or more silt.

Clay. Soil material that contains 40 per cent or more clay, less than 45 per cent sand, and less than 40 per cent silt.

Soil Type. 1: A subdivision of a soil series based on surface texture. At the present time in the United States a soil type is considered as a kind of phase and is not part of the soil classification system presently being used. See phase, soil. 2: In Europe, a class roughly equivalent to a great soil group.

Soil Variant. A kind of soil whose properties are believed to be sufficiently different from recognized series to justify a new series name but comprising such a limited geographic area that creation of a new series is not justifiable; based upon variations in subsurface horizons.

Sol Brun Acide. A former zonal group of soils developed under forest vegetation with thin A1 horizon, a paler A2 horizon which is poorly differentiated from the B2 horizon, a B2 horizon with uniform color from top to bottom, weak subangular blocky structure, and lacking evidence of silicate clay accumulation. The sola are strongly to very strongly acid and have low base status.

Solifluction. The slow downhill flowage or creep of soil and other loose materials that become saturated.

Solonchak. A former great soil group of the intrazonal order and halomorphic suborder, consisting of soils with a gray, thin, salty crust on the surface, with fine granular mulch immediately below, underlain with grayish, friable, salty soil; formed under subhumid to arid, hot or cool climates, under conditions of poor drainage, and under a sparse growth of halophytic grasses, shrubs, and some trees. See Table 2, page 554.

Solonetz. A former great soil group of the intrazonal order and halomorphic suborder, consisting of soils with a very thin, friable, surface soil underlain by a dark, hard, columnar layer, usually highly alkaline; formed under subhumid to arid, hot to cool climates, under better drainage than Solonchaks, and under a native vegetation of halophytic plants. See Table 2, page 554.

Solum (pl. sola). The upper part of a soil profile, above the parent material, in which the processes of soil formation are active. The solum in mature soils includes the A and B horizons. Usually the characteristics of the material in these horizons are quite unlike those of the underlying parent material. The living roots and other plant and animal life characteristic of the soil are largely confined to the solum.

Specific Gravity. The relative weight of a given volume of any kind of matter (volume occupied by solid phase, pore space excluded) compared with an equal volume of distilled water at a specified temperature. The average specific gravity for soil is about 2.65. Contrast with bulk density.

Spillway. An open or closed channel, or both, used to convey excess water from a reservoir. It may contain gates, either manually or automatically controlled, to regulate the discharge of excess water.

Splash Erosion. See erosion.

Spodic Horizon. See diagnostic horizons.

Spodosols. See soil classification.

Spoil. Soil or rock material excavated from a canal, ditch, basin, or similar construction.

Spoilbank. A pile of soil, subsoil, rock, or other material excavated from a drainage ditch, pond, or other cut.

Sprigging. The planting of a portion of the stem and root of grass.

Sprinkler Irrigation. Irrigation where water is applied by means of perforated pipes or nozzles operated under pressure so as to form a spray pattern.

Standard Deviation (statistics). A measure of the average variation of a series of observations or items of a population about their mean. In normally distributed sets of moderate size, the interval of the mean, plus or minus the standard deviation, includes about two-thirds of the items.

Standard Error of Estimate (statistics). An estimate of the standard deviation of means of samples drawn from a single population, often calculated from a single set of samples.

State Soil Conservation Committee, Commission, or Board. The state agency established by state soil conservation district enabling legislation to assist with the administration of the provisions of the state soil conservation districts law. The official title may vary from the above as new or amended state laws are made.

Stolon. A horizontal stem which grows along the surface of the soil and roots at the nodes.

Stone Line. A concentration of coarse fragments in soils. In cross section, the line may be marked only by scattered fragments or it may be a discrete layer of fragments. The fragments are more often pebbles or cobbles than stones. The line generally overlies material that was subjected to weathering, soil formation, and erosion before deposition of the overlying material.

Stones. See coarse fragments.

Stoniness. The relative proportion of stones in or on the soil, used in classifying soils. See coarse fragments; stony; very stony; stony land.

Stony. Containing sufficient stones to interfere with tillage but not to make intertilled crops impracticable. Stones may occupy 0.01 to 0.1 per cent of the surface. Stoniness is not a part of the soil textural class. The terms "stony" and "very stony" may modify the soil textural class name in the soil type, but this is simply a brief way of designating stony phases.

Stony Land. Areas containing sufficient stones to make the use of machinery impractical, usually 15 to 90 per cent of the surface is covered with stones. A miscellaneous land type. See stoniness; rubble land.

Storage Capacity. See available water-holding capacity.

Storm. In general, a disturbance of the atmosphere. The term may be qualified to emphasize a particular kind of meteorological disturbance, such as windstorm, sandstorm, rainstorm, or thunderstorm.

Stover. The dried, cured stems and leaves of tall, coarse grain crops, such as corn and sorghum, after the grain has been removed. Contrast with fodder and hay.

Streambanks. The usual boundaries, not the flood boundaries, of a stream channel. Right and left banks are named facing downstream.

Strip Cropping. Growing crops in a systematic arrangement of strips or bands which serve as barriers to wind and water erosion. See buffer strips; contour stripcropping; field strip cropping; filter strip; wind strip cropping.

Strip Grazing. A system whereby animals are confined to a small area of pasture for a short period of time, usually 1 to 2 days.

Stubble. The basal portion of plants remaining after the top portion has been harvested; also, the portion of the plants, principally grasses, remaining after grazing is completed.

Stubble Crops. 1: Crops that develop from the stubble of the previous crop.

2: Crops sown on grain stubble after harvest for turning under the following spring.

Stubble Mulch. The stubble of crops or crop residues left essentially in place on the land as a surface cover during fallow.

Subgrade. The soil prepared and compacted to support a structure or a pavement system.

Subgroup. See soil classification.

Subhumid. A term applied to regions or climates where moisture is normally less than under humid conditions but still sufficient for the production of many agricultural crops without irrigation. Natural vegetation is mostly tall grasses. Annual rainfall varies from 20 inches in cool regions to as much as 60 inches in hot areas. Contrast with humid.

Subirrigation. Applying irrigation water below the ground surface either by raising the water table within or near the root zone, or by using a buried perforated or porous pipe system that discharges water directly into the root zone.

Suborder. See soil classification.

Subsidence. A downward movement of the level of the ground surface caused by solution and collapse of underlying soluble deposits, rearrangement of particles upon removal of coal, reduction of fluid pressures within an aquifer or petroleum reservoir, or by decomposition of organic soils (Histosols).

Subsistence Farm. A low-income farm where the emphasis is on production for use by the operator and his family.

Subsoil. The B horizons of soils with distinct profiles. In soils with weak profile development, the subsoil can be defined as the soil below the plowed soil (or its equivalent of surface soil), in which roots normally grow. Although a common term, it cannot be defined accurately. It has been carried over from early days when "soil" was conceived only as the plowed soil and that under it as the "subsoil."

Subsoiling. The tillage of subsurface soil, without inversion, for the purpose of breaking up dense layers that restrict water movement and root penetration.

Substratum. Any layer lying beneath the solum.

Subsurface Irrigation. See subirrigation.

Subsurface Tillage. Tillage with specialized equipment which loosens and prepares a seedbed but does not invert the surface residual mulch. See mulch tillage.

Succession. The progressive development of vegetation toward its highest ecological expression, the climax; replacement of one plant community by another.

Summer Fallow. The tillage of uncropped land during the summer in order to control weeds and store moisture in the soil for the growth of a later crop.

Sump. Pit, tank, or reservoir in which water is collected for storage or withdrawal.

Supplemental Irrigation. Irrigation to insure or increase crop production in areas where rainfall normally supplies most of the moisture needed.

Supplemental Pasture. Additional pasture for use in adverse weather, usually annual forage crops for dry periods or winter.

Surface Irrigation. Irrigation where the soil surface is used as a conduit, as in furrow and border irrigation, as opposed to sprinkler irrigation or subirrigation.

Surface Soil. The uppermost part of the soil ordinarily moved in tillage or its equivalent in uncultivated soils, ranging in depth from about 5 to 8 inches. Frequently designated as the plow layer, the Ap layer, or the Ap horizon.

Swamp. An area saturated with water throughout much of the year but with the surface of the soil usually not deeply submerged, usually characterized by tree or shrub vegetation. See marsh; miscellaneous land type.

T

Talus. Fragments of rock and other soil material accumulated by gravity at the foot of cliffs or steep slopes.

Tank, Earth. A structure for impounding water, formed by an excavation and an earthen dam across a drainage area.

Taproot System. A plant root system dominated by a single large "taproot," normally growing straight down, from which most or all of the smaller roots spread out laterally, such as alfalfa. Contrast with fibrous root system.

Taxonomic Unit. See soil taxonomic unit.

Tensiometer. Instrument used for measuring the suction or negative pressure of soil water.

Terrace. An embankment or combination of an embankment and channel constructed across a slope to control erosion by diverting or storing surface runoff water instead of permitting it to flow down the slope at an erosive velocity. Terraces or terrace systems may be classified by their alignment, gradient, outlet, and cross section. Alignment is parallel or nonparallel. Gradient may be level, uniformly graded, or variably graded. Grade is often incorporated to permit paralleling the terraces. Outlets may be soil infiltration only, vegetated waterways, tile outlets, or combinations of these. Cross sections may be narrow base, broad base, bench, steep backslope, flat channel, or channel.

Terrace Interval. Distance measured either vertically or horizontally between corresponding points on two adjacent terraces.

Terrace Outlet Channel. Channel, usually having a vegetative cover, into which the flow from one or more terraces is discharged and conveyed from the field.

Terrace System. A series of terraces occupying a slope and discharging runoff into one or more outlet channels.

Textural Classification. See soil texture.

Texture. See soil texture.

Thermic. See soil temperature classes for family groupings.

Threshold Velocity. The minimum velocity at which wind will begin moving particles of sand or other soil material.

Tight Soil. A compact, relatively impervious and tenacious soil, or subsoil.

Tile, Drain. Pipe made of burned clay, concrete, or similar material, in short lengths, usually laid with open joints to collect and carry excess water from the soil. Perforated plastic or asphalt pipe may also be used.

Tile Drainage. Land drainage by means of a series of tile lines laid at a specified depth, grade, and spacing.

Till. 1: Unstratified glacial drift deposited directly by the ice and consisting of clay, sand, gravel, and boulders intermingled in any proportion. 2: To plow and prepare for seeding; to seed or cultivate the soil.

Tillage. The operation of implements through the soil to prepare seedbeds and root beds.

Tillage Pan. See pan, pressure or induced.

Tilth. The physical condition of soil as related to its ease of tillage, fitness as a seedbed, and impedance to seedling emergence and root penetration.

Topsoil. 1: Earthy material used as top-dressing for house lots, grounds for large buildings, gardens, road cuts, or similar areas. It has favorable characteristics for production of desired kinds of vegetation or can be made favorable. 2: The surface plow layer of a soil. 3: The original or present dark-colored upper soil that ranges from a mere fraction of an inch to two or three feet thick on different kinds of soil. 4: The original or present A horizon, varying widely among different kinds of soil. Applied to soils in the field, the term has no precise meaning unless defined as to depth or productivity in relation to a specific kind of soil.

Toxic Salt Reduction. Decreasing harmful concentrations of toxic salts in soils, usually by leaching, and with or without the addition of soil amendments such as sulfur or gypsum.

Trace Elements. See micronutrients.

Transpiration. The process by which water vapor is released to the atmosphere by the foliage or other parts of a living plant.

Tributary. Secondary or branch of a stream, drain, or other channel that contributes flow to the primary or main channel.

Truncated Soil Profile. Soil profile that has been cut down by accelerated erosion or by mechanical means. The profile may have lost part or all of the A horizon and sometimes the B horizon, leaving only the C horizon. Comparison of an eroded soil profile with a virgin profile of the same area, soil type, and slope conditions, indicates the degree of truncation.

Tundra. The treeless land in arctic and alpine regions, varying from bare area to various types of vegetation consisting of grasses, sedges, forbs, dwarf shrubs, mosses, and lichens.

Tundra Soils. 1: Soils characteristic of tundra regions. 2: A zonal great soil group consisting of soils with dark-brown peaty layers over grayish horizons mottled with rust and having continually frozen substrata, formed under frigid, humid climates with poor drainage. Native vegetation consists of lichens, moss, flowering plants, and shrubs.

U

Ultisols. See soil classification.

Umbric Epipedon. See diagnostic horizons.

Undifferentiated Soil Group (mapping unit). Two or more soils or land types that are mapped as one unit because their differences are not significant to the purpose of the survey or to soil management.

V

Valence. That property of an element that is measured in terms of the number of gram-atoms of hydrogen that 1 gram-atom of that element will combine with or displace; for example, the valence of oxygen in water, H_2O, is 2.

Value, Color. The relative lightness or intensity of color, approximately a function of the square root of the total amount of light. One of the three variables of color. See Munsell color system; hue; chroma.

Vegetation. Plants in general or the sum total of plant life in an area.

Vegetation Type. A plant community with distinguishable characteristics.

Venturi Tube. A closed conduit that gradually contracts to a throat, causing a reduction of pressure head by which the velocity through the throat may be determined.

Vertisols. See soil classification.

Very Coarse Sand. See soil separates; soil texture.

Very Fine. See particle-size classes for family groupings.

Very Fine Sand. See soil separates; soil texture.

Very Fine Sandy Loam. See soil separates; soil texture.

Very Stony. Containing sufficient stones to make tillage of intertilled crops impracticable. The soil can be worked for hay crops or improved pasture if other soil characteristics are favorable. Stones occupy approximately 0.1 to 3 per cent of the surface. See stony for discussion of phase names.

Volume Weight. See bulk density.

W

Water Conservation. The physical control, protection, management, and use of water resources in such a way as to maintain crop, grazing, and forest lands; vegetal cover; wildlife; and wildlife habitat for maximum sustained benefits to people, agriculture, industry, commerce, and other segments of the national economy.

Water Control (soil and water conservation). The physical control of water by such measures as conservation practices on the land, channel improvements, and installation of structures for water retardation and sediment detention (does not refer to legal control or water rights as here defined).

Waterlogged. Saturated with water. Soil condition where a high or perched water table is detrimental to plant growth, resulting from overirrigation, seepage, or inadequate drainage; the replacement of most of the soil air by water.

Water Management. Application of practices to obtain added benefits from precipitation, water, or water flow in any of a number of areas, such as irrigation, drainage, wildlife and recreation, water supply, watershed management, and water storage in soil for crop production. See irrigation water management; watershed management.

Water Quality Standards. Minimum requirements of purity of water for various uses; for example, water for agricultural use in irrigation systems should not exceed specific levels of sodium bicarbonates, pH, and total soluble salts.

Water Requirement (plant physiology). In a strict sense, the ratio of the number of units of water absorbed by the plant during the growing season to the number of units of dry matter produced by the plant during that time. More generally, the amount of water lost through transpiraion during the growing season, since the amount retained in the plant is very small compared to the amount evaporated from it. Water requirements vary with plants, climatic conditions, soil fertility, and soil moisture.

Water Resources. The supply of ground water and surface water in a given area.

Water Rights. The legal rights to the use of water. They consist of riparian rights and those acquired by appropriation and prescription. Riparian rights are those rights to use and control water by virtue of ownership of the bank or banks. Appropriated rights are those acquired by an individual to the exclusive use of water, based strictly on priority of appropriation and application of the water to beneficial use and without limitation of the place of use to riparian land. Prescribed rights are those to which legal title is acquired by long possession and use without protest of other parties.

Watershed Area. All land and water within the confines of a drainage divide or a water problem area, consisting in whole or in part of land needing drainage or irrigation.

Watershed Management. Use, regulation, and treatment of water and land resources of a watershed to accomplish stated objectives.

Watershed Planning. Formulation of a plan to use and treat water and land resources.

Watershed Protection and Flood Prevention Projects. A system of land treatment or soil conservation practices combined with structural measures installed to improve infiltration and reduce erosion of land within a drainage basin and to protect lands from floods.

Waterspreading. The application of water to lands for the purpose of increasing the growth of natural vegetation or to store it in the ground for subsequent withdrawal by pumps for irrigation.

Water Table. The upper surface of ground water or that level below which the soil is saturated with water.

Water Table, Perched. The surface of a local zone of saturation held above the main body of ground water by an impermeable layer or stratum, usually clay, and separated from the main body of groundwater by an unsaturated zone.

Water Use Efficiency. Crop production per unit of water used, irrespective of water source, expressed in units of water depth per unit weight of crop; for example, 10 inches of water per ton of alfalfa hay. This concept of utilization applies to both dryland and irrigated agriculture.

Waterway. A natural course or constructed channel for the flow of water. See grassed waterway.

Weathering. The group of processes such as chemical action of air and rainwater and of plants and bacteria, and the mechanical action of changes in temperature, whereby rocks, on exposure to the weather, change in character, disintegrate, decay, and finally crumble in the process of making parent material of soils.

Weed. A plant out of place.

Wheel-Track Planting. Plowing and planting in separate operations with the seed planted in the wheel tracks.

Wilting Coefficient. See wilting point.

Wilting Point (or permanent wilting point). The water content of soil on an oven-dry basis at which plants, specifically sunflower plants, wilt and fail to recover their turgidity when placed in a dark humid atmosphere. It is approximately the moisture content at 15-bar (15 atmospheres) tension.

Windbreak. 1: A living barrier of trees or combination of trees and shrubs located adjacent to farm or ranch headquarters and designed to protect people and livestock from cold or hot winds and/or drifting snow. 2: A narrow barrier of living trees or combination of trees and shrubs, usually from one to five rows, established within or around a field for the protection of land and crops. May also consist of narrow strips of annual crops, such as corn or sorghum.

Wind Erosion. The detachment and transportation of soil by wind.

Wind Strip Cropping. The production of crops in relatively narrow strips placed perpendicular to the direction of prevailing erosive winds.

Winter Irrigation. The irrigation of lands between growing seasons in order to store water in the soil for subsequent use by plants.

Woodland Suitability Groups of Soils. A woodland suitability group is made up of kinds of soil that are capable of producing similar kinds of wood crops, that need similar management to produce these crops when the existing vegetation is similar, and that have about the same potential productivity.

X

Xerophyte. A plant capable of surviving periods of prolonged moisture deficiency.

Z

Zone of Aeration. Subsurface zone above the water table in which the soil or permeable rock is not saturated with water.

NAME INDEX

SUBJECT INDEX

A

Acid soils, 277, 278, 280, 281
Actinomycetes, 192
Adams soil series, 108, 109, 124, 125
Adsorption and anionic exchange, 61
Agric endopedon, 137
Airborne substances, 481, 482
Air-dry weight, 212, 213
Alaska, climate and soils of, 90–93
Albic endopedon, 136
Alfalfa, boron deficiency symptoms in, 467
Alfisol in color, facing 100
Alfisols, 102, 103, 106–110, 140, 167, 168, 249–253, 257
Algae, 192, 193
Alkali soils, 381, 382
    reclamation of, 382, 383
Alluvial soil materials, 70, 75
Ammonium fixation, 64
Analysis, mechanical, 32, 33
Animal wastes, 475, 476
Anionic exchange, 61
Anthropic epipedon, 136
Apple, magnesium deficiency symptoms in, 467, 468
Argids, 102, 103, 108, 109
Argillic endopedon, 136
Argiudolls, 108, 109
Arguistolls, 102, 103
Aridisol in color, facing 100
Aridisols, 14, 102, 103, 108–112, 138, 168
Arid regions, soils in, 380–382
Available water, 211–213

B

Bacteria:
    autotrophic, 186, 187
    heterotrophic, 187, 188
    legume, 188, 189
    nitrogen fixing, nonsymbiotic, 190, 191
    nonnitrogen fixing, heterotrophic, 191
    symbiotic, 188, 189

Bean, manganese deficiency symptoms in, 466, 467
Biosphere and soil formation, 95, 96
Boralfs, 102, 103, 108, 109
Border irrigation, 400
Brussels sprouts, molybdenum deficiency symptoms in, 468
Buffer capacity, 64
Bulk density, 38–40
Bush fallow, 265–269

C

Calcium:
    in plant nutrition, 230
    soil test for, 436
Cambic endopedon, 136, 137
Carbonation, 84
Cationic exchange, 58, 59
Chelation of phosphorus, 65, 66
Chemical properties of soils, 52–67
Chemical transformation, 81–84
Chemical weathering 81–84
Chenango soil series, 164, 165
Chiseling, 335, 336, 340
Chroma, 46–48
Clay:
    colloidal, 53–58
    minerals, 53–58
    soils, nutrients in, 15
Climate and soil formation, 87–92
Clod, 39, 40
Coarse fragments, 36
Colby soil series, 108, 109, 113, 114
Colloidal clay, 53–58
Colloidal properties of soil, 52–67
Color, soil, 46–48
Compact horizons, root growth in, 343
Compost, 422
Consumptive use of water, 389, 390
Contour tillage, 353–355
Corn, deficiency symptoms in, 463, 464
Corrugation irrigation, 399
Cotton, deficiency symptoms in, 465, 466